THE ELECTROSTATIC LOUDSPEAKER DESIGN COOKBOOK

THE ELECTROSTATIC LOUDSPEAKER DESIGN COOKBOOK

First Edition

By

Roger R. Sanders

Audio Amateur Press

Publishers

Peterborough, New Hampshire

FIRST EDITION
1995
Copyright ©1995 by Roger R. Sanders

Distribution Agents:
Old Colony Sound Lab
Post Office Box 243
Peterborough, New Hampshire 03458-0243 USA

Library of Congress Card Catalog Number 93-071654

ISBN: 1-882580-00-1

ACKNOWLEDGMENTS

I wish to thank the many loudspeaker builders and experimenters who graciously contributed their experiences, ideas, and suggestions to the ESL Clearinghouse. I've presented many of their ideas for your use. I only regret that too many names and years have passed for me to list them all.

ESL Clearinghouse
P.O. Box 647
Halfway, OR 97834
Phone: (503) 742-7640

Mike Riddle wrote *FastCAD*, the most outstanding Computer Aided Design program I've seen. It's speed, power, and ease of use made it possible for me—a person with no artistic talent—to provide you with the hundreds of illustrations found in this book.

Evolution Computing
437 S. 48th Street
Tempe, AZ 85281
Phone: (602) 967-8622

I am not an electrical engineer or mathematician. The English engineer **Reg Williamson** graciously provided the simplified electrical formulae and designed some electronics for my electrostatic projects.

Finally, **Edward T. Dell**'s belief that amateurs can build speakers and electronics that are as good or better than commercial products is directly responsible for this book. His publication, *The Audio Amateur*, featured a home built electrostatic loudspeaker project by **David Hermeyer** that was the catalyst that started my career in the field of do-it-yourself audio. I doubt that I would have found the motivation, knowledge, or contacts needed to design and build speakers and electronics without his vision.

Audio Amateur Publications, Inc.
P.O. Box 576
Peterborough, NH 03458-0576
Phone: (603) 924-9464

DISCOVERY can be fun as well as intriguing. I remember the first time I discussed electrostatic loudspeakers, sometime in the late '50s, through a monthly electronics magazine. As a teenager new to electronics, I was discovering many aspects of this new hobby. These monthly periodicals, also new, were shaping my preferences toward gadgets unique and different. The more knobs and functions along with "bells, whistles, and flashing lights" a construction article had, the better and more intriguing its lure.

Similarly, when I spotted a construction article for a simple tweeter project called an electrostatic loudspeaker, my appetite was really whetted. Not only was it "simple," but uniquely different. An air of magic and mysticism accompanied a speaker without a motor. There was also definite intrigue in finding something so tantalizing and captivating to the psyche.

A decade passed before electrostatics became an integral part of my life and stereo system. At that time the electrostatic devices were the Radio Shack Model ESA-3s. Next, I auditioned the Infinity Servo Statics which I heard at the West Coast Hi-Fi Show. Oh boy! was I ever falling in love.

In 1974, I chanced to meet a customer in the local electronics store who told me about a journal dedicated to audio amateurs. The journal, *The Audio Amateur* (TAA), had published a complete construction article for a full-range electrostatic speaker and its companion 900W amplifier. Now this, fellow audiophiles, is the stuff of dreams.

I share this personal story with you because we all have had love affairs with people, animals, places, cars, boats, aircraft, etc. What about you? We are not that different, as evidenced by the fact that you are about to read this book. *You* have *discovered* and are now *intrigued* by the phenomenon of the electrostatic principle, and the possibility that a set of speakers can actually be created by your own hands.

Creation is defined by one dictionary as the act of forming out of nothing. More than likely your previous endeavors into speaker building were to "create" suitable boxes and installing various drivers already in existence. Nice, but think how much more satisfying it would be to take the basic elements and fabricate, in toto, that which is envisioned by your mind's eye. Herein lies our purpose.

Over the years much has been published about ESLs in patents, articles, and texts. Many of these tend to be rather technical, but ambiguous. It remained for the work and research of one man, David Hermeyer, and *The Audio Amateur* magazine, to design a practical home-built device to rival commercial offerings.

In the years following the 1972 publication of this project, many developments occurred. Among the first was the research and experimentation done by the author into the improvement of the original Hermeyer design. This research was published along with several follow-up projects. Additional original designs by Mr. Hermeyer and others, yours truly included, have also been published.

Your attention is also directed to *Speaker Builder* (SB), sister publication to *The Audio Amateur*. The intent by all of us has been to push the state-of-the-art to new limits and to make affordable to everyone a speaker we consider to be the finest available.

It is one thing to "roll your own" by following the printed instructions. It is quite another to understand and appreciate the "whys and wherefores" involved in the genesis of a design. As a builder you can definitely alter dimensions of a design to fit a new criterion, but is this designing?

Roger Sanders and I want to challenge you to do more. To *create*, you must possess certain innate knowledge; to fully appreciate your handiwork, it is also necessary to have this knowledge.

Roger Sanders has spent more than two years compiling and organizing the basic elements of this electrostatic book. In the chapters that follow, he presents a logical, systematic approach to the various aspects of design.

You will find information presented in short blocks rather than lengthy paragraphs. This is very beneficial: first, when learning the material, and later when the book is used as a reference. He has taken great pains to remove the suffix "ese" from the word "technical" and to add "ease" into each explanation.

After you delve into theory, methodology and architecture, you will find a construction project for several affordable speakers. We believe the Compact Integrated TL/ESL to be far superior to commercial offerings. But for all its design objectives and attributes, it is not the zenith. The evolution of this design took twenty one years. Its forerunners, constructed over the prior decade, are still very much the pinnacle of performance.

Does this mean nothing new can be learned or

gained? Absolutely not! It does attest to the fact that a good design concept is, if not timeless, at least enduring, maintaining its appeal far longer than most. By applying the knowledge presented in this book to your speaker design, it promises not only to be enduring but endearing well into the next century.

Please keep in mind, however, as you read the book and formulate a plan, that by their very nature these devices are elegant in their simplicity. The addition of knobs and lights for function is OK, but do not get caught up in the philosophy of "bells and whistles." As a design becomes cluttered with extra component parts, performance and elegance are placed in jeopardy.

It is our hope that you will accept the challenge and use this information, together with your unique ideas, to help advance the state of the art. Join with others who have gone before. The Electrostatic Loudspeaker User's Group was formed for the exchange of information and ideas for its worldwide membership. Your contribution and participation is most welcome. Roger Sanders and I welcome news about your project.

Whether you wish to learn more about the subject of ESLs in general, or to create a uniquely individual system, I sincerely hope that your initial curiosity and anticipation have been raised and that your time and effort will be amply rewarded.

Barry Waldron
Placerville, California

ABLE OF CONTENTS

TABLE OF CONTENTS

THE
ELECTROSTATIC
LOUDSPEAKER DESIGN
COOKBOOK

CHAPTER 1

OVERVIEW

You're smart, talented, energetic, and like fine music systems. You find conventional magnetic speaker systems disappointing. The smoothness, detail, low distortion, delicacy, and transparency of electrostatic loudspeakers thrill you. But they're expensive, don't truly meet your needs, and there are things about them you'd like to change.

You're fascinated with the idea of building electrostatics. Their construction seems simple. You know you could do it—if only you had some practical design and construction guidance.

What little information you've found on the subject was written in engineering jargon. It was full of math. Engineers wrote it for engineers. It was boring and hard to understand. It was incomplete. It just wasn't functional.

If the author presented a speaker design, it was hard to build and didn't fit your needs. You needed to know *why* he did what he did—but he didn't tell you.

You're frustrated. Your goal is tantalizingly close, but just out of reach. You are right on the edge of taking the plunge, but realize that you could make costly blunders. You'd prefer to learn from others' mistakes instead of your own.

If this sounds like you—sit back and relax. Your search is over.

This book provides the knowledge you need. Directed to the amateur, it's written in simple concepts instead of engineering terms.

Laymen often find the mathematical models and expressions commonly seen in engineering texts difficult to understand. A "how-to" book doesn't need them.

Since the emphasis is on the practical, the only formulae included are those that you can use to solve down-to-earth problems. The bibliography contains references for engineers.

The cold, detached, impersonal, and complex writing style typical of engineers troubles me. It's hard to read and comprehend. I think a friendly, relaxed, personal writing style gets the message across better.

An important feature is the use of **GUIDE-LINES**. Scattered throughout, and boldly displayed in shadowed boxes, these guidelines present critically important information in a simple, easy-to-grasp form.

I use guidelines for critical concepts, as a substitute for formulas, for defining values based on experience, and for essential parts of construction and setup. Although brief, they are a quick and handy source of information.

Because readers have diverse backgrounds, I've included a chapter that defines, explains, and gives examples of common engineering terms. Although the book starts with basics, it rapidly evolves to advanced topics. Technically skilled readers will find the depth they want.

I present the information sequentially. Later chapters assume a grasp of earlier ones.

Optimizing one design parameter often degrades another. When many are involved, the interrelationships are complex and confusing. Successful ESL (**E**lectro**S**tatic **L**oudspeaker) designers juggle design choices to meet their most important goals, while sacrificing minor ones.

You must understand these compromises. I've put much effort into making these interactions and compromises clear so your first design will be successful.

Listeners differ in their speaker system needs. Therefore, this book is not geared toward one specific speaker design. Instead, I give the information necessary to engineer and build whatever design you like.

If you want to build proven designs, you'll find several well-reasoned and thoroughly tested ESL projects with full details on how to assemble them. You will also discover *why* I made the design choices I did.

Not everybody wants to design ESLs. You may just want directions on how to build one. If so, you can skip the first part of this book and go directly to the construction section (*Chapter 8*). There you will find the help you need.

Associated electronics are an essential part of ESLs. I've devoted a comprehensive chapter on them. Included are parts sources and construction details.

Over the years, I've found that most amateurs have more problems and questions about the electronics than about the ESLs. To solve this problem, I've included a chapter on how to build electronics.

For reasons that will become clear later, many builders use conventional driver bass systems with their ESLs. It has been notoriously difficult to obtain a seamless match between ESLs and magnetic drivers, so that magnetic bass sounds electrostatic. I spent nearly as much effort solving this problem as I spent developing ESLs.

A chapter on magnetic woofer systems covers both theory and practice. This mainly involves woofer enclosures; I've made no attempt to delve into the theory and construction of conventional drivers.

As with ESLs, I present the principles associated with enclosures, so you can design and build woofer systems specialized to your needs. You will find thoroughly tested and proven designs with full construction details.

Mechanical and electrical devices sometimes fail. It is rare to find a "fix-it" section in a design text, but I have included a troubleshooting section to help you find and fix problems.

I used the experiences of manufacturers, engineers, and hundreds of home builders to guide you. While I've built and published many ESL projects, much practical knowledge came from the ESL Clearinghouse during its 18 years of life. Untold numbers of builders contacted me with their experiences, ideas, and questions.

This Clearinghouse yielded not only innovative new designs, but provided the labor for extensive building and testing that I haven't the time to do alone. The result has been the development of startlingly simple and effective ways to build ESLs.

A good example of this is *Chapter 11*. Although I invented the free-standing curved ESL, I do not have as much experience building them as some of my readers. To give you the best and most up-to-date information possible, I asked Barry McClune to share his experience and building techniques with you.

After spending years helping amateur builders, I believe I understand your needs. I've tried to anticipate and answer most of your questions. Still, an unusual problem or new idea periodically arises.

I am always looking for improvements to ESLs. Please feel free to contact me for additional support, or for sharing information.

Roger R. Sanders
P.O. Box 647
Halfway, Oregon 97834 U.S.A.
Phone: (503) 742-7640

However simple ESLs may seem, they involve complicated physics and electronics. Analyzing them demands the use of technical terminology. I regret having to trouble you with this already, but we must speak a common language. I promise to make it short, easy, and understandable.

For simplicity, I've left out subtle technical details in some of the descriptions. Bear with me. The explanations are good enough to get through this project.

VOLTAGE. Voltage is the force or pressure that pushes electric current through a conductor. It is measured in **Volts**. ESLs use thousands of volts. One thousand volts is also known as one **kilovolt,** which is abbreviated **kV**.

Polarity refers to the poles (north or south) of magnetic force. It also defines the sign (positive or negative) of a voltage.

Voltages may change their polarity, as in audio signals. This produces an alternating current waveform. *Figure 2-1* shows a **sine** wave and a **square** wave.

We must decide at what point on the wave to measure the voltage. Do we measure the peak or the average voltage? The standard measurement is **RMS**, which is the acronym for **R**oot **M**ean **S**quare. RMS voltage can be thought of as the average voltage of the waveform. Technically it is not quite the average, but is close enough for this discussion.

Occasionally, it is more useful to measure voltage at the top or peak of the wave. We call this measurement **peak-to-peak voltage,** and abbreviate it **P-P**.

CURRENT. The flow of electrons through a conductor is **current** and is measured in **amperes**. We usually abbreviate this to **amp**. ESLs only use a few thousandths of an amp. One thousandth of an amp is a milliamp, its abbreviation is **mA**.

Direct Current is the energy that flows in only one direction. It is abbreviated **DC**.

Alternating Current is energy that changes direction. Its abbreviation is **AC**. We call the rate at which it changes direction its **frequency**. The frequency may be a stable 60 cycles per second, as in your home electric power, or highly varied, as in music.

Power is the ability to do **work**, such as move air to make sound. The **watt** is a measurement of electrical power. Volts times amps equals watts. For example, 5A @ 10V = 50W.

RESISTANCE. Resistance restricts the flow of current. It only applies to DC circuits. Its abbreviation is **R**.

Resistance is measured in **ohms**. One ohm is the amount of resistance required to limit the current to one amp when pushed by one volt. We call 1,000 ohms a **kilohm** and abbreviate it to **k**. One million ohms is called a **Megohm** and is abbreviated to **M**. The symbol for resistance is the Greek letter omega = Ω.

The energy used to overcome resistance is dissipated as heat. Resistors used in electrical circuits have heat dissipation ratings measured in watts. If you dissipate 10W in a resistor rated at ½W, you will soon see smoke as it overheats and fails.

REACTANCE. Reactance is the resistance to the flow of alternating current. You can think of it as a special type of resistance that

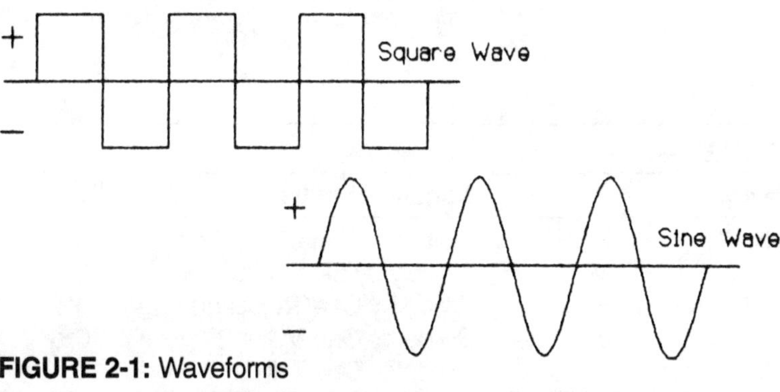

FIGURE 2-1: Waveforms

changes with frequency. There are two types of reactance: **capacitive** reactance and **inductive** reactance.

Capacitive reactance refers to the resistance that a capacitor presents to an alternating current. It *decreases* as the frequency of the AC increases. To put it another way, it is inversely proportional to frequency. It is abbreviated X_C

Inductive reactance refers to the resistance that an inductor presents to an alternating current. It *increases* with increasing frequency. In other words, it is directly proportional to frequency. Its abbreviation is X_L.

IMPEDANCE. Impedance is resistance in AC circuits. If that sounds like "reactance," that's because they are related. Reactance is one of the three factors that determines impedance. The three factors are capacitive reactance (caused by capacitors), inductive reactance (caused by inductors and transformers), and resistance (caused by resistors). It is the combination of all three that determines the total impedance in an AC circuit. Its abbreviation is **Z**.

Frequency refers to how rapidly the current changes direction. It's measured in cycles per second. Rather than use the long term "cycles per second," we use **hertz** and abbreviate it to **Hz**. One thousand hertz is a **kilohertz** and is abbreviated to **kHz**.

For example, the winding on a transformer may have only a few ohms of resistance (remember *resistance* is a DC measurement). When you measure it with an alternating current of 1kHz, its *impedance* may be several hundred ohms. The impedance may be several million ohms at a frequency of 100kHz.

OHM'S LAW. Ohm's Law is at once the easiest, simplest, and yet one of the most important of all electrical formulae. It's so handy that you should be familiar with it. It mathematically specifies the relationships between voltage, current, and resistance. The resistance

is measured in ohms. I've modified it to layman's terms:

CAPACITANCE. A **capacitor** is an electrical device capable of storing and discharging electrons. You can think of it as a special type of extremely fast-acting battery.

$$\boxed{\begin{array}{c} \textbf{OHM'S LAW} \\[1em] \text{Amps} = \dfrac{\text{Volts}}{\text{Resistance}} \\[1em] \text{Resistance} = \dfrac{\text{Volts}}{\text{Amps}} \\[1em] \text{Volts} = \text{Amps} \times \text{Resistance} \end{array}}$$

The number of electrons required to produce a given voltage on a capacitor defines its electrical **capacitance**. Again, this is analogous to a battery—the more electricity it can store, the greater is its capacitance.

Capacitance is measured in **farads**. A farad is too large for most of our electrical measurements. Rather than use decimals or fractions it's easier to use prefix modifiers.

For example, it's easier to say that a certain capacitor has a value of 20 picofarads (abbreviated pF) than it is to write 0.00000000002 farads (did I get the right number of zeros?). You will find most capacitors rated in either picofarads or microfarads (μF).

PREFIX MODIFIERS. Measurements have modifiers attached to them to make their use more practical. If you understand the modifiers, measurements are logical and easy to remember. Some common ones are listed in *Table 2-1* followed by their abbreviations and exponential expressions.

Prefix modifiers are very common. Besides the farad noted above, other examples include a thousand meters (kilometer), a thousandth of a meter (millimeter), a thousandth of a gram (milligram), a thousand grams (kilogram), a thousand ohms (kilohm), a millionth of a meter (micrometer or micron), and many more.

The English system of measurement is poor, particularly when dealing with small quantities. The metric system is superior, but is not accepted and therefore not well understood in the US.

ESLs use tiny measurements that are best

TABLE 2-1

Name		Number Modifier	Exp
Kilo	K	Multiply by One Thousand	10^3
Meg	M	Multiply by One Million	10^6
Milli	m	Divide by One Thousand	10^{-3}
Micro	μ	Divide by One Million	10^{-6}
Pico	p	Divide by One Trillion	10^{-12}

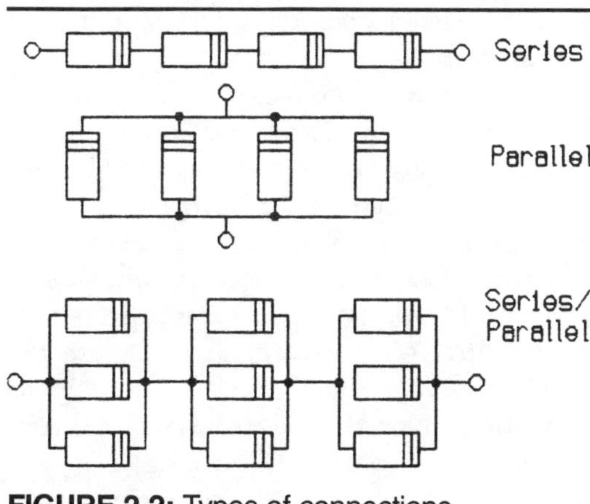

FIGURE 2-2: Types of connections.

communicated in millimeters or microns. I'm in a bind. Should I use the superior metric system known worldwide, but not (amazingly) used in the States? Or should I use ridiculous English measurements with which Americans are more familiar? Although I would prefer to use metric units, I'll use English in this text since it is published in the US.

Even so, you probably haven't heard of the English measurement **mil**. I will use it frequently when discussing small distances like the thicknesses of glue bonds, diaphragms, wire, sheet metal, and spacers. As suggested by the prefix modifiers above, a mil is one-thousandth of an inch (0.001″).

PARALLEL AND SERIES CONNECTIONS.
You can make connections to electronic parts in **series**, in **parallel**, or a combination of the two. This is easier to illustrate (*Fig. 2-2*) than to explain.

MASS AND INERTIA.
Scientists define mass and inertia in a roundabout manner that fails

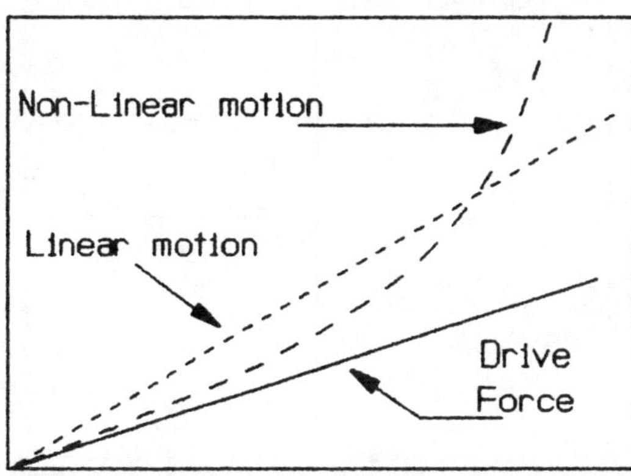

FIGURE 2-3: Types of motion.

to define either one: **mass** is that which has inertia. **Inertia** is a property of mass. This gets a bit more meaningful when we add that inertia is the tendency for mass to remain in motion (or at rest), unless acted upon by an outside force.

Weight is the force produced by mass in a gravitational field. Although many people use weight and mass interchangeably, weight is a **property** of mass—they are not the same. The weight of a speaker is not the problem—its inertia is.

Inertia ruins transient response. It prevents the speaker from responding instantly to the electrical drive signal. It causes the speaker to "remain in motion" and **overshoot** the intended waveform.

When mass (like a speaker cone) is combined with a spring (like the suspension in a speaker), it produces resonances and "ringing." **Resonance** is best visualized in a guitar string. The mass of the string vibrates at one particular frequency—its **resonant frequency**. If you increase the mass of the string or reduce the spring tension, the frequency falls. **Ringing** is a type of resonance where the speaker continues oscillating after the drive signal is removed.

AUDIO TERMINOLOGY.
Nonlinear distortion occurs when a linearly increasing voltage or other force results in exponential rather than linear motion. *Figure 2-3* graphically depicts **linear** and **nonlinear** motion.

Harmonic distortion occurs when the desired (fundamental) frequency has harmonic (mathematically related) frequencies inappropriately added to it. For example, if you want a pure 1kHz frequency tone (also known as a **sine wave**), but find that there is a 2kHz tone added to it, the 2kHz tone is harmonic distortion. It is harmonic distortion, because it is a multiple of 1kHz and isn't supposed to be there.

It is also the *second* harmonic of 1kHz because it is 1kHz × 2. What would the third harmonic be? Right . . . 3kHz. If the amplitude of the 2kHz tone is 20% of the fundamental, you have 20% second-harmonic distortion. Harmonic distortion measurements usually lump all harmonic components together and call it **T**otal **H**armonic **D**istortion (**THD**).

Intermodulation distortion (**IM distortion**) is where the desired frequency is modulated by another frequency. Imagine a cone speaker producing deep bass. If it also produces high frequencies, the high frequencies will have a subtle wavering sound to them, since their radiating surface is rapidly changing position. IM distortion is not only a problem in speakers—similar problems occur electrically within amplifiers.

Frequency response is the way an audio component's output relates to its input at different frequencies. A picture is worth a thousand words (*Fig. 2-4*).

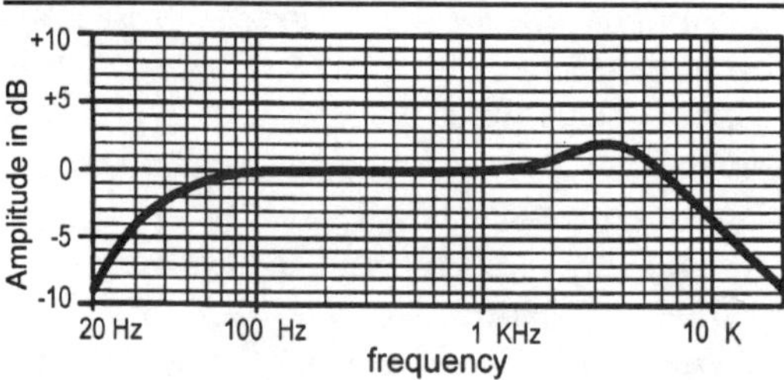

FIGURE 2-4: Frequency response graph.

In this hypothetical example of an amplifier, the ratio between the output and input is the same at all frequencies between 100Hz and 1kHz. At higher and lower frequencies the ratio changes, and the output either climbs or falls.

The 100Hz–1kHz range forms a ruler straight and flat line. We often call this region **flat**, or **linear**. The range from 1–4kHz has **nonlinear** or a **rising** frequency response. Below 20Hz and above 4kHz the frequency response is also nonlinear, because it is **falling**.

Bandwidth refers to the frequencies reproduced linearly by a component. In the above amplifier, the linear bandwidth is 100Hz–1kHz.

Damping is the process of stopping something. For example, the shock absorbers on a car (correctly called suspension dampers) stop the car from bouncing. Sound absorbing material in a room or speaker enclosure stops resonances. Note that damping is not the same as dampening which means to wet something. The two are commonly confused.

Transducers are devices that change one type of energy into another. In the case of speakers, microphones, and phono cartridges, they turn electricity into motion or vice versa.

Frequency response in transducers is similar to that of electrical components except that the ratio is between the input of one form of energy and the output of another.

Line-level voltage comes out of your tuner, CD player, preamplifier, or tape deck. It ranges up to about a volt at only a few milliamps of current. From a practical point of view, all com-

ponents usually operate at line level except phono cartridges and microphones, which have much lower output and require preamplifiers to bring them up to line level.

Voltage gain (usually simply called **gain**) refers to the increase or decrease in audio voltage as it passes through a circuit. The practical effect is to change the loudness of your music. For example, adjusting the volume control on a preamplifier changes the gain of the audio signal.

Audio power amplifiers increase the voltage and current of the line level signal so serious work can be done (like making loud music). A large power amplifier may produce several amps at up to as much as 200V.

Equalizers are frequency shaping electronic circuits. Just as there are many kinds of amplifiers, there are many types of equalizers. Some are adjustable, some are not. Equalizers can be **passive** or **active**. Passive types do not have amplification; active equalizers do. Amplification can be produced by using tubes, transistors, or integrated circuits—ICs are most commonly used in equalizers.

Many familiar audio components are equalizers even though we don't call them that. Have you ever heard of a crossover, graphic equalizer, phono preamp, or tone control? I thought so. These are all equalizers.

Filters are just different names for certain type of equalizers. *Figure 2-5* shows many different kinds of equalizers with their special names (if any) and their associated frequency response trends.

Crossover networks are equalizers that separate the audio spectrum into limited frequency bands suitable for driving woofers, tweeters,

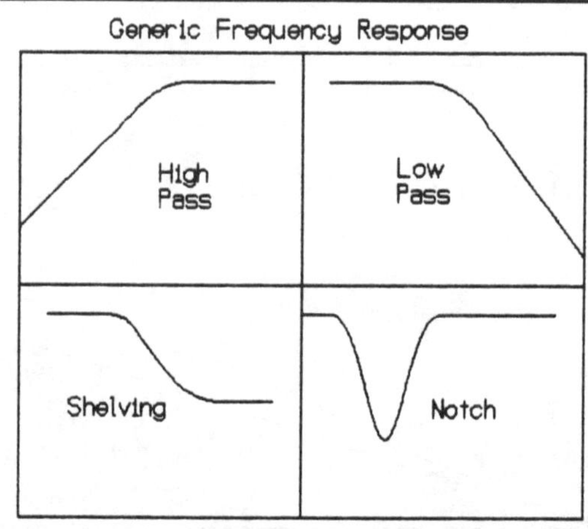

FIGURE 2-5: Various filter types.

and midranges. They can be either **high level** or **low level** as well as **passive** or **active**.

High level crossovers are always passive. You will find them in the common magnetic speaker system's enclosure where they intercept the signal from the power amplifier before it gets to the drivers. After splitting the signal into various frequency bands, they pass them on to the appropriate drivers.

Low level crossovers may be passive or active. The active types are more common and are often called **electronic crossovers**. They split the line level signal *before* it reaches the power amplifier. They perform better than high level crossovers and allow you to use a separate amplifier for each driver. Both significantly improve a speaker system's performance although they require multiple power amplifiers.

Biamp systems are two-way speaker systems using low level crossovers and two power amplifiers. A **triamp** system's format should now be obvious.

Simple equalizers like crossovers have slopes that are multiples of 6dB/octave. Each 6dB increment is called an order. First-order crossovers are 6dB/octave, second-order is 12dB/octave, third-order is 18dB/octave, and so on.

Phasing refers to the diaphragms or cones of two drivers moving in similar directions, when the same voltage polarity is applied to them. When setting up speakers, it is very important that the drivers in both channels move in the same direction.

If one moves one way when the other moves in reverse, the sound will have a diffuse and directionless quality. The image will *not* appear centered between the speakers, as it will in a correctly phased system. Phasing also applies to multiple drivers in a speaker. If the tweeter is moving toward you, the woofer should also.

White noise is a random mixture of all the frequencies in the audio spectrum. Subjectively, it sounds like a waterfall. White noise is an exceptionally good test for frequency balance.

Sound pressure level is a measurement of speaker **output** or **loudness**. Its abbreviation is **SPL**. It is measured in **decibels** which we abbreviate **dB**.

Decibels change exponentially, not linearly. Doubling the sound pressure level does not double the dB—it only increases it by three. To put it another way, a voltage increase of 1,000 is a decibel increase of only 60.

To give you an idea of SPLs, I've specified some recognizable sounds with their approximate decibel levels.

- 0dB The quietest sound a human with perfect hearing can detect under ideal conditions.
- 60dB A quiet office.
- 72dB A quiet car at highway speed.
- 90dB Loud music.
- 100dB "ear shattering" music levels. If sustained, permanent hearing damage results.
- 105dB A full symphony orchestra at maximum peak loudness (Row A).
- 120dB The threshold of pain.
- 140dB Ten yards from a jet engine at full power.

CHAPTER 3

THEORY OF OPERATION

ELECTROSTATIC FORCE. The ESL's operational principle is the attraction of electrically charged objects. You can experience this force by combing your hair on a dry day.

The friction from combing generates an electrostatic charge on the comb. You can feel this force gently tugging on the hair on your arm if you move the comb near your skin. The force on the comb can pick up dust or tiny bits of paper.

This charge is **static**—it does not move. In this respect, it is similar to a permanent magnet.

Atoms contain positively charged protons in their nuclei surrounded by orbiting negatively charged electrons. When protons are surrounded by an equal number of orbiting electrons, the atom is electrically neutral. When the number of electrons and protons are different, the atom has a net electrical charge.

Ionization is the production of electrically-charged atoms by removing electrons from, or adding electrons to, the outer shells of an atom. The net electrical charge of ionized atoms is the **electrostatic force**. When applied to ESLs, we often use the terms **force**, **charge**, and **voltage** interchangeably.

The strength of the charge depends upon the number of electrons missing or added. When there are excessive electrons present, the net charge is negative. When electrons are missing, the net charge is positive.

ELECTROSTATIC LOUDSPEAKER PARTS.

An ESL contains three types of parts:
- Diaphragm
- Stators
- Spacers

Diaphragms are the moving parts of the ESL. Most ESLs have only one diaphragm.

Stators are electrically-conductive stationary plates that drive the diaphragm. Typically, there are two, for reasons that will soon be apparent.

Spacers separate the diaphragm from the stators. The spacers are in direct physical contact with the stators and diaphragm. They must have exceptionally good electrical insulating qualities to keep the charge on the diaphragm separated from the charge on the stators.

DIAPHRAGMS. They are made of an extremely thin and lightweight plastic film placed under tension like the head of a drum. The diaphragm drives the air in the room to make sound.

The diaphragm's mass is equivalent to a layer of air about ¼-inch thick. Compared to the room's air mass, this is insignificant. You can consider an ESL massless for practical purposes.

The diaphragm must have an electrostatic charge of several kilovolts. This is the **polarizing** or **bias** voltage. A high voltage power supply produces the polarizing voltage.

Plastic film is an insulator. You must make one side of it conductive to accept the electrostatic charge. Many materials can produce a conductive coating—the most common method for amateurs is to rub powdered graphite on the diaphragm.

Remember that this charge is static—it does not move. If it weren't for tiny electron leakage paths that are present, the power supply could be discarded once you charge the diaphragm.

Dust and water vapor (as humidity) cause the leakage paths. In practice, if you remove the power supply, the speakers continue to play for several hours, but the sound gradually becomes fainter as the charge slowly escapes.

STATORS. This electrostatically-charged diaphragm doesn't move by itself—one or more stators drive it. A stator is an electrically-conductive, acoustically-transparent plate or grid. It is stationary and is placed close to, but does not touch, the diaphragm.

The stator must allow sound to pass through, while simultaneously presenting a strong electric field to the diaphragm. It therefore must have holes or slots in it. Their shape and size is important, as they influence ease of construction, cost, and speaker performance.

An audio power amplifier applies voltage to the stators, as determined by the music. Like magnetic forces, opposite electrostatic forces attract each other and like ones repel.

Let's say that the diaphragm has a positive charge, and at a given moment a stator has a negative charge. The diaphragm will be attracted to the stator. If they are both positive, they will be repelled.

The power amplifier's rapidly changing polarity causes the diaphragm to be attracted to and then repelled from, the stator. This action pushes the air and generates sound waves.

As demanded by the music, the power amplifier varies the frequency of the voltage supplied to the stators. Higher stator voltages interact more strongly with the static charge on the diaphragm. This creates greater force on the diaphragm, more diaphragm motion, larger amplitude air waves, and louder sound.

Electrically, music simply consists of changing frequencies and voltages. The frequency patterns define the *type* of sound (piano, flute, orchestra, and so forth), and the voltage controls the loudness.

Conventional power amplifiers lack the voltage necessary to drive ESLs to adequate output levels. A **step-up transformer** between the amplifier and ESL increases the amplifier's output to several thousand volts.

SPACERS. These precisely distance the diaphragm from the stator. This distance is critical. If it is too large, the speaker output will be very low or nonexistent. If the diaphragm touches the stator, it can't move to generate sound, and the static charge on the diaphragm will be lost to the stator—again reducing output.

The spacers must be excellent electrical insulators. We usually make spacers out of plastic like Plexiglas® or Lexan® because they are readily available, cheap, and easy to assemble.

CELLS & PANELS. When sandwiched together, these three parts comprise an electrostatic **cell** or **panel**. We call an ESL with only one stator **single-ended**. *Figure 3-1* is a cross-sectional view of such a cell.

Note that the stator/diaphragm assembly

CAPACITANCE DEFINING FACTORS

• Diaphragm-to-stator spacing

• Stator design

• Diaphragm area

just described is a capacitor. It stores electrons when charged just like any capacitor.

The capacitance is an important feature of ESLs, as it represents the electrical **load** seen by the power amplifier. The capacitance of the speaker presents many problems.

Electrostatic force decreases by the square of the distance between the stator and diaphragm. Changing the distance produces an exponential change in the force seen by the diaphragm. It is not linear. Therefore, single-ended ESLs suffer from nonlinear distortion unless the motion of the diaphragm is extremely small.

To avoid this, all modern ESLs use a double stator, **push/pull** drive system. This consists of a stator on each side of the diaphragm, as shown in *Fig. 3-2*. From this point on, I'll discuss only push/pull electrostatic speakers.

The power amplifier is connected to the stators so that as one stator is driven to a positive voltage, the other goes negative. This causes one stator to pull the diaphragm, while the other pushes it.

The nonlinear increasing force seen by the diaphragm as it approaches one stator is balanced by the nonlinear decreasing force seen by the diaphragm as it moves away from the other. The net effect is that the diaphragm "feels" a constant force and moves in a linear fashion.

Carefully note that the motion is linear *anywhere* between the two stators. The beauty of this is that the sound *quality* is not affected by construction tolerances, although sound *quantity* is. This is important, because it is impossible to build stators and diaphragms perfectly flat and uniformly equidistant from each other.

This is particularly fortunate for the amateur speaker builder, because it guarantees superb sound even with poorly built cells. I've seen incredibly distorted, warped, and generally poorly built ESLs whose sound quality is the same as faultlessly built ones. An additional advantage of the push/pull ESL is that the forces on the diaphragm are twice those of a single-ended ESL, which improves output.

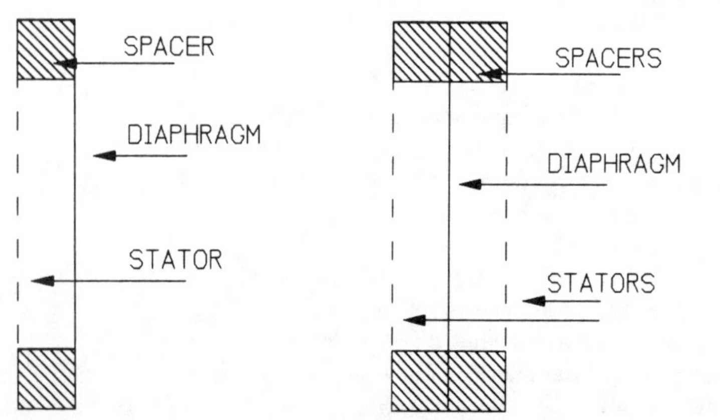

FIGURE 3-1: Single-ended ESL.   **FIGURE 3-2:** Push-/pull ESL.

CONSTANT CHARGE OPERATION. In a push/pull ESL, the charge on the diaphragm must remain motionless. If it moves as the diaphragm changes position between the stators, the forces seen by the diaphragm will change. This produces nonlinear distortion.

Figure 3-3 shows a simplified schematic diagram of the electrical drive system needed for an ESL. Note the electrical path from the diaphragm, through the polarizing power supply, through half the transformer's secondary winding, to the stators. The diaphragm's charge can follow this path and change value, as it bleeds to one stator or the other.

Unfortunately, as the diaphragm moves towards one stator, the capacitance between the diaphragm and stator changes. To compensate, the charge on the diaphragm tries to move. If it does, nonlinear distortion results.

You can solve this problem by removing the polarizing power supply after you energize the diaphragm, since this would disrupt the electron's path. This is impractical, because the charge slowly leaks from the diaphragm.

The solution to this problem is resistor R1, the **charging resistor**. It resists rapid movements of current through the polarizing power supply. Although it doesn't completely stop the current, it slows it enough so that at audio frequencies the charge on the diaphragm is essentially stable.

An added benefit is that this resistor limits the electron flow when you turn on the polarizing power supply. Large electron flows occur when you first charge the diaphragm. These flows tend to burn off the diaphragm coating, where the polarizing power supply connects to the diaphragm. Eventually, the diaphragm coating would fail and the speaker would quit. The charge resistor prevents this type of failure.

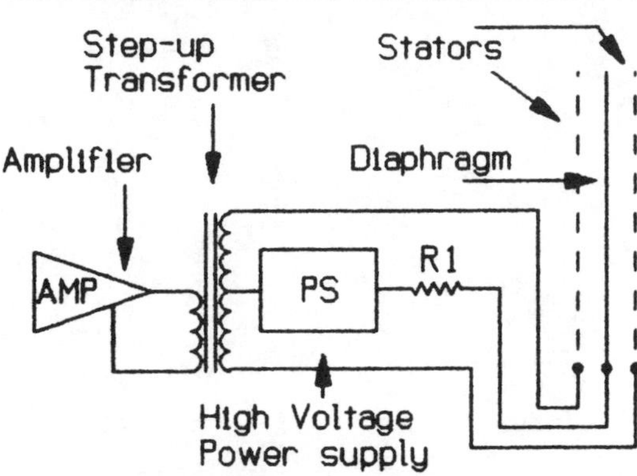

FIGURE 3-3: ESL drive schematic diagram.

> ## SMALL CELLS AND LOW FREQUENCIES DEMAND HIGH VALUE CHARGING RESISTORS

You can calculate the value of this resistor by keeping the time constant RC of the charging circuit large in comparison to one-half the frequency. If you're not sure what that means, relax; I'll show you another way. If you're an engineer—have fun.

The value cannot be excessively high, or the charging current can't hold the diaphragm's voltage constant under high leakage conditions. If it's too low, distortion will increase.

> ## *GUIDELINE #1:*
> ### Charging resistor value
> 20–200MΩ.
> Use the higher values for small cells and low frequencies.
> The value is not critical
> 20MΩ works well in large ESLs.

Small cells store fewer electrons at a given voltage than large ones. For a given current flow, small cells will change voltage more than large ones. Therefore, they need more resistance in the current path than do large cells under similar conditions.

Greater diaphragm voltage changes occur at low frequencies, because the current has more time to flow. Therefore, for a given size cell, lower frequencies need higher-value charging resistors than cells used only at high frequencies. You may be asking yourself, "So what value resistor do I need?" The upper limit seems to be around 200MΩ. A value this high causes a noticeable loss of output under high leakage (humid) conditions in large ESLs.

The lower limit is less clear, since it is a function of speaker size, the lowest frequency of operation, and your opinion of acceptable distortion levels. I consider 10MΩ marginal, and generally use 20MΩ in large speakers.

"A speaker" refers to all the cells driven by one transformer or amplifier channel. Usually this is the entire speaker for one channel. Your speaker may be made of many individual cells

> **An ESL with a
> high value charge resistor and
> high impedance diaphragm
> operates in the
> CONSTANT CHARGE mode.**

wired in parallel. You don't need a separate charging resistor for each cell, since they act as one large one.

Besides the charging resistor, the diaphragm itself needs to be high resistance to prevent migration of the electrostatic charge on its surface. Construction irregularities cause the diaphragm to be closer to some areas of the stator than others. The charge tends to move to the closest areas where the forces are the greatest.

Practical experience demonstrates that low resistance diaphragms work adequately despite theory, as long as they have high value charging resistors. Some builders use aluminized Mylar® film diaphragms. Such diaphragms have a resistance of 4Ω per inch compared to the $50k\Omega$–$1M\Omega$ per inch resistance typically found on graphite-coated diaphragms.

ESLs sometimes produce an electrical arc (spark) between the diaphragm and stator. If this arc has high current, it can generate enough heat to melt the diaphragm or even cause the ESL to burst into flames.

This rarely causes a problem in high resistance diaphragms because the diaphragm's resistance limits the current to a safe value. The electrons from only a small area of the diaphragm can discharge into the arc, so the current is small.

A low resistance diaphragm can discharge the entire diaphragm charge across the gap. This can generate enough heat in a large ESL to cause it to ignite. At the very least, it will burn a hole in the diaphragm. An additional disadvantage of aluminum-coated diaphragms is that they are more massive than other types, although they are still light enough to work well.

CHAPTER 4

FREQUENCY RESPONSE

The driving forces within ESLs are linear and accurate. Because low mass causes the speaker to drive the air directly, ESLs should have perfect frequency response.

Unfortunately, this is not so. They have serious problems at both high and low frequencies.

Figure 4-1 shows the general trend of the frequency response of a full-range ESL with a minimum dimension of 18″. *Figure 4-2* is a photo of the actual frequency response of the same speaker obtained from a storage oscilloscope trace and 1/3-octave pink noise fed into a real-time analyzer. Other objective tests show similar trends. The bottom trace in *Fig. 4-2* shows what you can and must achieve if you want accurate performance.

The causes of ESL frequency response errors seem to depend on whose text you read. I have yet to see a comprehensive, authoritative, and scientifically-valid discussion of the subject. I don't have all the answers either. Fortunately, we don't have to positively know the cause of the problems to deal with them.

In the discussion that follows, I base my comments on the measurable and reproducible frequency response of ESLs as shown above, and only address causes that *unquestionably* affect it.

HIGH FREQUENCY LIMITS. The mass of an ESL's diaphragm is the ultimate limit on high-frequency response; but this usually isn't a problem, because the limit is supersonic. For example, in ¼-mil Mylar film diaphragms, the diaphragm's mass only *starts* to roll off the high frequencies at 32kHz. It's reasonable to think of ESLs as having unlimited high-frequency response.

From a practical viewpoint, the audio drive voltage restricts the high-frequency response. This is a significant problem that will influence your design options. I examine this in detail in *Chapter 7*, so will not discuss it further here.

LOW FREQUENCY LIMITS. Two undeniable problems impair low frequency response: phase cancellation and fundamental resonance. Before analyzing them in detail, you must first understand that we normally operate ESLs as dipole radiators. This means they

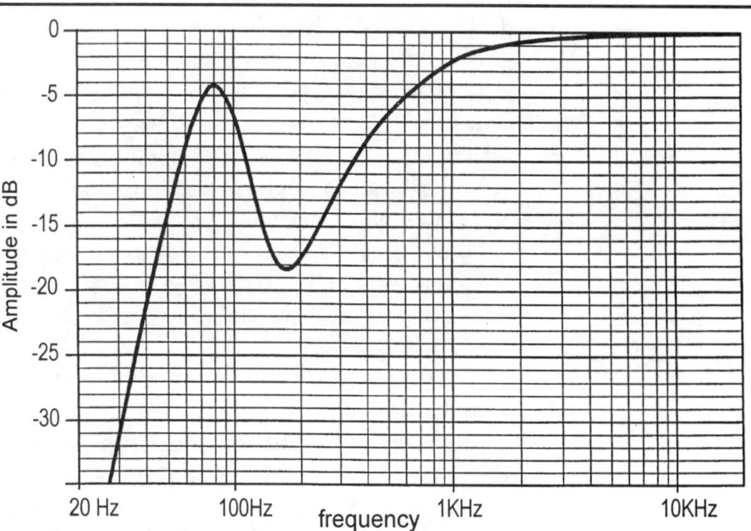

FIGURE 4-1: ESL frequency response trend.

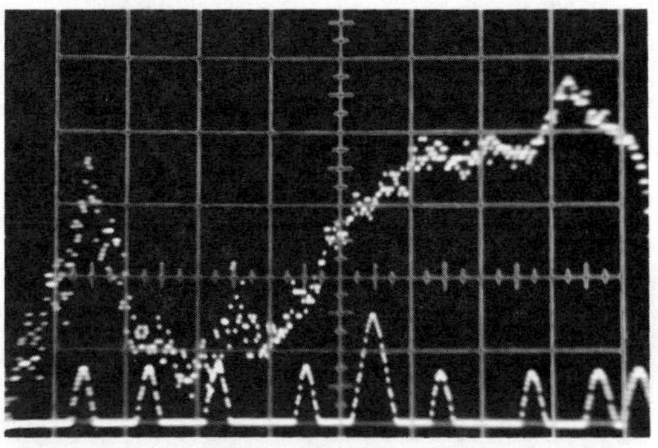

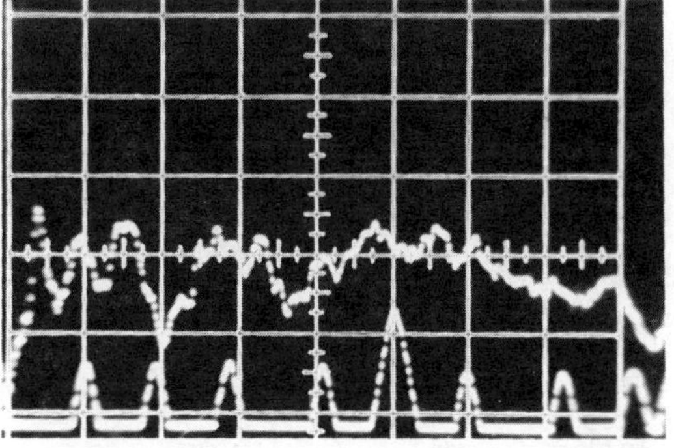

FIGURE 4-2: Measured ESL frequency response.

don't have enclosures. They are free to radiate sound both in front and behind them.

Avoid enclosures whenever possible, because they resonate and have diffraction effects—both impair performance. In cone-type magnetic speakers, enclosures usually can't be eliminated, which is one reason why they don't sound as clear as ESLs. Dipole operation is better, but is only practical with large speakers.

PHASE CANCELLATION. As a speaker's driven surface moves, it produces pressure on one of its sides and a vacuum on the other. These pressure waves are sound, and large waves are louder than small ones.

Unfortunately, dipoles allow the pressure generated on one side of the speaker to "leak" around the side of the driver to the vacuum on the other side. This reduces the size (amplitude) of the wave, which reduces output. This is phase cancellation.

The seriousness of the "leak" is a function of how far the air must travel to get to the other side (speaker size), and how much time it has to arrive (low frequencies). Phase cancellation losses begin when the wavelength of the sound is about one-quarter of the minimum dimension of the speaker. The output loss increases exponentially, with decreasing frequency.

Figure 4-3 shows the wavelength of sound versus frequency. *Figure 4-4* depicts the nonlinear progressive loss of bass typical of a dipole ESL with a minimum dimension of 18".

To picture this problem in your mind, imagine that the ESL is a canoe paddle, and that the air in the room is a bathtub full of water. If you move the paddle rapidly back and forth 6", you make large waves. This is analogous to an ESL reproducing loud high-frequency sound.

If you move the paddle slowly (bass in the ESL) the same 6", the water has time to travel around the edge of the paddle and cancel most

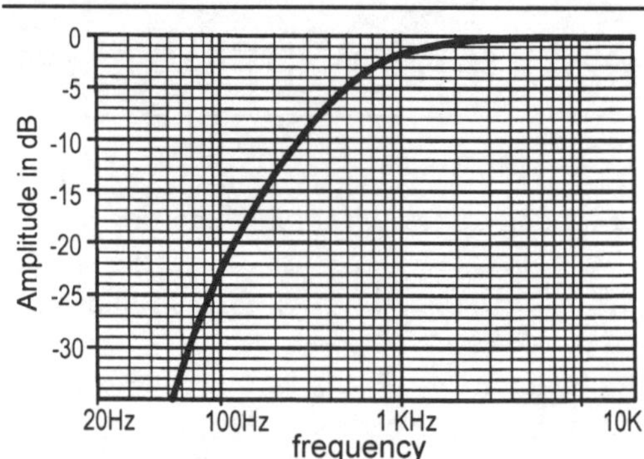

FIGURE 4-4: Trend of phase cancellation.

of the wave energy. The waves will be small. If you move the paddle *very* slowly (deep bass), there will be no waves at all.

Note carefully that the excursion of the paddle and the ESL is the same at both high and low frequencies. In other words, the ESL's diaphragm motion is linear. Despite this, the response falls with decreasing frequency.

This is not a flaw in the ESL itself. The problem is the fluid behavior of air. Regardless of the cause, phase cancellation is a major problem that must be corrected.

FUNDAMENTAL RESONANCE. An ESL's diaphragm must (unfortunately) be under tension. This makes the diaphragm act like a spring. The diaphragm's mass interacts with this spring and resonates.

This is the speaker's fundamental resonance. It is often an ESL's only resonance—but it is savage. It is usually 10–20dB in amplitude. Below it, the frequency response plummets. Review *Fig. 4-1*.

You can't even calculate or predict the resonant frequency with accuracy. Conventional resonance formulae calculate the mass of the driver against the spring rate. This doesn't work if you use the diaphragm's mass in the calculation. Its mass is so small that the resonance will appear to be supersonic.

The air mass "seen" by the diaphragm totally swamps the mass of the diaphragm itself. It is the mass of the air, plus the mass of the

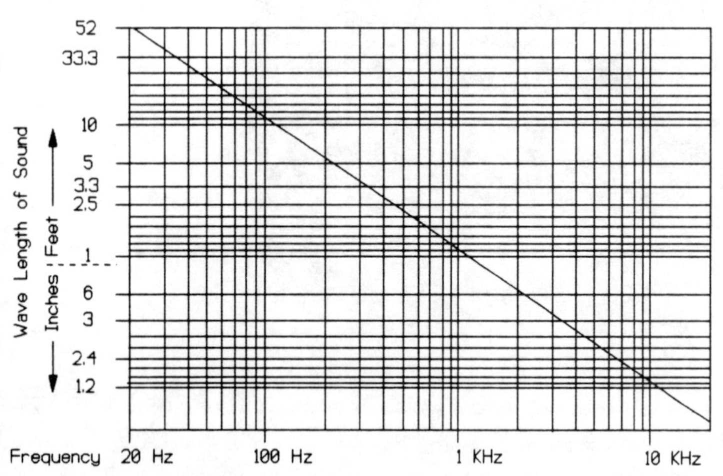

FIGURE 4-3: Wavelength versus frequency.

**OUTPUT CRASHES
BELOW
FUNDAMENTAL RESONANCE**

FACTORS DEFINING FUNDAMENTAL RESONANT FREQUENCY

- Air mass "seen" by diaphragm
- Area of diaphragm
- Room resonances
- Diaphragm mass
- Diaphragm tension
- Direction(s) of tension
- Distance between spacers
- Air temperature
- Elevation
- Humidity

diaphragm (the *effective* mass), that you need to use in your formula.

But how can you determine the air mass? It varies depending on temperature, humidity, elevation, barometric pressure, diaphragm area, room resonances, and room size.

Room size is significant. The mass varies depending on room volume. In a small room the diaphragm works with the room's entire air mass. As the room gets larger, you reach a point where the air's compressibility decouples it from the speaker, so that the diaphragm only "sees" part of the air in the room.

Room resonances, if close to the speaker's resonance, can "pull" the resonance away from what is calculated. How can you compute the mass with these unknown variables?

The diaphragm's spring rate is a major part of the resonance phenomena. It's difficult to predict, because it varies depending on the distance between supporting spacers, length, whether it's pulled in one or both planes, and its tension.

Even if you have a formula that deals with all these variables, you probably wouldn't find it useful, because you don't know the value of many of them. For example, what exactly is the diaphragm's tension? What is the mass of the air working with the diaphragm? Is this mass all the air in the room, or only part of it? Is a large room resonance affecting the speaker's resonance?

Incidentally, the air mass "sees" your speaker as a unit. It doesn't care whether you built it as one large cell, or many small ones. It is the total size that counts. You can't draw any conclusions about the system's resonance based on the behavior of one if its constituent cells.

Because of these problems, I consider the resonant frequency to be unpredictable. We can only estimate it. Still, we need to have at least some idea of what it will be, so we can deal with it in the design process.

I'll give you some idea of what to expect based on experience. Large ESLs (3–6-feet tall, 1–2-feet wide) with high diaphragm tensions, in moderate size rooms (12′ × 20′), will usually resonate somewhere between 50 and 150Hz.

The ESLs in this book have resonances in the 50–100Hz range. Smaller ESLs will have higher resonances, and huge ESLs will have lower ones, particularly if they have "floppy" low-tension diaphragms.

In summary, the massless nature of ESLs permits their use across the complete audio spectrum without crossovers. Phase cancellation and fundamental resonance impair the bass response, so that a full-range ESL sounds bright, thin, and anemic. It has no deep bass.

GUIDELINE #2:

Fundamental Resonant Frequency

100Hz (±50) for large ESLs in mid-sized rooms—higher for small ESLs and lower for very large ones.

CORRECTING FREQUENCY RESPONSE. Many designers leave the frequency response as is. In effect, they use fundamental resonance to compensate for phase cancellation.

This gives the speaker some bass, but a severe depression occurs in the frequency response above resonance that guts the midrange, and there is no deep bass. The sound is still bright and thin, and the output is severely compromised, because of the resonance.

This is not satisfactory. You don't have to sacrifice frequency response to get ESL sound quality. Let's explore practical ways to deal with these problems.

ENCLOSURES. Magnetic drivers are usually much smaller than ESLs. They suffer such severe phase cancellation problems that designers *must* use enclosures to deal with this.

Enclosures work by simply isolating the front of a driver from the back of it. This prevents the air from "leaking" around the side of the driver and cancelling the pressure waves.

While this solves the phase cancellation problem, it causes others—specifically resonances and diffraction. I discuss these topics in *Chapter 12* on transmission lines.

Magnetic woofers are relatively massive and have very low spring rates because their suspension systems are soft and "floppy." Their fundamental resonance frequency is low—much lower than the typical ESL. The frequencies of interest are usually above their resonance, so enclosing them to prevent phase cancellation works well to produce deep bass.

Putting an ESL in an enclosure doesn't work as well. The diaphragm is under high tension, so the spring rate is high. The mass of the diaphragm is low, so the speaker's effective mass depends on the air mass within the enclosure—which isn't much. The result is a relatively high fundamental resonance, with loss of the deeper bass below resonance.

If you put your ESL in an enclosure, you iso-

MURPHY'S LAW:

If anything can go wrong, it will—at the most inconvenient time

late one side from the room air. This reduces the mass loading on the diaphragm, which *raises* the fundamental resonance and further harms deep bass response. The enclosure does stop phase cancellation and the midrange "suck out," but the deep bass is a disaster.

The limited excursions of ESLs require large panels. Therefore, their enclosures must also be large. To complicate matters further, an ESL's enclosure needs to be extra large, so the diaphragm "sees" a large air mass.

ESLs need such enormous enclosures that for full-range ESLs, this usually involves closing off part of the room instead of building huge boxes.

You may be thinking, "Why doesn't he just reduce the diaphragm tension? Wouldn't that lower the resonant frequency and solve the problem?" You're right—you can control the resonance that way.

Unfortunately, Murphy's Law interferes. You know Murphy don't you? I'm sure you've met him—you just may not know him by name.

The problem with low diaphragm tension is that you must have high diaphragm tension for high output. Severe output loss is too great a price to pay for deep bass. What good is deep bass if you can't hear it? I cover this problem extensively in the next chapter.

You must treat the interior of the enclosure with sound absorbing material to minimize the reflection of bass energy back through the diaphragm. Such reflections cause aberrations in the frequency response and must be suppressed.

It is easy to absorb high frequencies. All you need is a thin layer of anything soft on the enclosure walls.

Stifling low frequencies is a challenge, because they contain a great deal of energy. The techniques used for sound absorption in large anechoic chambers work well, but are hard to do. Let's look at how the "pros" do it.

The basic idea is to trap the sound waves between large, tapered, soft, foam wedges so that it bounces back and forth between them. Each "bounce" converts some sound energy to heat, since it requires work to compress the

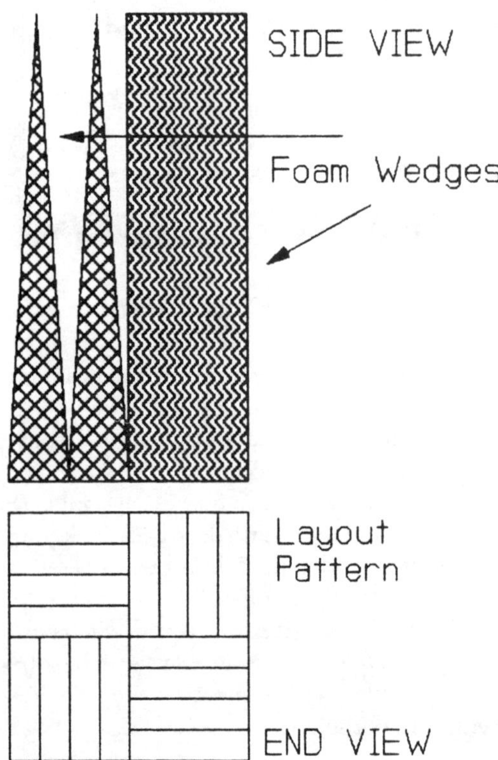

FIGURE 4-5: Anechoic chamber wedge design.

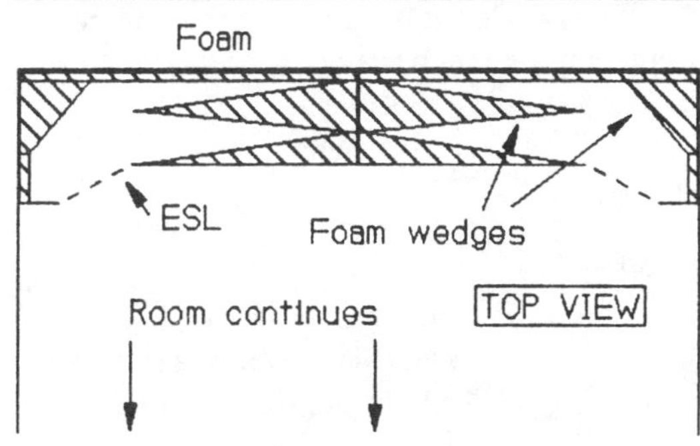

FIGURE 4-6: Room converted to ESL enclosure.

foam. If the wedges are large enough, you can stop any resonance.

Figure 4-5 shows the design and layout of foam wedges, as used in commercial anechoic chambers. Bigger wedges are better, but usually the room inside the enclosure is too small to use long enough wedges to do a good job.

Figure 4-6 shows one way to make a speaker enclosure for full-range ESLs. It uses a couple of huge foam wedges, rather than multitudes of smaller ones. The interior walls are made non-parallel with tapered foam sheets.

I outline a method for making large foam wedges and composite construction techniques at the end of *Chapter 12* on transmission line design. Please look there for details on how to make enclosures like the one shown in *Fig. 4-6.*

SPECIAL ESL ENCLOSURES. What follows are just thoughts and ideas about enclosures—"brain-storming," if you will. I've not tested or experimented with these ideas. I just throw them in for your interest and possible experimentation.

"Normal" speaker enclosures work well with ESL midranges or tweeters. Just be sure you control resonances and prevent energy from the woofer from escaping through the ESL.

However, it is unusual to find midrange and tweeter ESLs in enclosures, because enclosures cause resonances, diffraction, and take up space. ESLs sound better without them, and at higher frequencies, don't need them.

Although simple closed-box enclosures effectively deal with phase cancellation, they do not control the problem of fundamental resonance. But there are enclosures that can. One method is to add acoustic mass within the enclosure to reduce the resonant frequency. You can do this by enlarging the enclosure or replacing the air with a heavy gas.

I haven't explored this extensively, but sev-

eral industrial gases are much heavier than air. For example, sulfur hexaflouride gas is not only heavier, but has excellent electrical insulating properties. Even ordinary CO_2 is heavier than air, and is cheap and readily available from welding supply shops.

You can damp the resonant frequency and maintain response below this frequency either by using equalization (to be discussed shortly) or by modifying the enclosure, so it will generate a secondary acoustic resonant circuit—or both. One way is to make the enclosure deep and narrow. At wavelengths just under four times the depth of the enclosure, the resonance will be forced to the ¼-wavelength resonance of the depth of the enclosure.

You will need to add damping material to the enclosure to reduce the resonance. It is possible to decrease the low-frequency cutoff by over an octave using this technique.

Another possibility is to have a small enclosure immediately behind the ESL, connected to a larger enclosure by a port. By selecting suitable dimensions, you can develop resonant circuits that suppress the fundamental resonance while supporting the frequencies below it.

As you might expect, these enclosures are difficult to design and build—even if you have a room large enough for them. Appropriate design formulae for them are complicated and beyond the scope of this text.

I've never heard of anybody building enclosures more complicated than the closed-box type for ESLs. These usually have been disappointing. It appears that you must build complicated enclosures to deal effectively with the problems associated with bass and ESLs. The question is—are they worth the trouble when there are other, more practical ways of solving the problem?

EQUALIZATION. Using an equalizer to increase the diaphragm's excursion with decreasing frequency is a very effective technique for correcting phase cancellation (and other frequency response problems as well). All you have to do is build an equalizer whose electrical frequency response is a mirror image of the speaker's acoustic output.

This sounds simple, but before you can build an equalizer you must know the speaker's acoustic frequency response. Unfortunately, measuring speaker frequency response is neither easy nor practical for most amateurs. It is a notoriously difficult exercise which requires specialized equipment.

Another difficulty is even if you know the speaker's frequency response, it will almost surely not match the natural slopes of equalizers. Equalizers have slopes that are multiples of 6 such as 6-, 12-, or 18dB/octave. Speaker frequency response slopes will not only be different, but they won't be constant—phase cancellation changes with frequency, as does the slopes that produce fundamental resonance.

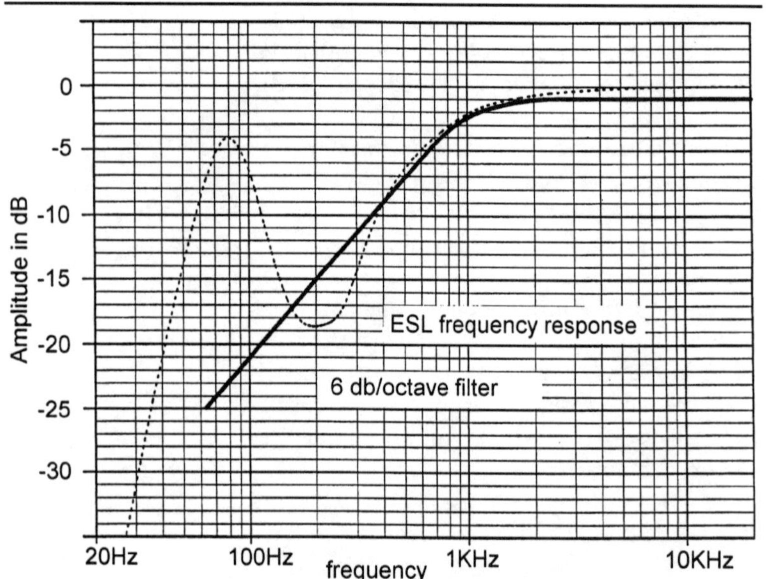

FIGURE 4-7: Comparison of ESL frequency response to equalization.

Fortunately, there are ways around these problems. If you use some logic and common sense, you can predict the frequencies likely to need equalization. You don't need to measure the speaker's frequency response to do this. Just look at the wavelength versus frequency graph (*Fig. 4-3)* to see where phase cancellation starts.

If you prefer, you can calculate the wavelength. The speed of sound varies depending on temperature, humidity, and elevation, but they don't make a significant difference for practical speaker design. I've rounded off the numbers, since they aren't critical. The following formulae are adequate for our needs:

$$\text{Wavelength (feet)} = \frac{1,083}{\text{frequency}}$$

$$\text{Wavelength (inches)} = \frac{13,000}{\text{frequency}}$$

Alternatively, you can directly calculate the frequency where the phase cancellation begins by the following formula:

$$F = \frac{52,000}{D_{min}}$$

Where:

F = Frequency where phase cancellation begins in hertz

D_{min} = Minimum dimension of the ESL in inches

Look at *Fig. 4-7.* The upper curve is the familiar full-range ESL frequency response. The lower one is that of a 6dB/octave equalizer. Note that the two curves match closely in the upper part of the phase cancellation range. If you limit the equalization to this range, and start it based on wavelength calculations, you can get nearly perfect equalization without measuring your speaker's frequency response.

For demonstration purposes, the graphs show the equalization following the speakers' response. In actuality, you would make the equalization a mirror image of the speakers' frequency response, so it would increase the output where the speaker's output is falling.

A major consideration with equalization is its effect on output. Equalization designed to correct phase cancellation increases output in the midrange compared to the high frequencies.

All speakers have a limit on maximum output. Equalization doesn't change this limit. Equalization decreases the higher frequencies to get the desired frequency balance—the midrange output remains the same and is limited by the speaker's output abilities. The net effect is that equalization reduces the system's output. Still, equalization is very effective and practical, if used in moderation.

Besides its ability to correct an ESL's phase cancellation problem, equalization has another use: it is a sensible way to tame fundamental resonance. A notch filter can put a dip in the audio drive voltage at fundamental resonance. This can reduce the fundamental resonance to tolerable levels.

Notch filters have other benefits as well. By reducing diaphragm excursion at resonance, they allow the use of smaller diaphragm-to-stator (D/S) spacings that increase output. Combining notch filters with other equalizers, or with other methods of controlling phase cancellation, is a practical way to flatten frequency response and increase output.

SEGMENTATION. So far, I've been talking about ESLs that are one big cell driven uniformly over its entire surface. An ESL may be broken into strips (segments), which can be independently driven at different frequencies.

These segments usually vary in size and performance. Small ones operate as tweeters and large ones as woofers, just like two- or three-way conventional magnetic speaker systems. *Figure 4-8* depicts such a design.

Segmentation controls phase cancellation by reducing the driven area as the frequency increases. Looking at it another way, the driven area increases as the frequency falls. The result is more bass in relation to the highs, which compensates for phase cancellation.

This was a common method of dealing with the problem in commercial ESLs in the '50s and '60s, although it is not used in current designs. The original QUAD and KLH 9 were segmented.

The KLH 9 used four large bass panels and one small tweeter panel. A crossover network split the incoming power amplifier signal into two parts, and fed the bass frequencies to the four bass panels and the highs to the tweeter.

This was a very crude attempt to offset phase cancellation, since there was only a single "step" in the frequency response. This single step was far from the smooth compensation needed for flat frequency response.

QUAD devised a three-way system using a midrange, tweeter, and a bass cell. This provided a closer approximation of the mirror image curve required, but with only two steps it was still far from perfection.

Segmented speakers tend to have discrete steps in their frequency response, where they shift to the various drivers. They do not produce the smooth response needed for accurate sound reproduction. *Figure 4-9* shows a general response trend of segmented ESLs.

Segmented ESLs have many other disadvantages as well. Besides only doing a fair job correcting frequency response, they have serious output limitations.

Reducing driven area reduces the output at high frequencies to match the weak bass. Since diaphragm excursion in the bass is the ultimate limit on output, the entire speaker is restricted to that output level.

Segmented ESLs need crossovers, introducing a whole new set of problems. If you must use them, it's best to keep the crossover point below 600Hz where their flaws can be made inaudible. As I'll discuss in detail in *Chapter 7*,

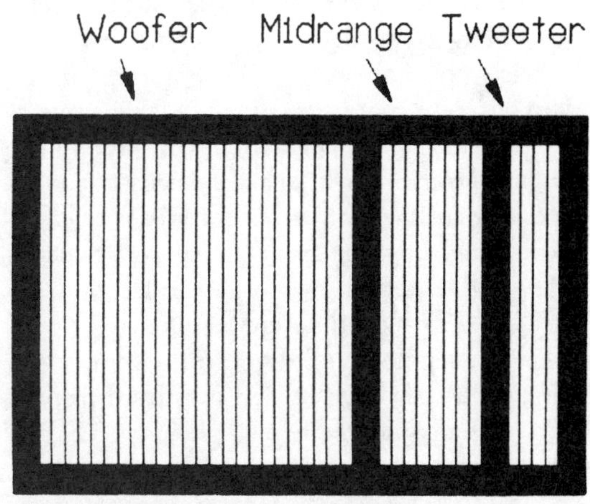

FIGURE 4-8: Segmented ESL.

even the best crossovers produce objectionable colorations in the critical midrange region. If you want a state-of-the-art speaker system, you must not use crossovers in the midrange. Phase cancellation begins at several thousand hertz, so it isn't practical to avoid crossovers in the midrange with a segmented ESL design.

Conventional crossovers are designed to drive resistive loads like magnetic loudspeakers. Designing crossovers to drive capacitive loads like ESLs is difficult and rarely done. Electrostatic high-level crossovers are not commercially available.

Although not perfect, segmentation can be made to work reasonably well subjectively. It has several advantages besides correcting frequency response.

One of its major advantages has to do with the step-up transformers necessary to drive ESLs with conventional amplifiers. The problem with step-up transformers is that it is difficult to get wide bandwidth, while simultaneously producing very high voltages. It is easier to make specialized ones which cover a more

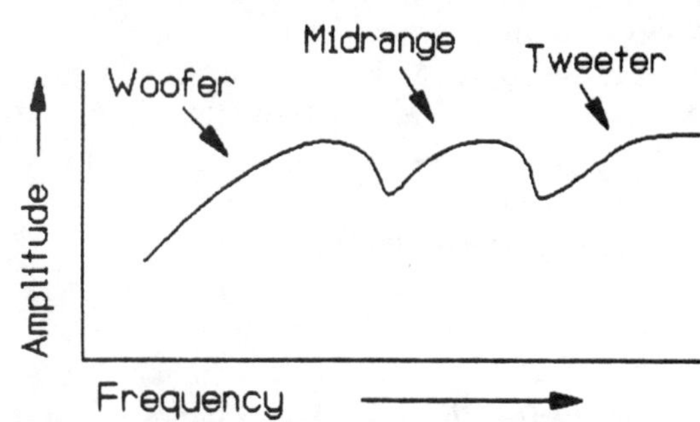

FIGURE 4-9: Frequency response trend of segmented ESLs.

limited frequency range. This serious problem can be largely resolved using segmented ESLs because you can use a step-up transformer with different characteristics for each segment.

Optimizing each segment for its frequency range is another advantage. As discussed in the next chapter, the requirements for bass cells are somewhat different from those for tweeter cells. Segmentation gives you the freedom to vary cell design for different frequencies.

Finally, segmented ESLs permit you to use a narrow-strip tweeter which improves dispersion, when compared to a single large cell. But don't expect miracles. Narrow ESL strip tweeters still have poor dispersion compared to quality magnetic designs.

HYBRID SYSTEMS. The advantages of ESLs are not needed in the bass, where they have severe problems with frequency response and output. It makes more sense to turn the job over to magnetic woofers that shine in the bass but fail in the higher frequencies, where ESLs excel. In short, specialize. Detail, smoothness, and delicacy are critically important at middle and high frequencies. Magnetic drivers cannot match ESLs in these respects—but these properties are of little importance in the bass.

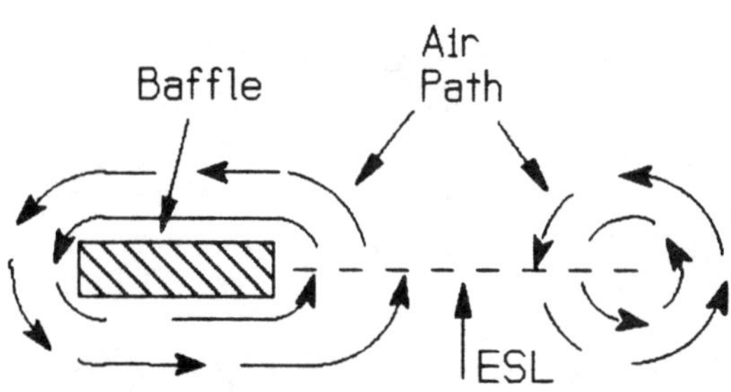

FIGURE 4-10: Baffle operation.

Woofers must provide high output, flat frequency response, and deep bass capabilities. Magnetic woofers out-perform ESLs in these categories. They are also smaller than ESLs of similar capability.

Magnetic drivers' tendency to overshoot and ring are insignificant at low frequencies, where music has slow rise and decay times. The ear is not sensitive to the relatively high harmonic distortion levels inherent in magnetic bass, so this also is not a problem. Using low frequency crossover points stops IM distortion, likely to be present in full-range ESL designs.

Specialization can provide superb frequency response and dramatically increased output. Although not easy, it *is* possible to build hybrid systems that sound as clean as a full-range ESL.

Disadvantages include such factors as complexity, size, and expense. Hybrids require crossovers, bi-amplification, woofer enclosures, extra work, and floor space.

If the crossover point is high enough, you can run the ESL in its linear frequency range, where there is no phase cancellation. In reality, you'll have to go lower than that to keep the crossover point below 600Hz.

A driver does not simply stop operating when it reaches the crossover point. Depending on the crossover slope, the woofer should ideally have good frequency response for at least a couple of octaves above it. Even the best magnetic woofers have difficulties above 2kHz. Therefore, about the highest practical crossover point for a high performance woofer system is 500Hz. This keeps the woofer within its best performance range and makes crossover problems inaudible.

An ESL with linear frequency response to 500Hz is absurdly large. A reasonable compromise is to use a smaller one and correct the mild phase cancellation with equalization, or a single step of segmentation, or perhaps by building a small enclosure.

Frankly, the biggest problem with hybrid systems is the typical audiophile's prejudice toward them. For example, you probably have a strong opinion that ESLs are better than magnetic drivers, or you wouldn't be reading this book. You may also believe that a full-range ESL is the only road to musical Valhalla.

Now here I am—supposedly an ESL expert—telling you to consider using *magnetic* woofers! I forgive you for questioning my sanity.

Allow me to make a confession. I'm not addicted to ESLs. Rather I'm addicted to superb sound reproduction. I'll do what I must to get my fix. That happens to be ESLs on top, but I also demand high output and deep bass. Full-range ESLs just can't do it.

It's difficult to match a magnetic woofer to an electrostatic tweeter. In my opinion, it has not been successfully done commercially, which is probably why hybrids have a poor reputation. But it *can* be done, as I'll describe further in *Chapter 12*.

Keep an open mind and carefully define your needs. If you listen exclusively to quiet chamber music and classical guitar in a small apartment, by all means make a full-range ESL. If your tastes lean toward full symphony orchestra, pipe organ, or rock reproduced at "Row A" concert hall levels, you should consider hybrids.

LARGE ESLs. Phase cancellation does not occur if the speaker is large enough. Also, a very large ESL will have a low fundamental resonance. They are simple, don't need crossovers, segmentation, magnetic woofers, bi-amplification, or equalization. Sounds pretty good, doesn't it?

Remember the rule: "If it sounds too good to be true, it probably is." The problem with large ESLs is the sheer size required for uncompromised performance. The bass frequency of 32Hz has a wavelength of around 32' and requires a speaker about four times that size to prevent phase cancellation. Where are you going to put it?

The speaker would have a huge capacitance, and will be impossible to drive with conventional power amplifiers in the usual way. You would have to use special direct-coupled power amplifiers or make the ESL in multiple strips, with a separate transformer and power amplifier for each strip.

The maximum reasonable size for a home ESL driven with conventional electronics is about 2-feet wide and 6-feet tall. Anything larger usually requires special electronics and a custom-built sound room.

In short, ESLs large enough to produce uncompromised sound are impractical.

BAFFLES. A baffle is a passive extension on the speaker that lengthens the path the sound wave travels (*Fig. 4-10*).

They don't contribute to output. They make an already large speaker even larger, but can help if used carefully and in moderation. For example, they are ideal for thin strip ESL tweeters where they can considerably extend the low frequency response without being overly large.

A baffle doesn't have to be planar. It may extend perpendicular to the face of an ESL. Think of it as an open-back box. This still lengthens the path the wave has to travel, but minimizes speaker width (at the expense of depth). Avoid making the baffles parallel, as this can generate resonances. *Figure 4-11* shows various baffle designs.

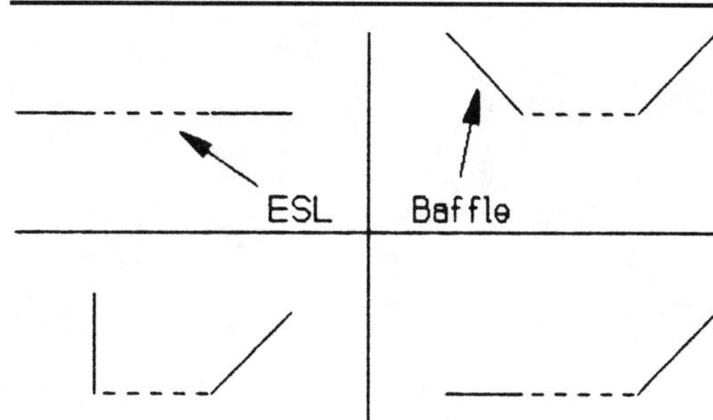

FIGURE 4-11: Baffle design.

Baffles needn't be massive or strong unless you want to use them for structural support. Their surfaces may be either hard or soft. You can reduce diffraction by rounding the edges.

DAMPING. Fundamental resonance can be tamed slightly with the use of damping materials on the ESL panel. This can take the form of various materials such as acoustic foam, thick grille cloths, or thin felt.

Damping materials, unfortunately, adversely affect the sound from the speaker. Grille cloths are particularly bad in this respect. If you use damping materials, be sure to test the speaker with and without them to see whether they impair the general sound quality. I discuss how to do this later in this book.

Depending on the material, the amplitude of the resonance, and the Q desired, the damping elements can either cover the entire speaker, or you may place them in strips. They can be placed on either or both sides of the speaker, but listeners usually prefer the rear, where they don't directly impair the sound.

Good output lies at the heart of ESL design. It's the key to success. Poor output is the main shortcoming of most ESLs.

Small, linear, easy-to-drive, full-range crossoverless ESLs are easy to make. Just pick a convenient size, measure the frequency response, and build a mirror-image equalizer. It would be inexpensive, simple, and have incredible sound quality. It would only have one fault—you could barely hear it!

The need for high output is the cause of every problem in ESL design. You will compromise all other design parameters to achieve it. Its importance cannot be overemphasized.

In this chapter I'll examine the factors that influence output. I explain how you can manage these to obtain the highest output, and how this affects other design considerations.

This topic is complex and involves many interrelationships. I've outlined the design parameters that influence output below and will discuss some of them in this chapter.

FACTORS INFLUENCING OUTPUT.

   I. Polarizing Voltage
      A. Diaphragm Tension
      B. Diaphragm Materials
      C. Diaphragm Thickness
  II. Spacer Ratio
 III. D/S Spacing
      A. Arcing
      B. Acoustic Coupling
      C. Attainable Drive Voltages
 IV. Uniformity of Construction
  V. Stator Insulation
 VI. Stator Design–Field Density
      A. Hole Size Limits
      B. Stator Thickness
      C. Percentage of Open Area
      D. Stator Deflection
      E. Ideal Stator Opening Size
      F. Corona
 VII. Audio Drive Voltage
VIII. Dimensions
 IX. Crossover Slopes
  X. Crossover Frequency
 XI. Shunt Capacitance
 XII. Directionality
XIII. Proximity

GUIDELINE #3:
Maximum Polarizing Voltage
50V/mil of D/S spacing.

POLARIZING VOLTAGE. The static voltage on the diaphragm is the polarizing (or bias) voltage. It interacts with the charge on the stators to move the diaphragm.

The higher the polarizing voltage, the stronger the electrostatic force. You should use as much polarizing voltage as possible for maximum output. You get more diaphragm motion from a given stator voltage (the audio drive voltage) as you increase the polarizing voltage.

What is the maximum polarizing voltage you will need for your speakers? It varies based on the factors that follow. If you optimize all of them, the guideline will get you very close.

Unfortunately, many barriers limit the maximum polarizing voltage. One barrier is arcing caused by the voltage breakdown of air. Air is a poor insulator. The several thousand volts found on ESL diaphragms can overpower its insulation qualities and cause a spark to form between the diaphragm and a stator. This is **arcing**. The arc temporarily reduces the polarizing voltage until the charge builds back through the charging resistor (usually a few seconds).

This is annoying, because the arc makes a snapping sound and momentarily reduces the music level. More importantly, it may damage the speaker.

An **arc** is a stream of electrons. Rapid movement of electrons through resistance generates heat. If there is enough current in the arc, it can melt a hole in the diaphragm, destroy the conductive coating, or cause it to burst into flames.

POLARIZING LIMITATIONS
- Voltage breakdown of air
- Diaphragm stability

Holes may be as small as a pinhole, or the size of a saucer. Surprisingly, they do not affect the sound quality or quantity until they become very large and numerous.

Although the ultimate limit on polarizing voltage is the voltage breakdown of air, this is not usually a problem. Before the polarizing voltage gets high enough to arc spontaneously, the diaphragm becomes unstable and collapses into a stator.

The diaphragm is under tension like a drum head. This tension produces a **restoring force** that holds the diaphragm approximately midway between the two stators. Theoretically, if the diaphragm is equidistant from both stators, it will be attracted equally to them. You should need little or no tension to keep it there despite how much voltage you apply.

Theory is one thing, real world materials and construction techniques are another. In reality, the diaphragm will not be exactly equidistant between the stators. It always will be attracted more strongly to one stator than the other.

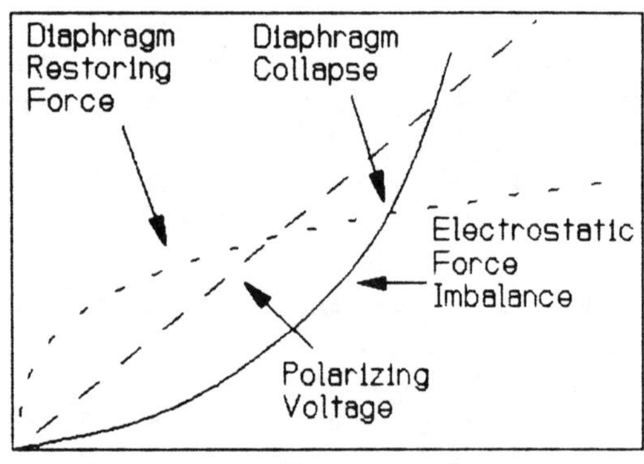

FIGURE 5-1: Forces leading to diaphragm collapse.

This is like having a nail hanging on a string between two magnets. If it is exactly the same distance from each magnet, it should hang between them. If it is slightly closer to one than the other, it will jump to the closer magnet. In reality, you will find it impossible to keep the nail centered—it will always jump to a magnet.

Even if you could build a perfectly symmetrical cell, you would still need restoring force. After all, the amplifier moves the diaphragm off-center to play music. When it does, the diaphragm will crash even in the theoretically perfect cell.

As you increase the polarizing voltage, the diaphragm is more and more strongly attracted to the favored stator until it comes into contact with it. When this happens, the diaphragm can't move and the music will almost disappear. You'll hear a hissing or frying sound as the diaphragm's charge bleeds away.

Usually, the diaphragm remains lightly stuck to the stator until you remove the polarizing voltage. When the charge dissipates, the diaphragm will pop back to its center position. Sometimes you can blow the diaphragm away from the stator with a quick, hard puff.

If the diaphragm doesn't stick, it oscillates. This happens when the diaphragm touches the stator but quickly discharges enough voltage so it doesn't stick. The voltage reduction allows the diaphragm to snap back to its middle position temporarily.

The charge builds to instability again, the diaphragm touches the stator, discharges, and snaps back. This oscillation may repeat several times per second to once every few seconds.

I've described diaphragm instability as though it were a linear event. By that I mean that as you increase the voltage, the diaphragm gradually moves closer and closer to the stator until they touch

In reality, it is nearly an all-or-nothing phenomenon. There is a definite **instability threshold** where just a few volts out of several thousand results in diaphragm collapse—like "the straw that broke the camel's back."

Recall that electrostatic forces are nonlinear. The force decreases by the square of the distance. As the diaphragm moves *slightly* closer to the stator, the electrostatic attraction increases by a large amount.

The diaphragm is simultaneously moving away from the other stator, so the attraction falls by a similar amount. This large change of force with only slight diaphragm motion further pulls the diaphragm toward the favored stator, although the voltage remains unchanged.

Fortunately, the diaphragm's restoring force increases geometrically as the diaphragm moves away from center. At first, the restoring force increases faster than the electrostatic force imbalance and the diaphragm remains stable.

When you increase the polarizing voltage enough, the restoring force doesn't increase as fast as the accelerating electrostatic imbalance. At this point, the diaphragm suddenly collapses into the stator (*Fig. 5-1*).

A fine line exists between maximum allowable polarizing voltage and diaphragm instability and collapse. I call this the instability threshold.

The diaphragm does not remain centered when it is under the influence of the polarizing voltage. Books on ESLs seem to ignore this, and view the diaphragm as though it remains centered between the stators except when driven by the stators. But this is not so. Close observation reveals that polarizing voltage causes the diaphragm to bow (*Fig. 5-2*).

The explanation for this behavior is simple: a perfectly straight diaphragm lacks restoring force. The diaphragm must be at least slightly bowed to generate restoring force.

This is similar to two men trying to pull a long rope straight. The harder they pull, the straighter the rope. But as they get it close to straight, increasing the tension on the rope has less and less effect. Although the weight of the rope is small, pulling it straight against the rope's weight requires an infinite amount of tension. It's impossible to make it straight—they can only get close. *Figure 5-3* shows the force relationships in graphical form.

The same is true of an ESL diaphragm. When you apply voltage, the electrostatic force moves it slightly off-center to generate restoring force.

You can't easily measure this, but you can see it by watching the reflection of a bright light in the diaphragm as you increase the polarizing voltage. The reflection moves and magnifies or reduces your image depending on the direction it bows. If you suddenly remove the voltage by shorting the diaphragm to a stator, the bowing is obvious, because the image "jumps" as the diaphragm snaps back to center.

This diaphragm offset does not affect linearity or sound quality. Recall that the diaphragm behaves linearly anywhere within the force field generated between the two stators.

You probably are saying, "OK, that's nice to know, but get to the point. What do I have to do to get the highest possible polarizing voltage?"

Although many factors theoretically have an influence on this, those that count raise the diaphragm's instability threshold. Some influence speaker output in other ways, also.

DIAPHRAGM TENSION. Increasing the diaphragm's tension increases the restoring force. This has a tremendous effect on output. If you want high output, you absolutely

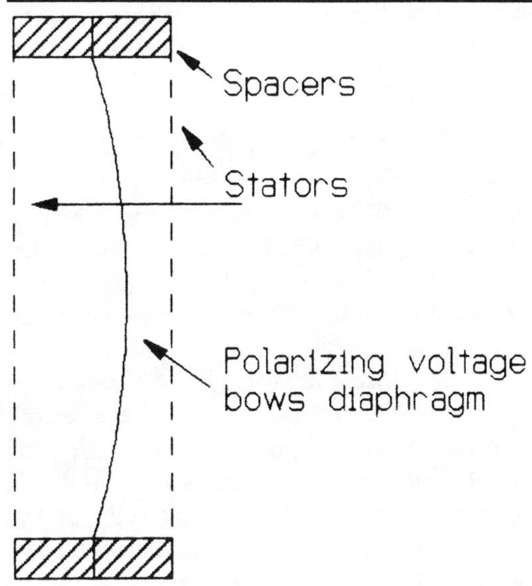

FIGURE 5-2: Diaphragm bowing.

must maximize diaphragm tension. The higher the tension, the higher the instability threshold, the higher can be the polarizing voltage, and the higher the output.

LOW DIAPHRAGM TENSIONS. High diaphragm tensions raise the fundamental resonance frequency. Since output plummets below resonance, some designers deliberately use low tensions to lower the resonant frequency and improve bass response. Their intent is valid, but they incur severe output penalties.

A little change in diaphragm tension makes a large difference in output, but it only makes a small change in the fundamental resonance frequency. Since full-range ESLs already have

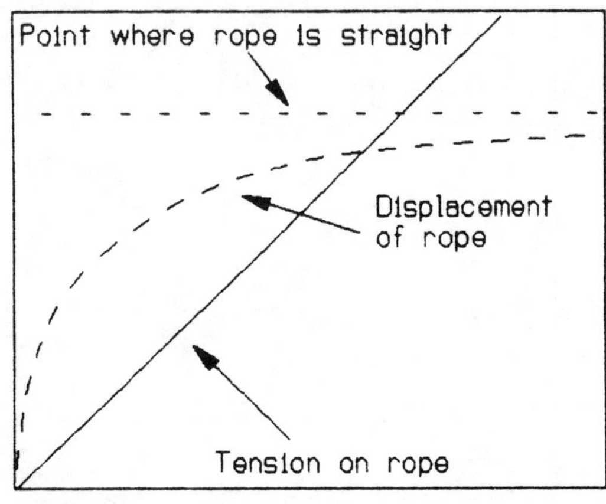

FIGURE 5-3: Displacement forces.

serious output problems, further losses due to low diaphragm tension aren't a good idea. What good is bass you can't hear?

VARIED-DIAPHRAGM TENSIONS. Some builders vary the tension across the diaphragm. By constructing many separate cells with different diaphragm tensions, they hope to tame the huge fundamental resonance by splitting it into a series of smaller ones at several frequencies.

At first glance, this seems like a good idea—but it doesn't work. The air mass "sees" all the little cells as one big one, so the builder still has only one large resonant peak. The spring rate is the average tension of the cells. In one large cell, the resonance depends on the average tension across the whole cell.

Using varied-diaphragm tension not only fails to solve any problems, it damages performance in several ways. Only one big resonance remains, but its frequency is higher than if the lowest tension had been used throughout. Since the lowest tension defines the instability threshold, the output is lower than it would be if the average tension had been used throughout. In summary, varied-tension ESLs compromise both output and bass response and gain you nothing in return.

Variable tension should be avoided unless the various cells are far enough apart for the air to see them individually. Unfortunately, if the air sees them as individual cells, it also will treat them as individual cells, with severe problems of phase cancellation and imaging. Phase cancellation is less of a problem when all the cells appear to the air as one large panel. Imaging is a major problem that I discuss in the next chapter.

Some believe high diaphragm tensions prevent the diaphragm from moving freely, which should *reduce* output. Again, their logic is sound, but it doesn't work that way in practice. Moving the instability threshold upwards increases output much more than higher tension reduces it.

DIAPHRAGM TENSION LIMITS. So what limits diaphragm tension? In practice, the limiting factor is the diaphragm material's tensile strength and ability to maintain high tension over time.

The diaphragm is made of a very thin plastic film, typically ¼–½-mil thick. Such thin films must be exceptionally strong to withstand high tensions. Diaphragm material choices are critical to ESL performance.

DIAPHRAGM MATERIALS. DuPont's discovery and development of polyester film revolutionized electrostatic speaker performance. They call their film Mylar, and at press time it remains the material of choice.

Since DuPont's patents have expired, many competing polyester films are on the market. Through the ESL Clearinghouse, I have reports on most of them. I heard so many horror stories about imported polyester films from builders that I asked David Hartwick (an ESL builder) to do a scientific study.

Using identical test jigs, he mounted diaphragms of various types of films—mostly polyester—and tested them for maximum tension, and tension retention. His results, my experiences, and reports I've received from others reveal that Mylar is capable of highest tensions.

I consider all polyester films inferior to DuPont's Mylar. Few can attain even half the tension of Mylar, and several sag to nearly zero tension within a few days. Apparently, there's polyester film and there's DuPont's Mylar. My strong recommendation is to demand Mylar and accept no substitute.

A new, nonpolyester, ½-mil plastic material named Clysar® also did well in these tests. It matched ½-mil Mylar, but I have no reports on its long-term performance. If it is available

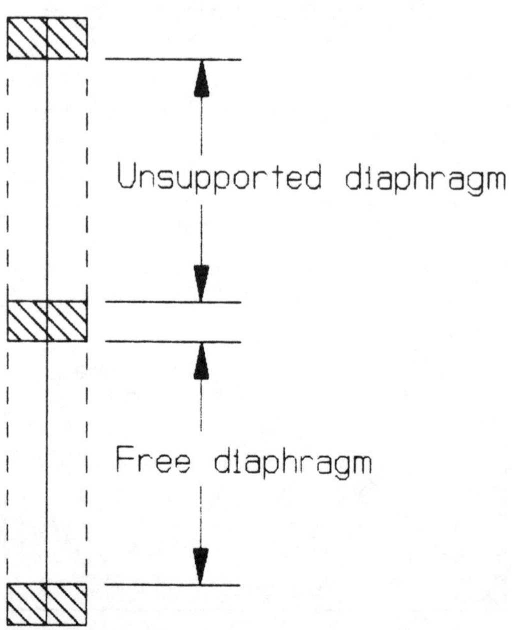

FIGURE 5-4: Unsupported diaphragm length.

from several manufacturers, I have no reports on any differences among them.

Long-term stability and durability are important issues. Mylar diaphragms last many years—some have lasted 12 years under highly adverse conditions (dust, insects, metal chips, and carbon fibre contamination). I have no long-term data on Clysar, so you are on your own if you use it.

Thin-film Mylar comes in two forms: Type "S" and Type "C." They are the same formulation, but Type "C" is used in capacitors, and its thickness tolerance is tightly controlled. ESLs don't need that level of precision. You needn't spend the extra money for Type "C," although it works fine and you may use it if you wish.

DuPont measures Mylar thickness in hundred-thousandths-of-an-inch increments. What ESL builders call half-mil is actually 0.00048". If you order Type "S," specify "48S Mylar."

DIAPHRAGM THICKNESS. Every material has an inherent tensile strength. The thicker it is, the stronger it is. In other words, a film 2-mil thick is capable of sustaining twice the tension of the same type of film that is 1-mil thick. Thicker films support more tension and more output.

As always, you must compromise. The mass of ½-mil Mylar starts to roll off the high frequencies at around 16–20kHz. Thicker, more massive films start to roll off at lower frequencies.

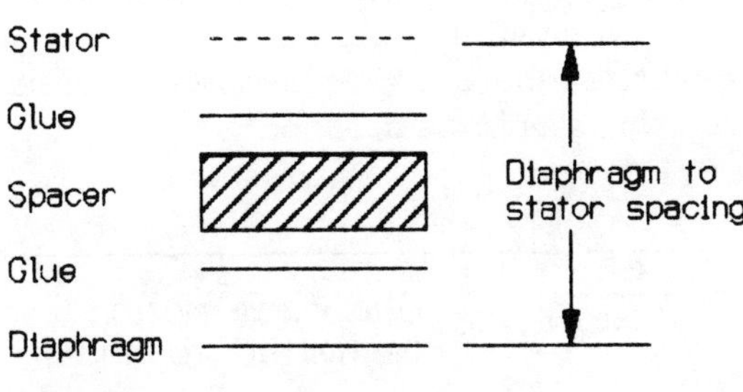

FIGURE 5-5: Diaphragm-to-stator spacing defined.

Films thicker than ½ mil are inappropriate for ESLs designed to operate as tweeters. If you are making large electrostatic woofers, it might make sense to use thicker films for higher tensions and higher output. The downside of this is that the higher tensions would raise the frequency of the fundamental resonance.

DuPont used to make "tensilized" Mylar, which had twice the tensile strength of standard Mylar. It was practical to make diaphragms only ¼-mil thick with it. Regrettably, it is no longer available, and DuPont has removed the equipment they used to make it. If you should happen to stumble across some—grab it. It's the best diaphragm material ever produced.

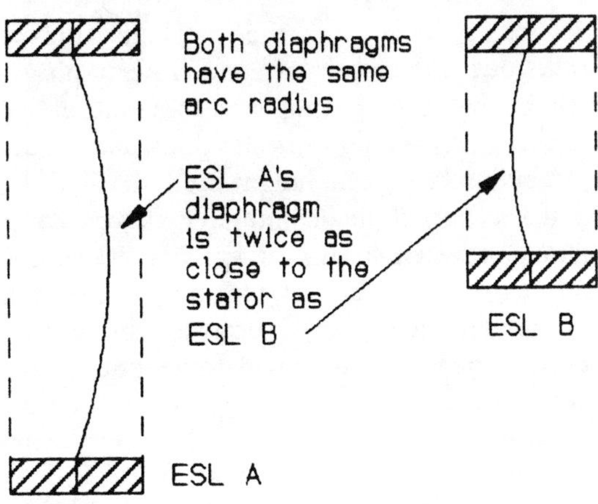

FIGURE 5-6: Effect of spacer distance on diaphragm stability.

The stators must be strong enough to resist the diaphragm tension. A large diaphragm produces a surprising amount of tension on the perimeter spacers. Most stators are adequately strong, so this isn't a problem, but you'll need to keep this factor in mind when designing stators.

SPACER RATIO. The diaphragm is separated from the stators by spacers. The distance from one spacer to the next determines the free (unsupported) diaphragm length (*Fig. 5-4*).

The spacers have a certain thickness which, along with glue bond thickness, defines the distance between the diaphragm and a stator. This distance is the D/S spacing (*Fig. 5-5*).

Spacer ratio is the term I use to describe the relationship between the distance between spacers and the D/S spacing. The spacer ratio has a large influence on diaphragm stability.

Diaphragm restoring force increases as the bow in the diaphragm increases. The diaphragm can bow more with low spacer ratios that increase the restoring force, thereby increasing stability. In other words, for a given D/S distance, putting spacers close together improves diaphragm stability and output (*Fig. 5-6*).

GUIDELINE #4:

Spacer Ratio

60:1 to 100:1

Close spacing raises the instability threshold, but adversely affects other things. Although close spacing increases output by raising the instability threshold, it *reduces* output by inhibiting diaphragm motion.

If a section of the diaphragm is clamped and glued between a pair of spacers, it cannot move to produce sound. Not only is the area of the trapped diaphragm between the spacers useless, but the diaphragm for about ¼" on each side of the spacer will have its motion restricted. The limitation varies from zero motion right at the edge of the spacer to free motion about ¼" from the spacer.

There is some confusion about the diaphragm's shape as the stator displaces it. Many believe the diaphragm moves in an arc. This is not so—at least at midrange and high frequencies. If you watch the diaphragm with a stroboscope, you will see that the diaphragm moves as an essentially flat plate with elastic edges.

The elastic edge width varies depending on excursion, but is generally about ¼-inch wide around the perimeter of the free diaphragm area. This pistonic motion is superimposed on the slight arc of the diaphragm caused by the polarizing voltage. The diaphragm has a complex shape, which I've simplified in *Fig. 5-7* for clarity.

If the spacers are ½-inch wide, and a ¼-inch wide strip on each side of the spacer is nonproductive, you have a 1-inch strip of useless speaker. To use an extreme example, if the spacer is a strip, and the strips are 2" apart, only about half the diaphragm can move freely to produce music.

Although this is undesirable, it isn't as bad as it sounds. Losing half the potential diaphragm area is an output loss of only 3dB. This is barely audible, subjectively. Furthermore, the instability threshold is tremendously increased, so you can use very high polarizing voltages. This will recover most of the output loss.

Proof of this is in narrow-strip tweeters. These may have a driven width of less than an inch, and have relatively large D/S spacing. The result is a very low spacer ratio—yet the output is perfectly adequate.

To summarize, the spacer ratio demands careful compromise. Low ratios enhance diaphragm stability at the expense of restricted diaphragm motion, reduced active diaphragm area, increased stray capacitance, and reduced output. Large ratios destabilize the diaphragm, reducing polarizing voltage and output. Extremes either way reduce output.

Over the years there has been much experimentation with spacer ratios, so I can provide you with excellent guidelines.

Most builders find 100:1 to be an excellent compromise, where both output and diaphragm stability are very high. Thus, for a D/S spacing of 70 mil (the thickness of 1/16" spacers plus two layers of adhesive), the distance between spacers should not exceed 7" (70 mil × 100). As little as 4.2" would still be excellent.

Larger D/S spacing requires a correspondingly greater distance between spacers. For example, 135-mil D/S spacing (1/8" spacers plus glue) would do well with spacers from about 8–13.5" apart.

For reasons that will soon become clear, you will usually choose the D/S spacing first, then decide the spacer distances using the spacer ratio. ESL tweeters are the exception. These often are built as narrow strips perhaps only an inch wide. Here, the distance between spacers is defined by the width of the cell, which you select first. You would then reduce the D/S spacing as much as possible to get close to the ideal spacer ratio. Probably, you can't reach the ideal ratio because of acoustic coupling. I'll discuss this shortly.

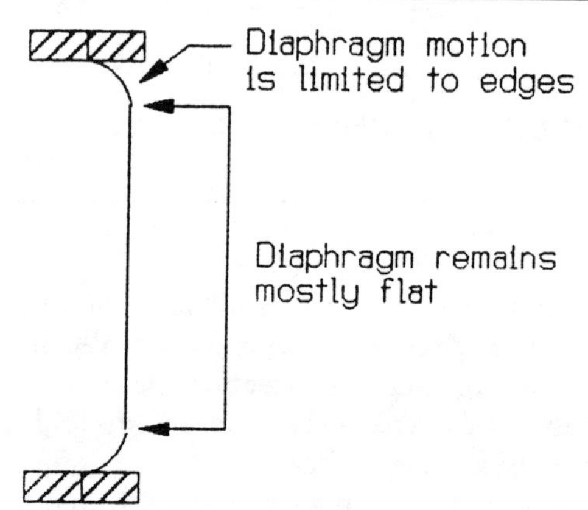

FIGURE 5-7: Diaphragm shape.

D/S SPACING. This is the second most important of the design parameters (the first being phase cancellation). Both excessively small and large spacing kill output. Unfortunately, you must choose a value based on many conflicting demands.

The thickness of the spacers *plus* the number and thickness of the glue layers defines the D/S spacing. Most ESL designs require two adhesive bonds, one between the diaphragm and spacer and the other between the stator and spacer. Depending on how you apply them, they may range from 2–20 mil in thickness.

The D/S spacing is so critical that these glue layers can make a significant difference in performance. Consider the effect the adhesive layers have on an ESL when using a common spacer thickness like 1/16-inch acrylic plastic ("Plexiglas"). This material is around 63-mil thick, and produces a D/S spacing of 67 mil if the adhesive layers are 2-mil thick.

The same spacers can form a D/S spacing of 100 mil, if you use very thick adhesive layers. This variable can adversely affect your design, if ignored. Also, it can be the cause of considerable construction nonuniformity—another problem to be discussed shortly.

D/S spacing places absolute limits on excursion, because the stators form a wall on each side of the diaphragm. The spacing must allow the diaphragm to move far enough to produce the SPLs required without striking the stators.

At the same time, the electrostatic force is weak and only travels short distances. Its force decreases by the square of the distance. Therefore, the stators *must* be close to the diaphragm for high output. This is nearly an insoluble problem, and requires careful compromise for adequate output.

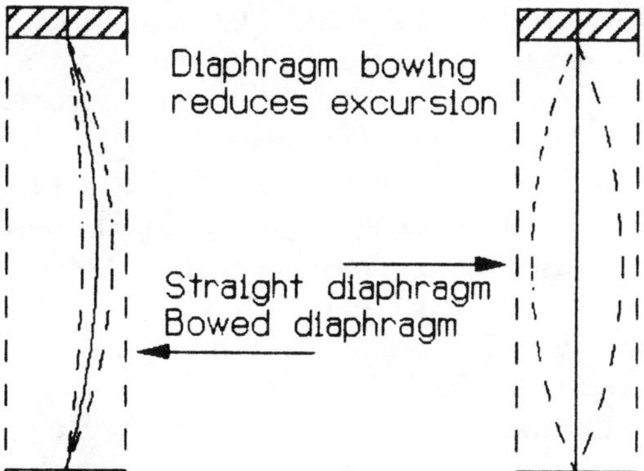

The energy spectrum of music is such that about 90% of the energy lies below 500Hz. The remaining 10% rolls off at 6–12dB/octave above that. This means tweeters don't have to produce much output. They may be physically small and have minimal excursion.

Woofers must have high output, so must make large excursions, particularly when considering the effects of phase cancellation. Fundamental resonance demands even more excursion.

DIAPHRAGM BOWING. The diaphragm can't use the total D/S spacing. As discussed previously, the polarizing voltage forces the diaphragm to bow slightly. This reduces excursion by the amount of diaphragm offset. *Figure 5-8* shows a highly exaggerated view of this. It also shows the diaphragm as if it were moving in an arc for clarity, although, as previously shown in *Fig. 5-7*, it has a pistonic motion.

ARCING. Large diaphragm excursions require thousands of volts on the stators. As the diaphragm moves closer to a stator, the air gap decreases, the voltage overcomes the air's insulating qualities, and the speaker arcs. In practice, this reduces the diaphragm's excursion, because it can't use the entire distance between stators.

ACOUSTIC COUPLING. There is a surprising and unexpected problem with bass and ESLs. I call it acoustic coupling. Large bass pressure waves produced by the woofer cause the midrange/tweeter ESL's diaphragm to move. These bass pressure waves can drive the diaphragm into the stators though the ESL itself is not producing the bass.

Here's why: the essentially massless ESL diaphragm has a large area, and is intimately coupled to the room's air mass. When the air mass moves, it carries the ESL diaphragm with it.

This can be a major problem if you want to produce concert hall output levels in a room with large bass resonances. I have driven ESLs into stators with powerful woofers, when the ESLs themselves were turned off! Acoustic coupling can be a serious problem in high output systems.

Acoustic coupling depends on room acoustics, the ESL's dimensions, the type of music to be reproduced, and your output

FIGURE 5-8: Diaphragm bowing reduces excursion.

requirements. It is easy to design an ESL to reproduce quiet classical guitar in a small apartment. It is challenging to design one to reproduce hard rock at ear shattering levels in a large square room.

Putting the speaker in a low-resonance room is a big help. The only other way to deal with this problem is to use enough D/S spacing to accommodate the effect, plus the excursion needed for the ESL's usual reproduction, plus that needed to accommodate diaphragm bowing.

For example, an ESL tweeter may produce adequate SPLs with a 10-mil D/S spacing. When mated with a powerful magnetic woofer (or an ESL woofer), acoustic coupling may cause the tweeter diaphragm to hit the stators at moderate SPLs. You probably will have to use 30-mil spacing to deal with acoustic coupling, although this will reduce the spacer ratio below optimum.

Large area midrange/tweeters (1' or more) are more prone to acoustic coupling than narrow strip tweeters (1" or so). Large diaphragms are more closely coupled to the air, and have larger unsupported diaphragm areas. This allows them to move more easily with the air mass.

From a practical standpoint, acoustic coupling usually determines the minimum D/S spacing. You should not base your spacing on the excursion calculated to reach a particular output, because acoustic coupling probably will be more important.

ESL woofers don't experience acoustic coupling, because they are the woofer—no other lower frequency driver causes pressure waves that affect them. Room and fundamental resonances will decide their excursion requirements.

So why use small D/S spacing anyway? Why not just use wide spacing? Remember that electrostatic forces are weak and only travel short distances. The advantage of small spacing is that a given set of voltages will have a more powerful effect on the diaphragm. This means that if all else is equal, an ESL with small D/S spacing will produce a much higher output than a similar ESL with large spacing.

This is a *big deal*, since the electrostatic force decreases by the square of the D/S spacing. A small change in spacing makes a large difference in output.

Small spacing makes the speaker more arc-resistant. Not only are lower voltages required for a given output, but if arcing does occur it is of much smaller magnitude, and is less likely to damage the speaker.

Many designers mistakenly suppose that small D/S spacing makes the speaker more likely to arc, because the parts are closer together. It doesn't work that way.

The voltages are reduced by the square of the distance. So when you reduce the D/S spacing, you reduce the voltages required for a given output. The result is an arc-resistant ESL.

ATTAINABLE DRIVE VOLTAGES. As the spacing increases, the voltages required quickly become unattainable or cause arcing. Your chosen output level requires that you square the voltage, if you double the D/S spacing.

This is discussed in greater detail later in this chapter and in the chapter on electronics. Suffice to say here that the limitation on drive voltage puts an absolute limit on D/S spacing.

SUMMARY. Acoustic coupling defines the minimum D/S spacing. The audio drive voltage and bandwidth limit the maximum D/S spacing.

Calculations for D/S spacing fail to optimize this critical parameter. Fortunately, practical experience with thousands of speakers has led to simple and reliable guidelines. ESL types and D/S spacing fall into three broad categories:

1. **Tweeters** are usually built as narrow strips about 1-inch wide but range between ½–3″. They may have other shapes such as circles, ellipses, or squares.

Many builders use baffles to delay phase cancellation and extend the low frequency response. Typical crossover frequencies are in the 4–6kHz region.

The usual D/S spacing is 40 mil, but may range downward to a minimum of 30 mil for very narrow designs. Although 20-mil spacing has been used successfully, it requires precision construction, and the tweeter becomes more susceptible to acoustic coupling.

Although the spacer ratio is low, it is still easy to obtain adequate output in 30–40-mil designs with minimal audio drive voltages. Smaller spacing is unnecessary and increases the risk of crashing the diaphragm due to acoustic coupling.

2. **Midrange/tweeters** are usually built as large panels up to 2-feet wide. Using equalization, they are crossed over at or slightly below 500Hz to magnetic woofers. They also are used in segmented systems as the midrange driver. They need medium to large excursions

to handle heavy acoustic coupling. Also, they must generate large amounts of equalized lower midrange energy.

In hybrid systems, they can produce truly impressive outputs—well above 100dB. If you need only moderate SPLs, you can operate this type of ESL full range, particularly if you suppress the fundamental resonance.

For this group of designs, 70–90-mil spacing is the norm. This is the best all-around compromise because you can get high outputs from wide-bandwidth, audio drive voltages. Arcing is not a problem and insulation is unnecessary.

3. **Woofers** often appear outwardly similar to the midrange designs described above. The key difference is the D/S spacing which ranges from 120–260 mil.

It is difficult to get enough audio drive voltage for these spacings. A 135-mil width is the maximum limit for high output with currently available audio drive systems. Even so, you will pay a penalty in frequency bandwidth. Also, you may need stator insulation.

Such large spacing allows enough excursion to handle acoustic coupling and 90+dB outputs—*if* you can find adequate audio drive voltages (more than 10kV).

You can use spacing greater than 140 mil to obtain higher outputs, but this will require very specialized audio drive systems. It is unlikely that you can get high-frequency response in a full-range system using this spacing. You must replace the air between the stators with a gas like sulfur hexafluoride that has better insulating qualities than air.

Large D/S spacing both requires and allows high polarizing voltages. With larger spacing also comes a higher instability threshold, because the greater distance between diaphragm and stator reduces the electrostatic force that destabilizes the diaphragm.

This reduction in force requires a higher polarizing voltage to compensate. Unfortunately, the output is not improved by the increased polarizing voltage, because the increased distance reduces its effectiveness.

D/S SPACING AND EFFICIENCY.

The D/S spacing has a powerful influence on output and apparent efficiency. You can look at efficiency, output, and D/S spacing in two ways:

- Pick a spacing that is wide enough to permit a given output level.
- Pick a spacing that will maximize the output from a given drive voltage.

The two are not the same. For example, you may want to produce 110dB output levels and find that it requires a D/S spacing of 130 mil and a drive voltage of 10kV. Your amplifier, when combined with the best step-up transformer you can find, will only deliver 5kV.

This combination will not meet your goal of 110dB. Realistically, it will only produce about 90dB. By reducing the D/S spacing, you can increase the *apparent* efficiency of the system and get more output. By decreasing the spacing to 70 mil, the output will climb substantially—probably to around 100dB. You still won't reach 110dB, but you'll come closer than if you used wider spacing.

The lesson here is to be realistic. If you don't have a monster amplifier, don't expect to get high output by using wide D/S spacings—in fact, it will lower it. You will get higher output by using *small* spacings with modest amplifiers.

You will pay a penalty if you wish later to upgrade to a more powerful amplifier. If you use a small D/S spacing to maximize output with a small amplifier, the speaker will arc or run out of excursion if you switch to a powerful amplifier later.

UNIFORMITY OF CONSTRUCTION.

The highest instability threshold, and therefore the highest output, is obtained when the D/S spacing is the same everywhere within a cell. In other words, your stators should be perfectly flat and at identical distances from each side of the diaphragm.

GUIDELINE #5:

D/S Spacing

Tweeters	30–40 mil
Midrange/tweeters	70–90 mil
Woofers	120–260 mil

While it is neither possible nor necessary to build perfect ESLs, you should avoid gross distortions in your stators such as warped or bent perforated metal or twisted and bent wire. These are obvious. A more subtle and very common problem results from nonuniform glue bonds.

I've seen glue bonds that range from 3–17 mil in the same cell! When you consider that there are usually two glue bonds in the D/S spacing, the nonuniformity could be double that, or about 30 mil total variation. That is almost half a 70-mil D/S spacing. If you add warped stators to variable glue bonds . . . well, let's just say your speaker is not likely to have high output.

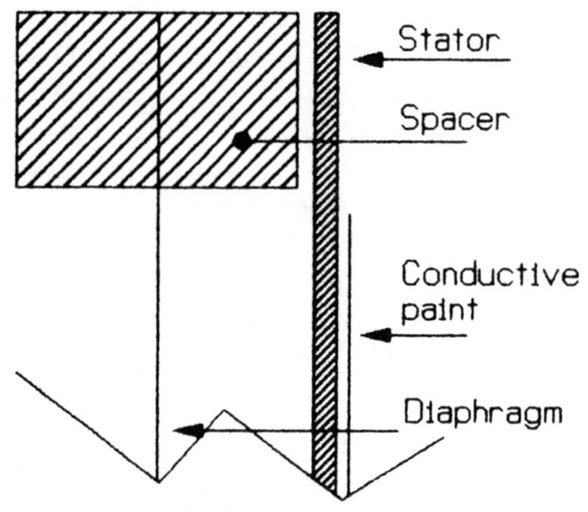

FIGURE 5-9: Construction for perfect insulation.

You make uniform glue bonds by using a thin *uniform* glue bead (1/16–1/8-inch wide), and press *hard* to squeeze the adhesive into a very thin film. Press every couple of inches, everywhere there is adhesive, to get a very thin film. You can't just slop down globs of glue, put some books on the assembly, and walk away expecting thin uniform glue bonds.

Using the "weakest link in the chain" analogy, arcing and diaphragm instability occur where the stator is closest to the diaphragm. If a section of stator is warped or bowed inward half the D/S spacing, then the whole stator might just as well be made with that spacing, because that's how it's going to behave.

Although the speaker's output will be adversely affected by poor construction uniformity, the *quality* of the sound will not be impaired, because the diaphragm behaves linearly anywhere between the two stators. The lesson here is that accurate, uniform construction produces maximum output.

STATOR INSULATION. Most designers insulate their stators presumably to prevent arcing. Although this has been taken as gospel since ESLs have been in use, it usually isn't necessary. Arcing will not damage an uninsulated cell if:

- It has a high-resistance diaphragm.
- The D/S spacing is less than 100 mil.

The higher voltages associated with larger D/S spacing pack more "punch." They may damage the diaphragm.

Low-resistance diaphragms (less than 1kΩ/inch) require perfect insulation. An arc will usually damage the diaphragm coating and may cause diaphragm ignition.

Insulation reduces output. It inhibits the effectiveness of voltages within the speaker. Its thickness must be added to the D/S spacing to maintain diaphragm excursion. Most stator insulation is ineffective—the speakers arc anyway. Common insulations include various clear paints and high voltage insulating varnish called Glyptol or "GLPT." I've never seen an ESL insulated this way that wouldn't arc.

Two insulating techniques, however, do work well. One is to make the stators out of perforated plastic and paint a conductive surface on the *outside* (*Fig. 5-9*). This separates the conductive surface from the diaphragm by a thick plastic insulator and is highly effective. This technique has three problems.

1. The insulator is very thick, often as thick as the D/S spacing. This drastically increases the D/S spacing for a given diaphragm excursion and reduces output significantly.

2. It is difficult to spray on the conductive coating without having some of it get inside the perforations. You bypass the insulation if the conductive material migrates to the inside of the stator.

3. Plastic stators require expensive custom perforation. Not only is this costly, it's difficult to arrange and time consuming.

Teflon® is a very good insulator. Its dielectric strength is exceptional. It doesn't require a very thick coating for Teflon to be effective. You can make tensioned wire stators with Teflon-coated wire that are very arc-resistant.

Teflon's disadvantages include cost, difficult construction, and limited stator designs. The only practical method of using it is in tensioned wire construction (to be discussed shortly).

Although the thickness of the Teflon adds to the D/S spacing, the increase is moderate because Teflon insulation is relatively thin. This type of stator works very well in large-D/S-spaced cells where the thickness of the insulation is a small percentage of the total.

STATOR DESIGN—FIELD DENSITY. Stator design has a very large influence on how effective a given set of voltages is in pro-

WARNING

Because of the risk of fire, speaker damage and high distortion—***I discourage the use of low-resistance diaphragms.***

ducing output. Several factors are determinative. They include:

- The ratio between stator thickness and hole/slot size
- The percentage of open area
- The ratio of hole/slot size to D/S spacing
- The cross-section shape of the holes/slot
- Shunt capacitance

Recall that electrostatic forces travel only short distances. If the stator has large perforations, the electrostatic charge in the open areas will be weak or nonexistent.

The **field density** is the average electrostatic force produced by the stator. It is the average of the open and conductive (closed) areas of the stator. High field densities produce the highest output. The stator design you select determines field density.

A look at the logical extremes of stator design can help us to better understand this concept. A solid conductor stator has the highest possible field density, but won't pass air to produce sound. A stator with no conductive areas has no electrostatic field to drive the diaphragm so it, too, will not produce sound. Either extreme produces zero output.

Another way of looking at field density is to envision a stator made of parallel wires. At a certain voltage, one wire exerts a specific force on the diaphragm. It should come as no surprise that for a given area, ten wires would exert twice as much force as five wires. In other words, for a given area and set of voltages, more wires closer together produce more output.

The same is true of perforated sheet stators. The ideal stator would have small holes or slots, so that the open areas are very close to conducting areas. This produces high field density, more force on the diaphragm, and high output.

HOLE SIZE LIMITS. There are limits to how small the holes or slots can be. If the stator is a solid metal sheet with round holes in it, the metal thickness defines the smallest hole size allowed. The air must see the perforations as "holes," not as "tunnels."

If the metal is thick and the holes are small, the openings appear to the air as tunnels. The air mass in the holes causes them to form Helmholtz resonators with distressing effects on sound quality. Also, it is difficult to perforate small holes in thick metal. Therefore, the diameter of the holes should be at least twice the thickness of the stator.

STATOR THICKNESS. The minimum practical thickness of the metal- or plastic-sheet stator is based on the distance between the supporting structures. If the distance between supports is too great, the sheet can sag or warp. The resulting lack of uniformity in D/S spacing reduces output.

The perforating process can cause warping stresses within sheet metal. You can't prevent the warping, but you can control it very well. You can use braces to hold the sheet flat. This is very easy to do by laminating the sheet to the insulating spacers that separate the diaphragm from the stator. Just lay the bow in the metal in the same direction as the center spacers. This way the spacers will support the metal and hold it flat. If you lay the bow across the spacers, it can sag between them.

Thinner sheets require placing the spacers closer together. At some point, the requirement to control warpage in thin stators can compromise the spacer ratio, although I have yet to see this in practice. This limitation on stator thickness can be circumvented by using external ribbing or bracing. It's much more work, but does allow extremely thin stators and small holes.

Stiff-wire stators have similar problems. They must be thick enough not to sag between spacers or be deflected by the electrostatic drive forces. Generally, wire is round and as such will not form a Helmholtz resonator because the slots formed between the wires will, in effect, have tapered walls (*Fig. 5-10*).

GUIDELINE #6:

Hole Size Limit

Perforations should be at least twice the stator thickness.

GUIDELINE #7:

Open Area Percentage
More than 40%

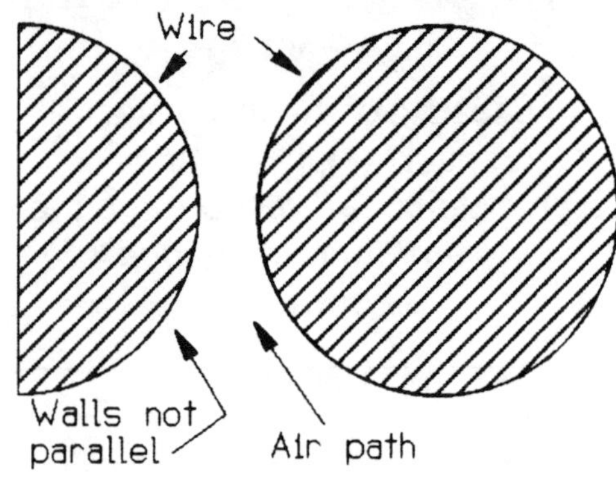

FIGURE 5-10: Nonparallel "holes formed by wire.

PERCENTAGE OF OPEN AREA. Holes in stators should be at least 40% of the total area to pass sound freely, but there is no reason why you can't use larger percentages. You can only use high percentages of open area if the conductors are close enough together to have high field density.

Although a higher percentage of open area is better (assuming a high field density), there is a point of diminishing returns. Eighty percent does not sound better than 50%, but construction difficulties abound.

About the only practical way to make high open area percentage stators is by using very small diameter wires under tension so you can keep them straight. For example, you might use tensioned 10-mil wires spaced 60 mil apart in a cell that uses 70-mil D/S spacing. It would work well, but be difficult to build, and the output increase compared to a well-designed simple cell would probably be inaudible.

STATOR DEFLECTION. If you use very thin metal or small diameter wires, the driving forces within the speaker may cause them to bend. Flexing will impair linearity and output.

Recall that electrostatic forces are very weak. The stator would have to be *very* flimsy for the electrostatic force to bend it. Also, the stiffness of the stator material is not the only thing resisting the electrostatic force. The stator's mass at audio frequencies has significant inertia and strongly resists deflection.

In theory, it is possible to build a stator that is so low in mass and stiffness that flexing is a problem, but I've never found it to be so, even when using very light perforated aluminum stators only 20-mil thick. In short, deflection is unlikely to be a practical problem, but keep it in mind if you decide to design something radical.

IDEAL PERFORATION SIZE. What is the ideal hole or slot size for maximum output? The opening should be no larger than the D/S spacing. For example, 135-mil spacing 1/8"

spacers plus adhesive) should have perforations no larger than 1/8".

This is a reasonable guideline, but leaving aside the issue of diminishing returns, there is no reason not to use even smaller stator openings. In other words, smaller is always better because the field density and output is higher, but at some point construction difficulties outweigh the performance benefits.

This rule also applies to wire stators. The center-to-center distance between the wires should be no more than the D/S spacing. Note that I said "center-to-center" distance, not "opening between the wires." Perforated metal and wire stators behave differently: wire is round while perforated metal has almost square edges (*Fig. 5-11*).

Ideally, we would like the D/S spacing to be the same everywhere. This is not possible if we are to have open areas. Because wire is round, it only matches the D/S spacing where the diaphragm is closest to the wire—the edge of the wire tangent to the diaphragm. The wire curvature effectively increases the D/S spacing.

Figure 5-12 shows 1/16-inch welding rod as the wire in a stator with 1/16-inch (70-mil) D/S spacing. I've overlaid the outline of 20-mil perforated metal so you can compare the two.

GUIDELINE #8:

Stator Opening Size
Perforations should be no larger than the D/S spacing.

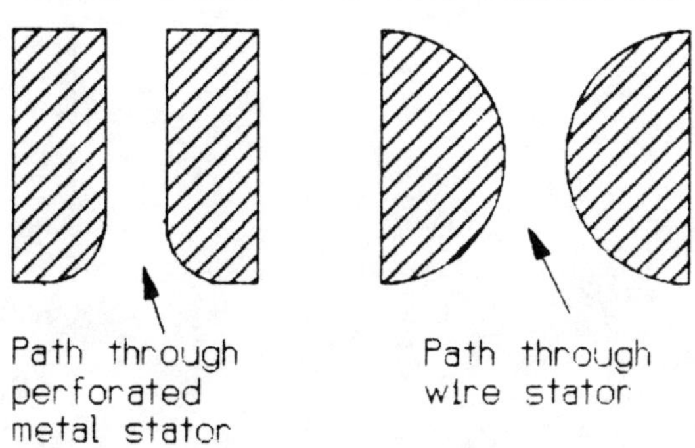

FIGURE 5-11: Air path differences between wire and perforated metal.

You can readily see that distance "C" from the edge of the perforated metal is closer to the desired D/S spacing "A" than is "B".

This is not a major problem, but it's worthwhile keeping it in mind when you make wire stators. You can minimize this effect by using small wire, just be sure you have at least 40% open area.

CORONA. Because of the effective increase in D/S spacing caused by round wire, you might think that perforated stators are better. This is not necessarily true. The somewhat sharp edge at the brink of a perforation tends to develop a corona—something that doesn't occur with round wire.

A corona is a cloud of ionized gases (air in this case) that tends to form around a sharp edge or point that is at high voltage. Once ionized, the gas becomes a conductor, not an insulator as air normally is. The result is that the speaker is prone to arc wherever a corona forms. The development of coronas forces you to reduce the polarizing voltage, which in turn reduces output.

The holes in perforated metal have sharp edges on one side and rounded ones on the other. It is very important to be *sure* the rounded ones face the diaphragm! Even so, in ESL designs that use very high voltages, you may find that wire stators are a better choice than perforated metal.

AUDIO DRIVE VOLTAGE. That the audio drive voltage affects output is hardly a surprise since it is the power amplifier that drives the speakers to produce sound. You need to understand the limitations and problems in obtaining high drive voltages. There is no point in designing and building an excellent

ESL only to discover that the necessary drive voltages are unattainable. I analyze audio drive systems in *Chapter 7*. At this point I'm only going to give an overview.

The following guidelines outline the voltages required for various D/S spacing. These are very close to the maximum voltages tolerable for a given D/S spacing. Higher voltages will cause arcing. If you can obtain higher voltages than specified for a given D/S spacing, increase the D/S spacing rather than trying to stop arcing by improving insulation.

AC voltages, such as audio signals, are usually measured as RMS voltages (*Fig. 5-13*). When discussing power amplifiers driving ESLs, it is more useful to talk in terms of peak voltages.

Peak-to-peak (P-P) voltages represent the absolute maximum voltages the system will see. These P-P voltages are what count when discussing arcing. It also makes design calculations easier. I use P-P voltages when discussing ESL audio drive voltages in this book.

Tweeters usually operate at around 1.2kV (I'm talking about audio drive voltages now, not polarizing voltages). They usually use a D/S spacing of 30–40 mil. The output requirements for tweeters are modest. They use small D/S dimensions and present a low capacitance load to the power amplifier. For these reasons, tweeters are easy to drive.

Midrange/tweeters can use a maximum

GUIDELINE #9:
High Field Density Stators

- Use thin stators with tiny openings
- Openings should be no larger than the D/S spacing.
- Maintain at least 40% open area.
- Make openings at least twice the stator's thickness.

GUIDELINE #10:
Audio Drive Voltage
100V/mil of D/S spacing.

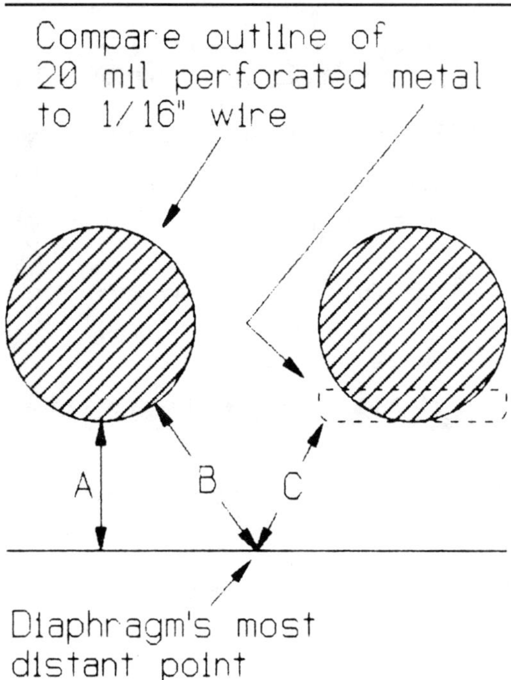

FIGURE 5-12: Effect of stator design on D/S spacing.

of 8kV when used with a 70-mil D/S spacing. It can be challenging to get such voltages if you also want wide bandwidth—and usually you do. Still, it *is* possible, and you can get very high SPLs this way.

Eight kilovolts is about all you can obtain with off-the-shelf parts. To exceed this voltage requires serious effort and money. The difficulty in getting higher drive voltages is the reason ESLs in general, and ESL woofers in particular, have trouble producing high output.

Woofers need large D/S spacing, commonly 135 mil, but some designs use as much as 250 mil. Such wide spacing demands extremely high voltages that are very hard to achieve.

We usually reverse the design process because of this. Instead of specifying a D/S spacing and then building a drive system to operate it, it is wiser to determine the maximum drive voltages that you can obtain, and at what frequency bandwidth. Then pick a D/S spacing that will maximize output. Woofer systems need all the voltage they can get—20+kV is not unreasonable.

There is a trade-off between voltage and bandwidth. High voltages and wide bandwidth are difficult to obtain simultaneously. From a practical standpoint, it usually isn't possible to get a full-range, crossoverless audio drive system for voltages much above 5kV.

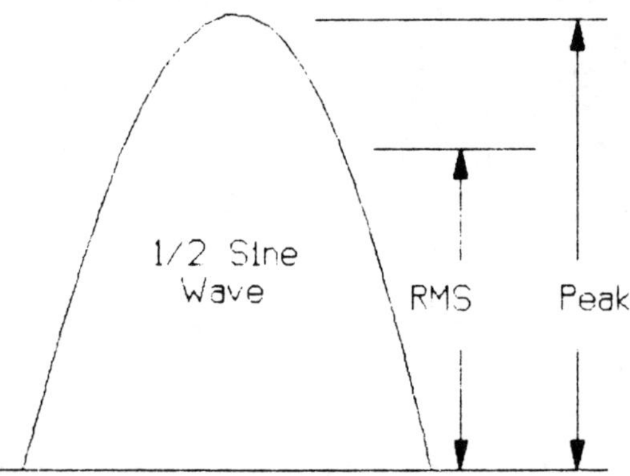

FIGURE 5-13: Different ways to measure AC voltage.

If you manage to get extremely high audio drive voltages, the speaker will be prone to arcing. You must insulate your stators to prevent this, while more modest voltages don't require stator insulation. In extremely high voltage systems, no amount of insulation will prevent arcing as the insulating quality of air simply is not good enough.

The Dayton-Wright ESL is an example using a D/S spacing of 250 mil and voltages above 20kV. Arcing was such a problem that D-W replaced the air between the stators with sulfur hexafluoride gas that has 6 times the insulating qualities of air. This solved the arcing problem. Although the gas replacement was an innovative solution to a major problem, the speaker couldn't produce high frequencies. The high audio voltages could only be generated at low frequencies.

No commercial power amplifier could produce the current required to drive the speakers at high frequencies. And no step-up transformer had a linear frequency response across the entire audio bandwidth at those voltages.

The speaker was massive for an ESL. Besides the heavy, sulfur hexafluoride gas, two additional diaphragms had to be driven. The diaphragms were passive covers that trapped the gas within the speakers. Since these covers were not driven, their added mass degraded high frequency response.

Because of these high frequency limitations, D-W was forced to use a piezoelectric tweeter and crossover network, which is no match for an ESL. The crossover network was in the critical midrange, resulting in flawed sonics.

One frustrated Dayton-Wright owner I know corrected the high frequency problems by having custom 1.5kW power amplifiers built. He equalized the input signal to compensate for the poor high frequency response of the ESL, and removed the piezo tweeters.

Although D-W speakers can produce high output in the midbass, they still cannot produce deep bass because of severe phase cancellation. Adding a 24″ woofer for the bottom octave, he reported good results with this system initially, but eventually the speakers failed.

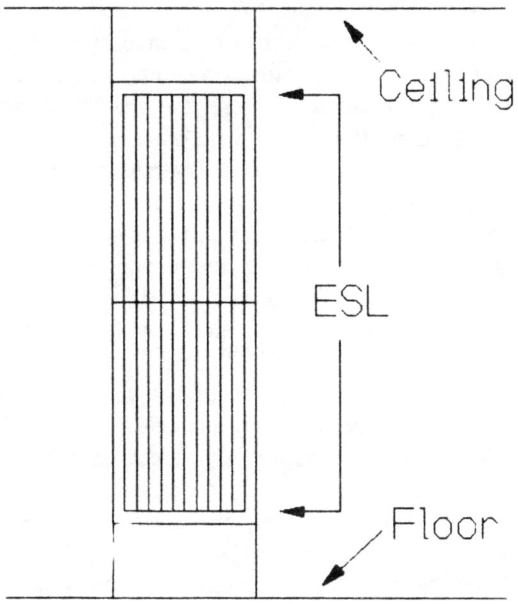

FIGURE 5-14: Modified line source.

DIMENSIONS. A large ESL has higher output than a small one, all else being equal. Not only does increasing the radiating area increase output, but it also reduces phase cancellation.

But not all else is equal. With larger cells comes greater capacitance and this increased load makes it more difficult to get adequate drive voltages and/or bandwidth. Reducing cell capacitance to solve this problem doesn't work because all factors that reduce capacitance also reduce output.

For example, increasing the D/S spacing reduces capacitance, but also reduces output for a given voltage. Increasing the drive voltage to compensate for increased D/S spacing costs you bandwidth, money, arcing, all three, or is impossible. Reducing the field density by using larger openings in the stator also reduces capacitance, but again at the expense of output.

When discussing dimensions, I am usually talking about width, because it is the *minimum* dimension of the ESL that counts. The mini-

mum dimension normally is its width because most ESLs are tall and narrow.

A tall narrow speaker is a **line source**. This is a popular shape because the speaker can be heard equally well whether you are standing or sitting. Imaging and detail are excellent because dipole line sources minimize room acoustics. More on this later.

Line sources have higher outputs than point sources of similar area because the sound radiation can expand in only one plane—horizontally (in a vertical line source). The vertical radiation is trapped between floor and ceiling and retains most of its energy.

With ESLs, the term "point source" is relative since the radiating area of even a small one is far larger than a tiny spot. Thus, a round or square ESL is *not* a line source—a tall skinny one is.

Picking the ideal ratio between height and width is not critical, but here are some things to consider:

- For a given area, output and imaging is better in a tall narrow line source.
- A line source minimizes floor space.
- Usually a line source is more aesthetically pleasing than a point source.
- You can hear a line source when both sitting and standing.
- For a given area, increasing width at the expense of height extends the low-frequency response by reducing phase cancellation.

The following observations may help you choose dimensions. If an ESL is to be heard in both sitting and standing positions, its minimum height is about 4′. This requires precise aiming, and if on the floor, it must be angled upward to "see" your head when you are standing. Otherwise, support it about 2′ off the floor. But this still is far short of the "perfect" line source that would stretch from floor to ceiling. A better compromise is to increase the height to 6′ the typical 8-foot high room.

Point sources are acceptable only if you sit when listening, and if you consider the other advantages of a line source of minimal importance. Doubtless ESLs can be made to operate adequately in both cases, but most listeners prefer line sources.

The width of the cell is at the heart of this decision. It must be wide enough to deal with phase cancellation, but not so wide that the power amplifier can't drive it. Also, width, in my view, is aesthetically undesirable.

Still, there is a way to drive ESLs of very

high capacitance: break the cell into more than one section and drive each section with its own power amplifier. Other than complexity and cost, the main difficulty here is that no two power amplifiers behave identically. By itself, an amplifier may sound flawless, but even slight differences in two amplifiers driving the same speaker may produce unpleasant effects.

A very practical compromise is to use a line source—but "cheat" on the extremes. Cut 1' off the top and bottom of an 8' line source. The resulting 6' speaker (in an 8-foot high room) acts like a floor-to-ceiling line source, but has 25% less driven area. The systems in this book use this trick and work very well (*Fig. 5-14*).

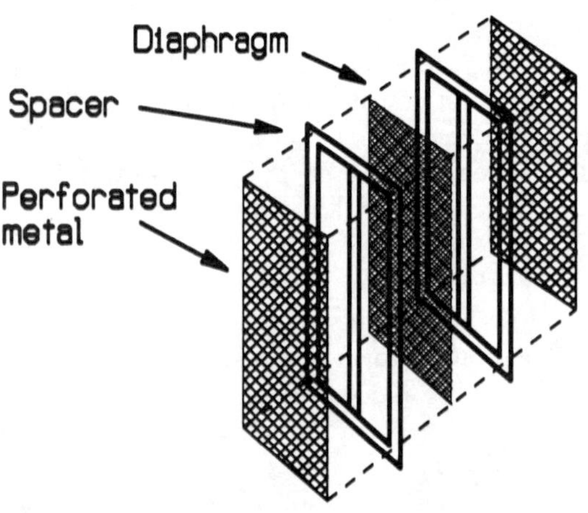

FIGURE 5-15: Simple ESL construction.

CROSSOVER FREQUENCY.

Hybrid and segmented speakers require crossover networks. The frequencies where the crossovers operate have a large influence on output. The higher the crossover frequency, the less the losses due to phase cancellation. The diaphragm excursion is less so you can use smaller D/S spacings. Smaller spacing improves output at a given drive voltage.

In short, the higher the crossover frequency, the higher the output. Unfortunately, high crossover frequencies exact penalties. These include more magnetic driver energy in the midrange, less detail in the sound, reduced image quality, poor frequency response, unpleasant phase effects, and a general deterioration of sound quality. More on this later.

CROSSOVER SLOPES.

Both the crossover frequency and its slope influence output. I discuss the advantages and disadvantages of various slopes in *Chapter 7*. Here, I only want to say that a speaker doesn't simply stop at its crossover point. It continues for several octaves beyond it.

SHUNT CAPACITANCE.

Shunt capacitance (also known as **stray** capacitance) increases the load on the power amplifier without contributing output. Shunt capacitance is any capacitance in the system that does not involve a moving part of the diaphragm. The simple rule is: *minimize it.*

Shunt capacitance has two causes: conductive stator overlying immobile diaphragm, and connecting wires. Where spacers trap the diaphragm, the diaphragm cannot move. If there is conductive stator over a spacer, the drive voltage sees it just like the rest of the speaker. It has capacitance—but the diaphragm can't move and produce output. This is the main cause of shunt capacitance.

Building an ESL with zero shunt capacitance is difficult (but possible). You cannot make simple stator designs without some shunt capacitance, typically about 10% of the total. Most builders trade this modest amount of shunt capacitance for building ease.

Simple stators usually are made of one large sheet of perforated metal with spacers glued to it (*Fig. 5-15*). This technique uses the spacers that produce the D/S spacing as the stator supporting structure. "Killing two birds with one stone" this way is a highly effective, economical, and easy way to build ESLs. Though these stators have considerable shunt capacitance, its effect can be minimized by keeping the spacers narrow.

How narrow? This depends on how narrow you can cut the spacers and how accurately you can assemble them. Narrow, stiff, and thin plastic strips are prone to shattering when cut or broken off the parent sheet. It is difficult to make them narrower than 3/8", although ¼" is possible.

The spacers on opposite stators must face each other in the completed cell to minimize shunt capacitance. It does no good to use narrow spacers then have them "miss" each other so you end up with a wide area of immobile diaphragm. Position spacers accurately so they match. Narrow spacers require precise positioning.

The spacers around the perimeter also contribute shunt capacitance. Reduce it by minimizing the amount of perforated metal or wire that overlays them.

A quarter inch of overlap is a practical figure. There is no reason for it to be more than ½". Less than ¼" may result in unreliable adhesion between metal and spacers.

You make zero shunt capacitance stators by not putting conductive stator over spacers. There are many ways of doing this—none of them as easy as the simple stator just described. The various techniques usually involve using some type of external ribbing to support the stator so it doesn't rely on the spacers for support.

Although perforated metal may be used to make zero shunt capacitance stators, most builders use wire for these designs. Perforated metal always is somewhat warped and needs more support than wire. There are compromise methods as well. It is not difficult to avoid placing stator over center spacers, but the edges are a problem. *Figure 5-16* shows a relatively easy construction technique where only the ends of a stiff-wire stator have stray capacitance.

Another means to build zero stray capacitance wire cells is to use a tensioned wire stator. In such a design, small wire such as magnet wire or Teflon-coated wire is wound back and forth across a frame to form the wire grid. The wires can be wrapped so that they don't overlap any center spacers. You then can mount the grid so that it doesn't overlap any perimeter spacers.

If the wire is thin and held at high tension, it will be straight. The beauty of tensioned wire stators is that they don't require external ribbing—*if* the perimeter frame is perfectly straight, *and* the wires are at high tension, *and* the unsupported length isn't excessive.

You can make external ribbing from stiff plastic or metal. Wood is not acceptable because it is not dimensionally stable. It is hygroscopic (absorbs water from the air) and may warp. Also, depending on your design, the ribs may need to be made of insulating material, and wood is a conductor at high voltages.

You may find it practical to buy fluorescent light diffusers and glue them to your stator. These are flat grids typically ½" thick with ½" square openings. These diffusers are flat, but they aren't very strong and they have more ribs than needed. They typically have a rib every ½", when one every 4" plenty.

A clever variation of the diffuser grid is to take something *really* flimsy—aluminum window screen—and glue it directly to the grid. This can form a high field density stator with zero shunt capacitance.

There are so many variations on stator designs that I can't describe them all in detail here. Let your imagination be your guide. If you find a better way, please call the ESL clearinghouse and describe your adventure.

CONNECTING WIRES. These wires between the ESL and the power amplifier can be a source of shunt capacitance. They should not be close together, but separated by ¼" or more.

What this means from a practical viewpoint is that the wires should be laid loosely on the floor rather than pulled tightly into a bundle. Under no circumstances should they be braided together.

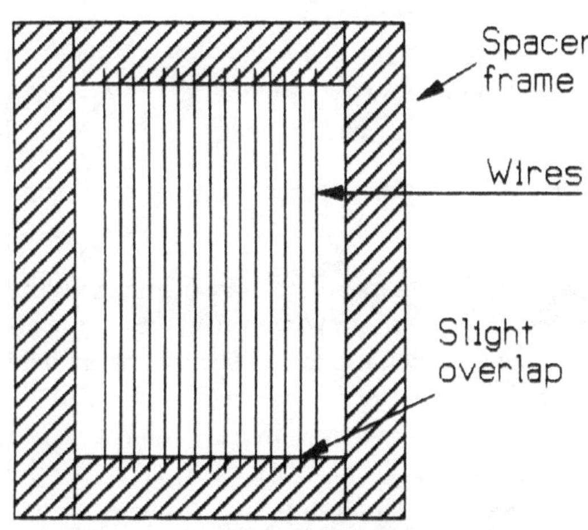

FIGURE 5-16: Stray capcitance in wire stator.

In summary, minimize shunt capacitance but keep a reasonable perspective. While everything involving output counts, shunt capacitance is a minor problem. It's OK to balance ease of construction against the relatively small negative effects of stray capacitance.

DIRECTIONALITY. Directionality is so important I have devoted the entire next chapter to it. I only mention it here to point out that if all else is equal, a highly directional speaker will have more output than one with a wide dispersion.

Of course, not all else is equal and the issue of directionality is hotly debated. Speakers can be built either way—just keep in mind the effects of directionality on output.

Why directional speakers seem to have higher output is not well understood. If you measure, in a room, two identical ESL reproduction systems which differ only in their dis-

WAYS TO MAXIMIZE OUTPUT

- Minimize shunt capacitance.
- Use uniform thin glue bonds to maintain D/S spacing.
- Make high field density stators.
- Be certain the diaphragm is conductive everywhere so there is output from the entire diaphragm.
- Maximize diaphragm tension.
- Build accurately.
- Maximize polarizing voltage.
- Use powerful amplifiers and high step-up ratio transformers to maximize the audio drive voltage.
- Use highly directional ESLs to confine energy to the first-arrival wavefront.
- Listen near your speakers.

persion characteristics, they will measure the same. But, when you listen to the two speakers, the directional one will seem substantially louder. This is not imaginary. It can be easily proved in a hybrid system that you must turn up the woofer level to match the ESL in the directional system compared to the wide dispersion one. This appears to be a psychoacoustic phenomenon.

I believe the most likely explanation for this is that the percentage of energy in the first-arrival wavefront from the two speakers differs. In a highly directional speaker, all the sound energy is radiated directly to you. In a wide dispersion one, most of the sound is directed toward the room walls. The sound bounces around and after a short delay, you hear it. Only a small percentage of the sound is radiated directly at the listening location as the first arrival wavefront. The rest arrives later because it has to travel further.

I think our brains only respond to the first-arrival wavefront when considering loudness. Since there is less energy in the first-arrival wave-front of a wide dispersion speaker, the brain perceives the sound as less loud than from a directional design. The SPL meter does not recognize this short time delay, so it correctly indicates that both speakers have equal output.

But make no mistake about it—directional speakers are subjectively much louder. How much louder? Since I can't measure the difference objectively, I can't give you a factual answer. Still, you deserve an estimate. I would hazard a guess of perhaps 6dB.

PROXIMITY. The closer you are to the speakers, the louder they sound. This is because less of the sound energy is lost to air friction, dispersion, and absorption by the room surfaces.

Unlike magnetic speakers, ESLs sound the same regardless of how close you are to them. This makes it possible to have different stage presentations based on the percentage of room acoustics present in the image.

If the speakers are distant, the performers sound as though they are in your room. If you are close to the speakers, the performers sound more like they are in the concert hall where the recording took place. In other words, a distant perspective is dominated by the listening room acoustics while room acoustics are absent in a close listening position.

Close listening locations have higher SPLs, improved detail resolution, more precise instrument location, and greater intimacy with the performance. The effect is like listening to good headphones but without their bizarre imaging. The only negative thing about close listening is that the close visual perspective tends to override what you hear and the sound may seem compressed into a smaller space. Close your eyes and the soundstage opens up. In time, this psychoacoustic phenomenon disappears.

I have set up ESLs only, 5′ from my listening location with 5′ separating them. The imaging and sound is remarkable. Close listening is a radical idea that is available only from ESLs. It deserves serious consideration.

DETAILS. Finally, pay attention to the details. Each individual factor may not seem to be much, but when you combine them, their sum can be impressive. This reminds me of the first rule of lightweight backpacking: *pay attention to the ounces and the pounds will take care of themselves.*

CHAPTER 6:
DISPERSION

Speakers have directional qualities. They may spread sound widely or radiate it in a tight beam. This sound directionality is called **dispersion**.

I've just described dispersion as though all frequencies behave in the same way. This usually is not so. Bass is always widely dispersed while the high frequencies may not be. Therefore, when I talk about speaker directionality characteristics, I'm really talking about high frequency dispersion.

Furthermore, I'm talking about **horizontal dispersion** Vertical dispersion usually is not an issue, although I will discuss it briefly later.

Pros and cons apply to both wide- and narrow-dispersion patterns. You wouldn't think so reading commercial speaker specifications. Most manufacturers don't even mention dispersion, and those who do only rave about the width of their dispersion. Can audiophiles be blamed for assuming that wide dispersion is "good?"

In my view, wide dispersion is "bad"—at least if you want what I consider to be the finest possible sound reproduction. The laws of physics strongly favor highly-directional speakers.

You can build ESLs with either wide- or narrow-dispersion patterns. You probably are unfamiliar with the advantages of narrow-dispersion speakers, so I want to spend a little time outlining their benefits. I'll explain the advantages and disadvantages of various-dispersion patterns and the reasons for this behavior.

Highly-directional (narrow-dispersion) speakers have better imaging, detail, and output than wide-dispersion ones. The penalty for this outstanding performance is that they sound best at one critical location—called the **focal point**, **sweet spot**, or **focus**. This location is equidistant from both speakers with the speakers' sound beams aimed directly at that spot (*Fig. 6-1*).

ROOM ACOUSTICS. Room acoustics degrade a speaker's sound. What is it about the room and the way it affects sound that causes problems?

Consider a wide-dispersion speaker. It sends a tiny percentage of its sound directly to you, while simultaneously bouncing the majority of its content off nearby walls and objects. The reflected sound (**room acoustics**) must travel a longer distance than does the direct sound so it reaches you after a *very slight* time delay (*Fig. 6-2*).

In home-sized rooms, this delay is too short for your brain to identify the "late" sound as an echo. The walls also selectively absorb some of the energy in the reflected sound, which alters its frequency response. Usually the high frequencies are reduced by room effects.

Keep in mind there is not just one late sound, but hundreds or thousands. Each comes from a slightly different location and has a different time delay and frequency response characteristic.

The flawed frequency response of the room acoustics damages the frequency response of the direct sound by mixing with it so you hear

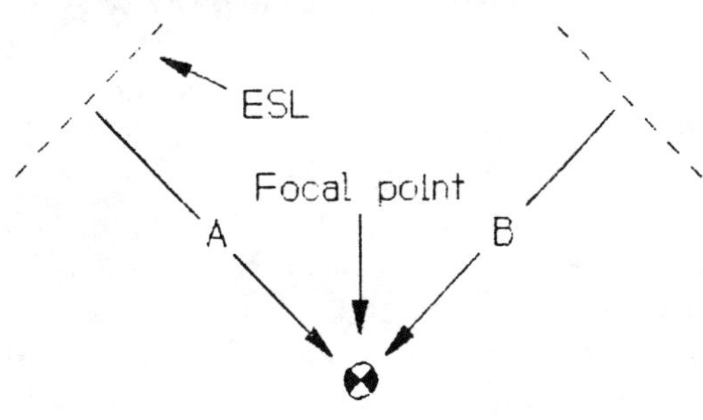

FIGURE 6-1: Proper speaker geometry.

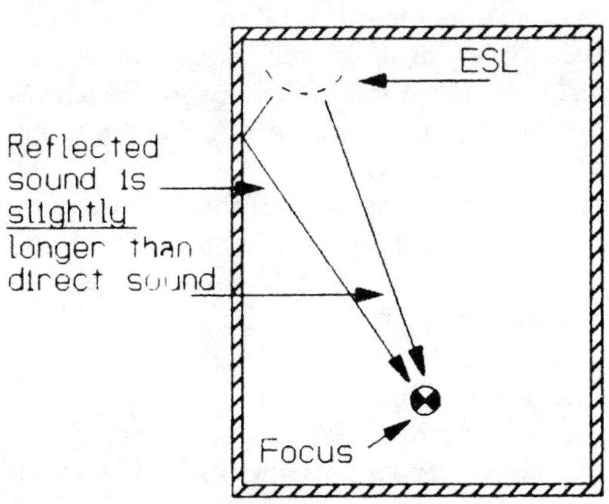

FIGURE 6-2: Room reflections delay sound.

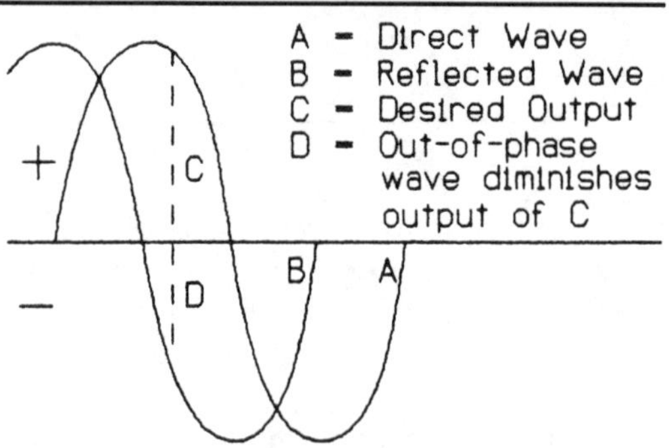

FIGURE 6-3: Delayed sound affects frequency response.

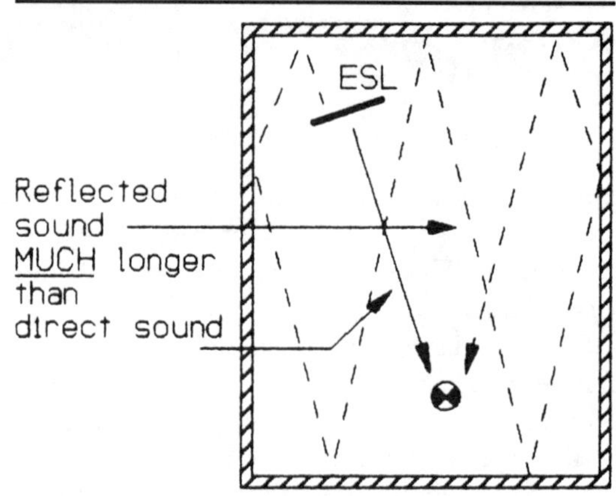

FIGURE 6-4: Long delay of narrow-dispersion speaker.

an *average* of the frequency responses. Also, the direct sound is adversely affected by attenuation and augmentation of individual frequencies caused by phase interactions with the delayed sound (*Fig. 6-3*).

For a speaker placed against a wall, the first reflections will be delayed by about 2mS, which hardly matters at low frequencies. The duration of each cycle of a 40Hz sound wave is 25mS, so at that frequency a reflection with a delay of 2mS is essentially in phase with the direct sound from the speaker. The direct sound and the reflection will reinforce each other, strengthening that frequency.

At 260Hz the duration of each cycle is only 4.2mS. A reflection of a 260Hz sound will arrive almost exactly a half cycle late. It will be negative when the direct sound is positive. The direct and reflected sound waves will cancel and reduce the sound level at that frequency. The audible effect of late reflections is an altered frequency response with peaks and valleys in it. This is highly undesirable.

Another problem with room acoustics is that they ruin the detail in the sound. To get crisp, sharply etched detail in the sound, you need a crisp, clean first-arrival wave front. The slightly delayed reflections in room acoustics "smear" this wave front, destroying the detail. This problem with slightly delayed room acoustics is a tremendous problem if you want clear, well-defined sound. Most audiophiles seem unaware of the problem—and their system's sound quality suffers as a result.

Remember your first headphone experience? They may have had faults, but surely you were impressed by their clarity and detail. Why do even poor headphones have so much detail? The reason is that there are no room acoustics in headphones. Therefore, they sound remarkably clear. Wouldn't it be won-

derful if we could do the same thing with speakers? Well, you can. The trick is to use highly-directional speakers.

Highly-directional speakers minimize room acoustics by sending their sound directly to you instead of elsewhere. They get it to you *before* exciting room acoustics.

Compare *Fig. 6-4* to *Fig. 6-2*. Note that the sound path for reflected sound is much longer in the highly-directional speaker. Therefore the delay time between the direct sound and the delayed sound is significantly greater.

Also the reflections are heavily attenuated because of their long paths and multiple reflections. Thus, after a relatively long delay, you hear very weak room acoustics.

Your brain can easily tell the difference between the direct sound and this faint, highly delayed sound. It hears the room acoustics of the narrow-dispersion speaker as faint echoes and ignores them. Because they are so weak, they don't affect the frequency response. The result is that highly-directional speakers have a much clearer and detailed sound than do wide-dispersion speakers.

As an aside, note carefully that your listening location can also have a strong influence on room acoustics. For best sound, your listening chair should not be against a wall! The reflected sound from the wall immediately behind you produces a severe, short delay room reflection that degrades the sound (*Fig. 6-5*).

IMAGING. Narrow-dispersion speakers project a more realistic and believable image than wide-dispersion ones. Their sonic images have often been compared to the visual images produced by laser holographs. They have great

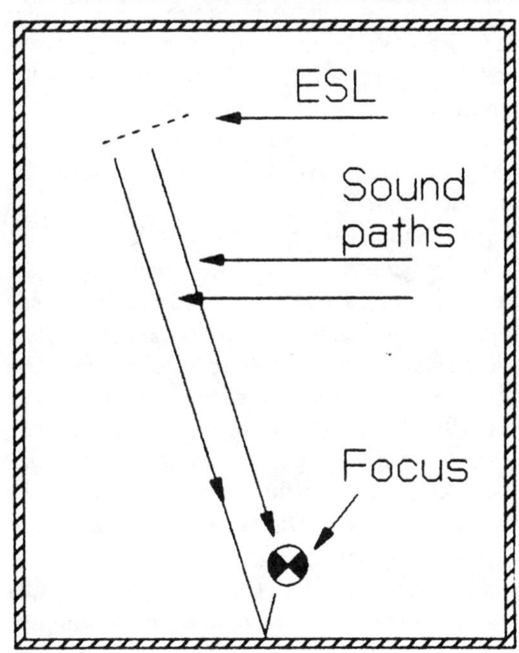

FIGURE 6-5: Adverse effect of listening near rear wall.

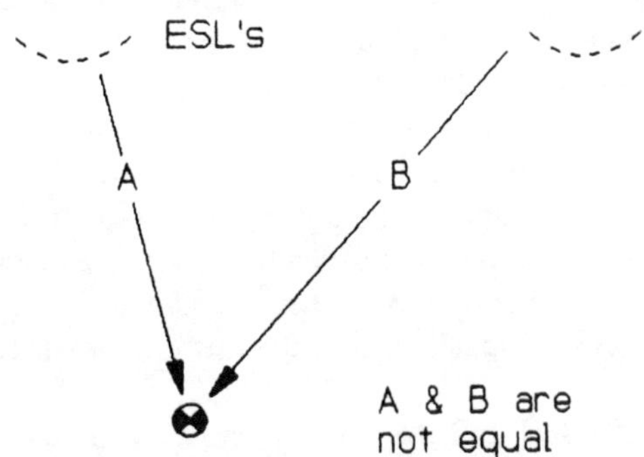

FIGURE 6-6: Different arrival times.

detail and three dimensionality (if the recording has been done in true stereo). This may sound like advertising hype—until you hear a highly-directional speaker.

Highly-directional speakers are phase coherent. This characteristic is difficult to describe in words, but you'll have no difficulty recognizing it when you hear it. The sound just seems more organized and intelligible.

The sound is more realistic than wide-dispersion types, because it does not seem to come from the speakers. The image floats in the room without seeming to involve the speakers.

In comparison, wide-dispersion speakers create an image which is diffuse and spatially ill-defined. The sound seems to come from the speakers. Your mind finds the image artificial and two (rather than three) dimensional. When further confused by all the room reflections, the image simply isn't precise, clear, realistic, and detailed.

As discussed in *Chapter 5* narrow-dispersion speakers have more output than wide-dispersion ones. Although the cause is psychoacoustic, the effect is real. Because ESLs have low output, this is an important consideration.

The supposed advantage of wide-dispersion speakers is they sound good everywhere. You are not confined to sitting in the sweet spot. This is false.

Even in wide-dispersion speakers, sound must arrive from both channels simultaneously for accurate stereo imaging. Although you can hear high frequencies from good wide-dispersion speakers anywhere in the room, they still sound best at the focal point—just as do highly-directional types.

If you listen to them elsewhere, phase and image coherency are further degraded, because the sound from each speaker arrives at different times (*Fig. 6-6*).

Wide-dispersion speakers get away with it, because their phase coherency and imaging is so compromised at the focal point that performance elsewhere isn't much worse. But make no mistake about it, it is worse. If you want the best sound from them, you must listen at the sweet spot.

The point I'm trying to make is that both wide- and narrow-dispersion speakers sound best at the focal point. If you want the best sound quality possible, you must listen there.

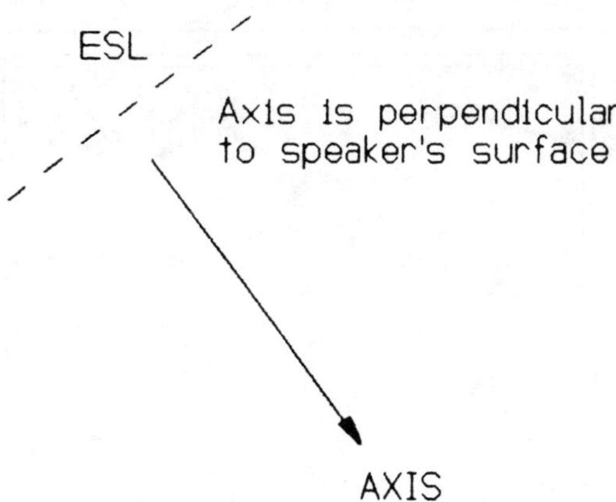

FIGURE 6-7: Axis of planar speaker is perpendicular to speaker's surface.

Since you must listen there with either type of speaker for best results, why sacrifice performance for wide dispersion?

I hope I've persuaded you to consider narrow-dispersion ESLs. I present wide- and narrow-dispersion designs in this book, and it doesn't matter to me which you choose. I just want to be sure you understand that highly-directional speakers have the best performance.

PLANAR ESLs. They exist on one plane (i.e., they are flat), and radiate sound perpendicularly to their diaphragm. This is called the axis of the speaker (*Fig. 6-7*).

The sound from a planar radiating surface will be tightly focused whenever the wavelength of sound is smaller than the minimum dimension of the speaker. Please review *Fig. 6-8*, the wavelength versus frequency graph.

A large planar ESL essentially has zero dispersion—it's like a laser beam. It does a superb job of imaging and minimizing room acoustics. You must experience it to appreciate it.

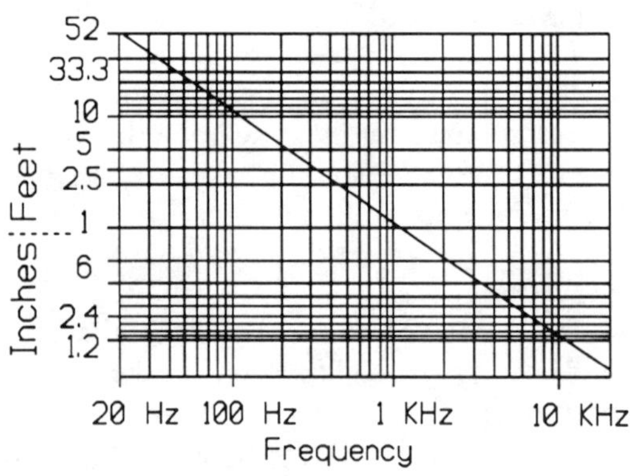

FIGURE 6-8: Wavelengths of various frequencies.

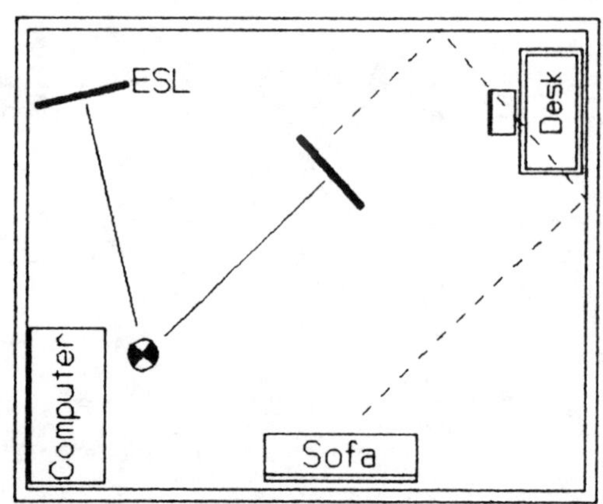

FIGURE 6-9: Good casual listening locations.

Many audiophiles are interested in planar ESLs because of their superior performance, but they worry about the sound for casual listening when they are not at the focal point. One of the most common questions asked about planar ESLs is, "What do they sound like outside focus?"

Although the sound at focus is very different from the sound outside focus, it is not objectionable. The best way I can describe it is to say it takes on a directionless and ethereal quality.

The sound is difficult to localize, because there is no direct sound to give you any clues. It is all reflected sound from many directions.

Also, because the high frequencies are partially absorbed by the room, the frequency balance changes. The sound is softer, more mellow, and distant.

The difference between the sound at focus and elsewhere is much like the difference between Row A and Row Z in a large concert hall. Like the concert hall experience, the sound can be enjoyed at both locations, however strong preferences abound.

Make no mistake, for serious listening you surely will prefer the sound at focus—just as you probably prefer close seats in a concert hall. For background music/casual listening, the sound outside focus is acceptable, much like listening in the back row of a concert hall.

You can reduce the difference between the two locations for casual listening. One way is to position yourself so you are in one speaker's beam. Although the stereo image is compromised, the high frequencies are not. For casual listening imaging is not important, and the high frequency beam makes the speaker sound more intimate than when you are not in any beam.

Since you probably will use your ESLs as dipole radiators, there will also be a beam from the rear of the ESL. If you have a hard surface behind the speaker, you may be able to direct the rear beam into your listening location.

Often it is very easy to position a desk or sofa or computer center where it can take advantage of one of the four beams available. *Figure 6-9* shows several possibilities. Note that all three locations are in at least one direct or reflected beam. This technique works very well.

Another way to deal with the outside focus sound is to use a rear-wave beam splitter, which I'll discuss shortly.

If your speakers are in a different room from where you casually listen, the differences between narrow- and wide-dispersion speakers disappear, because you are listening to the reflected sound from both types. For example, your speakers might be in the den or family room, but you like to relax and read the paper in the living room. Under these rather distant listening conditions, wide- and narrow-dispersion speakers sound the same.

My point is that a planar, laser-beam-directional ESL is acceptable outside focus. You undoubtedly will prefer the sound at focus for serious listening, but for casual listening you probably will find the sound outside focus acceptable. If not, beam splitters or careful positioning of furniture can usually resolve the issue to your satisfaction.

One problem with a planar ESL can't be resolved: you can't have a dozen friends over and all listen at once at the sweet spot. You can't have four people on the sofa in front of the speakers and have them all at focus. Of course, you can't do so with wide-dispersion speakers either, but the differences are less obvious.

If your listening chair is out in the room, you can easily accommodate two listeners at focus. One sits in the chair and the other stands just behind him. Since *serious* listening by more than two people is rare, this works well.

I often have someone over to hear my system. I give them the listening chair at focus, and I stand behind them.

If two audiophiles come together to listen, I have them share the listening chair and the standing position, while I stay out of the way. After all, I can hear the system anytime.

More than two listeners usually requires that they take turns in the focal point. It is awkward but possible for a third to sit on the floor in front of the focal point.

Serious listening sessions are usually short. In my experience, the standing/sitting routine has not been a problem.

WIDE-DISPERSION ESLs. You can use many techniques to achieve dispersion in an ESL, but you must have a curved wave front. A pulsating cylinder is a good example (*Fig. 6-10*).

MULTIPLE-PANEL CYLINDRICAL ESLs. In an attempt to duplicate a pulsating cylinder, designers build small panels and place them around the surface of an imaginary cylinder. These designs can vary from a shallow arc to a complete cylinder (*Fig. 6-11*).

The resulting speaker produces several discrete beams going in many directions. Unfortunately, the beams don't blend and

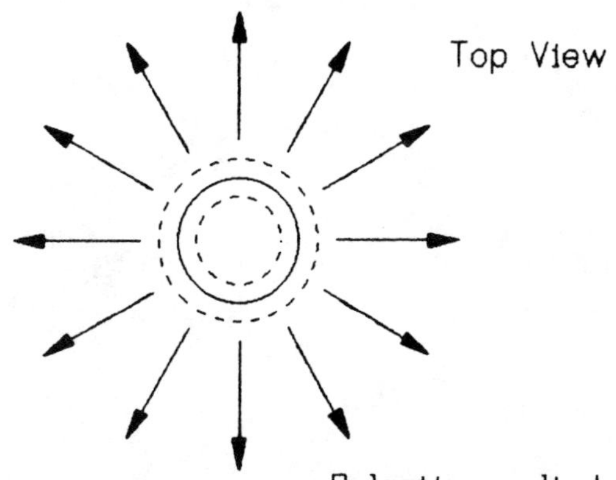

FIGURE 6-10: Wide-dispersion, cylindrical speaker.

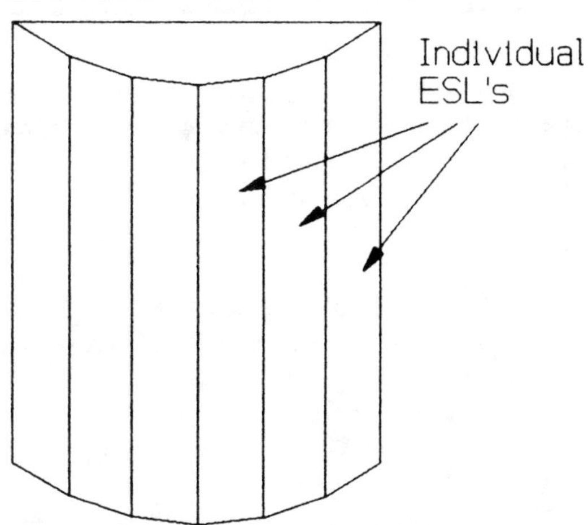

FIGURE 6-11: Wide-dispersion using many angled ESLs.

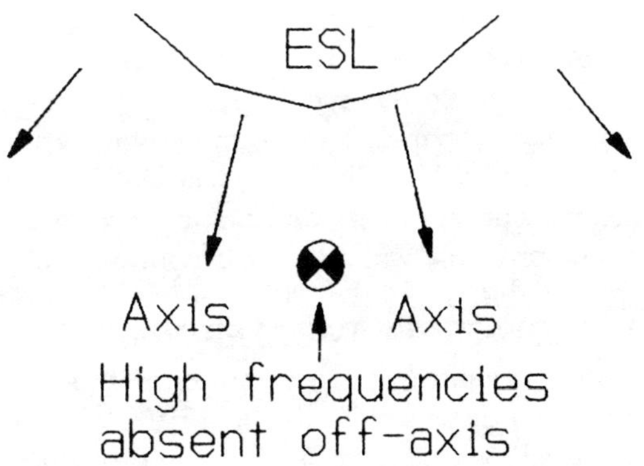

FIGURE 6-12: Cause of venetian blind effect.

smoothly disperse the sound. Remember, planar ESLs are extremely directional. Each panel acts as a separate driver (*Fig. 6-12*).

The sonic effects of these multiple beams are disturbing. The flaw is so characteristic it has a special name: the **venetian blind effect**.

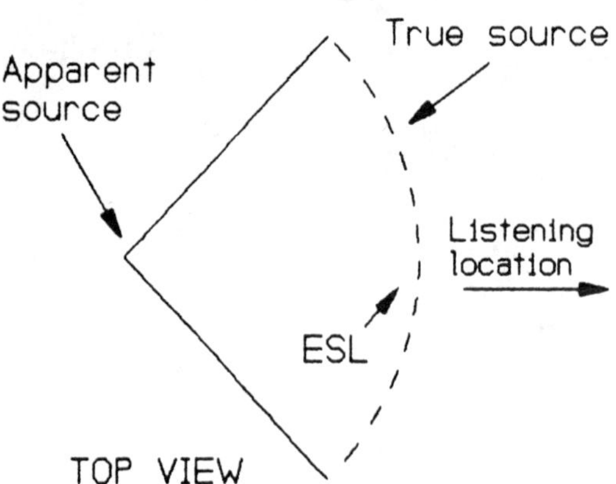

FIGURE 6-13: Wide dispersion causes apparant source to recede.

When you move around in front of such a speaker, the high frequencies come and go as though the sound is coming through a series of slots. This produces image-phase anomalies which vary depending upon location. The image is seriously compromised.

This highly undesirable effect can be minimized, but not completely cured. Reducing the angle between the cells to less than 4° helps. Wide dispersion requires many panels. To keep the speaker width reasonable, the cells must be very narrow.

This type of speaker moves the apparent-sound image away from you, which may or may not be a problem, depending on your preferred listening location and the degree of curvature. Shallow curves move the image further back than steep ones (*Fig. 6-13*).

Besides flawed sonics, there are many other disadvantages to this type of speaker. Making many narrow panels is much more work than making one large cell. There will likely be a high percentage of stray capacitance. The spacer ratio probably will be less than optimum.

NARROW-STRIP TWEETERS. In a full-range, segmented ESL, a common method of improving dispersion is to use a narrow-strip tweeter. Regrettably, this doesn't improve the dispersion enough to be satisfactory.

Recall that the sound from a planar-radiating surface will be tightly focused whenever the wavelength of sound is smaller than the minimum dimension of the speaker. The wavelength of a 20kHz wave is about ½". The driver must be smaller than that to produce dispersion. You can think of the longer wavelength as "spilling" over the edge of a narrow-planar ESL and forming the necessary curved wave front.

This is not an all-or-none phenomenon. As the frequency falls, the dispersion gradually increases. The practical implication of this is that a planar driver must be smaller than the wavelength of sound it produces to generate significant dispersion. A narrow-strip ESL tweeter would have to be narrower than ½" to adequately disperse a 20kHz tone.

The edge of an ESL diaphragm is clamped between spacers around its periphery. The spacers will be *at least* ½-inch wide, so even if you had a ¼-inch diaphragm, the width of the cell would be 1¼"—and this assumes that you don't have a mounting frame around the cell.

Also, the tweeter edges must be free of baffles, other ESL panels, or other obstructions in order to produce wide dispersion. If the sides of the strip tweeter have extensions, the air cannot "spill over" and form a curved wave front.

This is a serious dilemma, because these tweeters have phase cancellation problems and need baffles to extend their frequency response into the midrange. They are often flanked by midrange ESLs as a convenient way to mount, drive, and baffle them. The resulting narrow-strip tweeters usually have disappointing dispersion and fail to meet their designer's goals.

Magnetic tweeters have the same problem. They solve it by using a small dome which forces the wave into a curved shape.

ELECTRICAL DELAY LINES. Dispersion is produced by using a series of narrow-strip cells. The audio drive to each of these is delayed slightly more than the previous one.

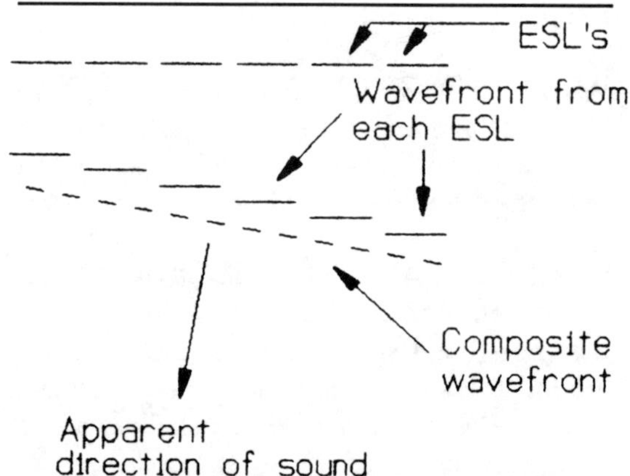

FIGURE 6-14: Time delay between cells produces angled wavefront.

If the delay between each segment is the same, the delay produces an angled wave front. This gives the impression that the sound is coming from the direction defined by the angle (*Fig. 6-14*).

By progressively increasing the amount of delay from one segment to the next, a curved rather than flat (but angled) wave front is formed (*Fig. 6-15*).

The delay is usually obtained by using an inductor in series with the stators of each cell. If you drive the ESL at its center with the delay propagating outwardly in both directions, a more or less curved wave front will be produced (*Fig. 6-16*).

QUAD Model 63 uses a clever variation, in which the various panels are concentric and the delay propagation spreads outwardly from the center of the speaker. This approximates a pulsating sphere, which some believe is the ideal speaker form.[1]

The disadvantages of delay lines are many. The obvious complexity of design and construction is one. The delay line does not produce a smooth dispersion pattern since it is executed in a series of steps. The inductors must handle high voltages. The load presented to the amplifier is more complex than a simple capacitor. Few use this technique.

CURVED ESLs. If you build the ESL in the shape of an arc, it will disperse the sound over the area "seen" by the arc as being perpendicular to the surface of the ESL (*Fig. 6-17*).

The dispersion from this type of ESL is smooth and free from venetian blind effect. As you move about the room, the sound quality is unchanging except for balance and imaging

1. Williamson, Reg., "The Quad 63, " *SB* 1/82, pp. 10–18.

due to unequal-length sound paths to each speaker. This technique is simple and you can use it with large cells, so other design parameters are not affected.

The idea of a curved ESL is not new, but the apparent impossibility of holding a free-tensioned diaphragm in the shape of a curve prevented its use until recently. The problem with forming ESLs into an arc is the diaphragm collapses into the inner stator, because diaphragm tension tends to flatten it like a drumhead.

Most curved designs use supporting structures to prevent collapse. These seriously impair diaphragm motion and other design parameters, such as the spacer ratio and D/S spacing. The results are inadequate output and poor frequency response.

In 1979 I solved this problem with a very simple design which prevented diaphragm collapse by selective stretching. It was first published in *Speaker Builder*, 2/1980.

Diaphragms collapse, because we normally stretch them in *all* directions. If we stretch one in only one direction where the tension holds it in a straight line only, it will not collapse.

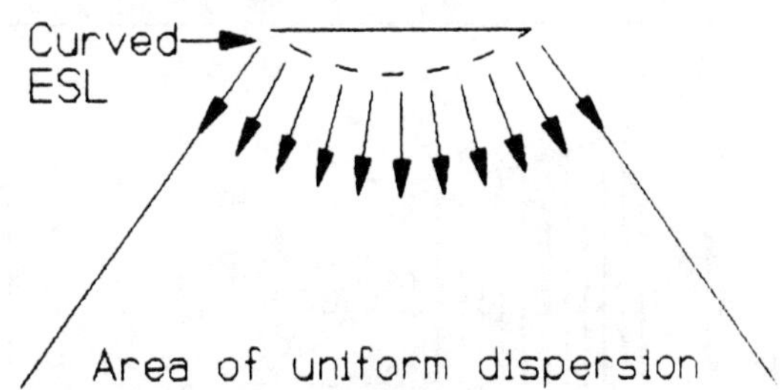

FIGURE 6-17: Wide dispersion produced by ESL.

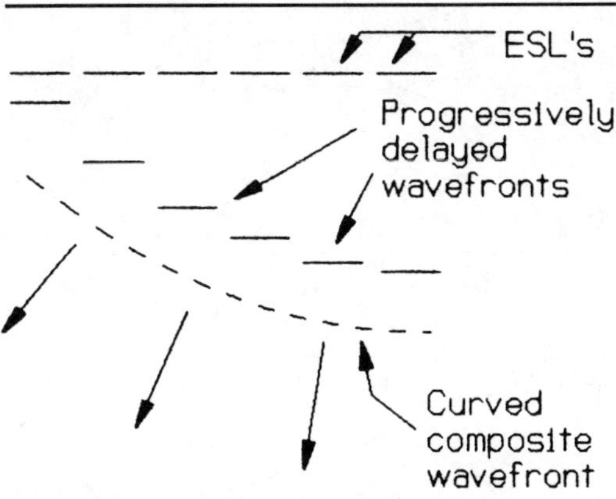

FIGURE 6-15: Apparent-curved wavefront.

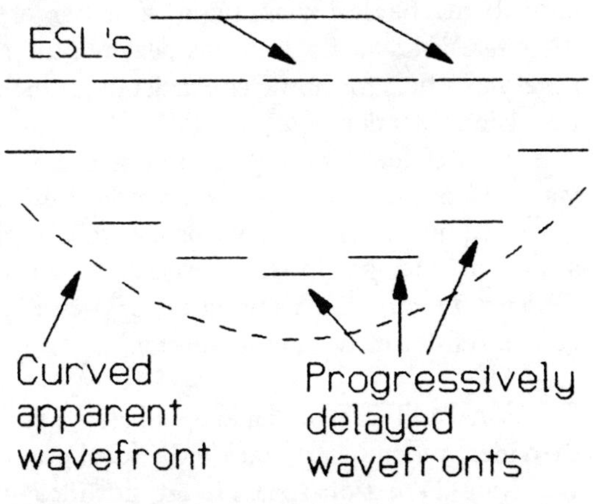

FIGURE 6-16: Curved wavefront.

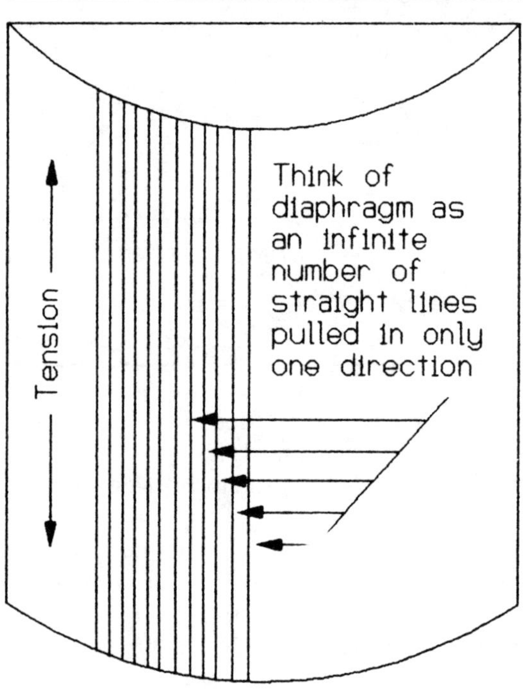

FIGURE 6-18: Curved diaphragm production.

You can better visualize this if you think of the diaphragm as an infinite number of straight lines lying side by side. If these are stretched only in the direction of the cylinder axis which forms the curved ESL, the lines will stay straight even though they may be placed in an arc formed by the cylinder circumference (*Fig. 6-18*).

You need no special support structures to build these. The technique is the same as with a planar—except it must be built upon a curved surface.

The required special construction parts like curved tables or special jigs, and possibly a curved, mechanical diaphragm stretcher are the main disadvantages to this design. Still, it is very practical for home construction, which I explain in detail in *Chapter 10*.

Barry McClune developed some very innovative ideas for building curved cells which don't require a curved table or a mechanical stretcher. He graciously agreed to write *Chapter 11* on his techniques of building curved cells, which you may prefer.

BEAM SPLITTERS. Planar speakers are less than ideal when heard outside the focal point, and wide-dispersion speakers are not ideal—period. What we need is a speaker which has perfect imaging for serious listening and wide dispersion for casual listening—and its wide dispersion must not harm the image at the focal point.

A planar dipole sounds poorly off-axis mainly because the highs are lost. As long as you are in the main- or rear-reflected beam, the sound is acceptable, but these beams only cover very narrow areas. The rear beam from a dipole radiator does not harm the image, because it has long delay times and is mostly decayed by the time it reaches the focal point.

The perfect speaker would have a multitude of rear beams going in many directions, so you would be in a beam no matter where you were in the room. We need only be careful to avoid directing any of the rear-reflected beams into the focal point, since that would smear the sound and reduce detail.

The rear-wave beam splitter spreads the rear radiation from a dipole across a wide area. It also reduces the sound energy in the beam, which reduces its ability to degrade the focal point.

Beam splitters take the form of some type of angled or curved surface behind the speaker which intercepts the rear wave and redirects it, the way a bumper on a pool table redirects the ball's travel. *Figure 6-19* shows various beam splitters.

Remember that great care must be taken to avoid directing any of the rear wave into the focal point. For this reason, curved surfaces should have a gap in them where they are parallel or nearly parallel to the diaphragm. Straight or stepped beam splitters should not have a flat surface parallel to the diaphragm. Rather, they should have a point or corner facing the diaphragm (*Fig. 6-20*).

Since planar speakers always face inward

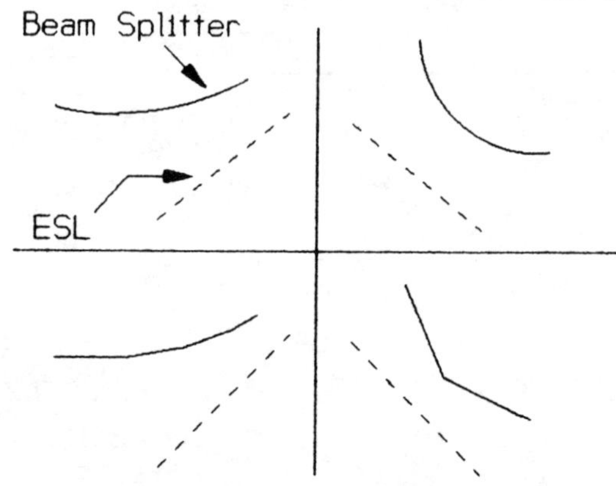

FIGURE 6-19: Various beam splitters.

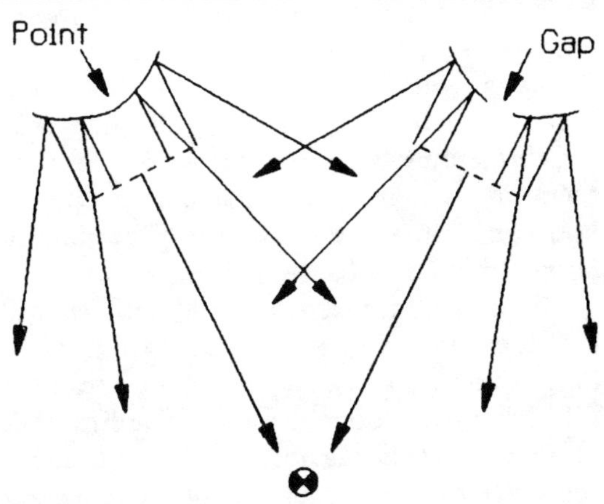

FIGURE 6-20: Avoid directing rear wave toward focus.

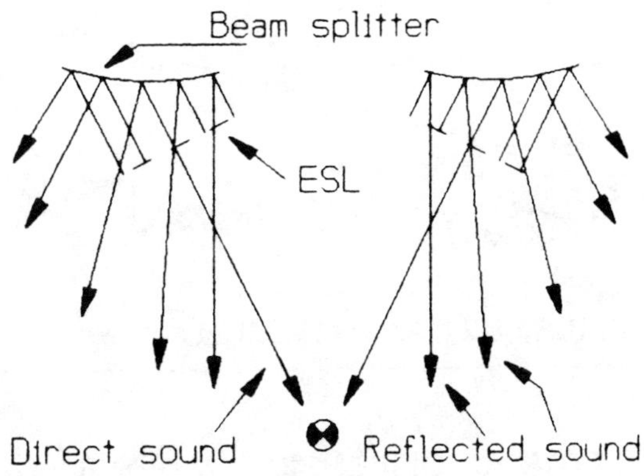

FIGURE 6-21: Ideal beam splitter pattern.

toward focus, beam splitters should be asymmetrical. Additionally, there is usually no need to listen to the sound in the triangle between the speakers and the focal point. The beam splitter can be simplified by directing the sound away from focus towards the outside of the speakers (*Fig. 6-21*).

Beam splitters can take any form which can redirect the rear beam. They needn't be strong or difficult to build, but must have a hard, highly-reflective surface to bounce the wave back into the room. If they are soft, they will absorb the high frequencies we are trying to redirect.

They can be made of plastic, fiberglass, or any other hard, thin material such as Masonite® or Formica®. Even cardboard works well. The beam splitter may be placed at any point behind the speaker where it can intersect the rear beam.

Generally, beam splitters are placed between the wall and the speaker, but if the wall is hard and highly reflective, the beam splitter can intersect the beam after it leaves the wall (*Fig. 6-22*).

Beam splitters may also be integrated into the basic speaker design rather than as a separate "add-on." The integrated ESL/TL designs described later use a combination of both techniques to achieve beam splitting without the use of a separate beam splitter panel. It does so by using a vertical column at an angle to the ESL. This column is the woofer's transmission line (*Fig. 6-23*).

The main problem with effective beam splitters is they are large and visually obtrusive. They may take considerable work to build, if you want them to be aesthetically appealing. They also take up floor space.

Rear-wave beam splitters are an idea I developed in 1989 and first published in *Speaker Builder* 1-3/1990. It is a promising concept which works well, and a fertile area for more study and experimentation.

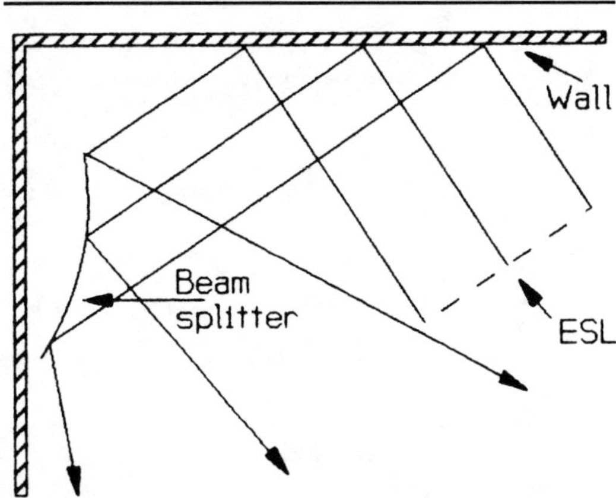

FIGURE 6-22: Using wall with beam splitter.

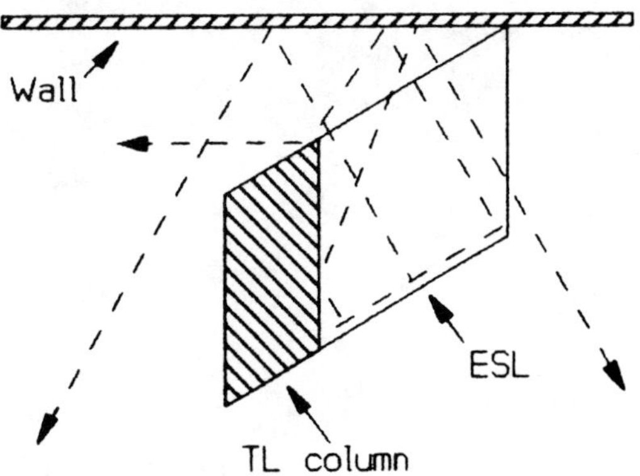

FIGURE 6-23: Using speaker enclosure as beam splitter.

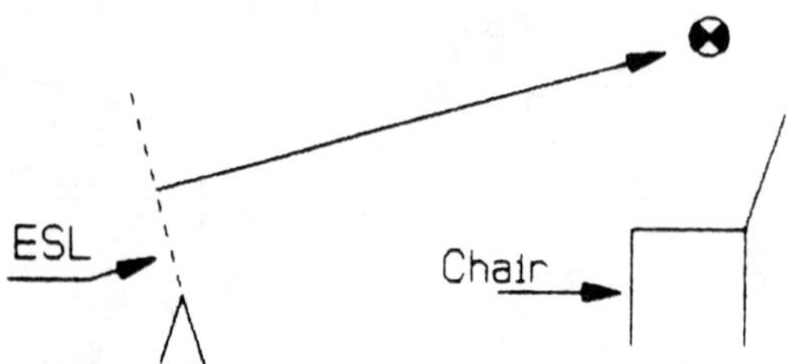

FIGURE 6-24: Aim small ESL at ear level.

VERTICAL DISPERSION. Vertical dispersion is less of an issue than horizontal dispersion, because we usually only need to hear our speakers when sitting down. Of course, it is nice to hear them when both sitting and standing, but this is neither essential nor a critical issue in most audiophiles' minds.

Still, excessive vertical dispersion generates room acoustics just as horizontal dispersion does. This should be avoided for optimum performance.

Excess vertical dispersion is usually not a problem in ESLs, because even curved ones are highly directional by nature. In magnetic speakers, particularly those with small dome tweeters, it is a significant cause of image smear.

The problem in ESLs is more likely to be a lack of vertical dispersion. If you build cells which are only 2- or 3-feet tall, you can hear them only when sitting (or standing), but not both. You can aim them to either location by tipping them back as needed (*Fig. 6-24*).

The minimum height of an ESL which will give adequate coverage in the sitting and standing position is about 4'. If you put the bottom of it 2–3' from the floor, it can be heard in both positions. Usually it is not difficult to put such a cell on long legs or use a magnetic-woofer enclosure to elevate it to the necessary height.

A tall line source is a better solution. If you build a line source which goes almost completely from floor to ceiling, you can hear it when standing or sitting. Such a line source will be extremely directional in the vertical position (because the dimension will be much greater than the wavelengths it's reproducing); it will not excite room acoustics. It also will maximize output. This is one of those rare occasions when you can "have your cake and eat it too." Murphy's Law finally gives us a break.

CHAPTER 7:
ASSOCIATED ELECTRONICS

Many speaker builders become so involved with their speakers they give the associated electronics only passing thought. The electronics not only are equally important, but usually cost more than the speakers.

You don't need to know much of this chapter to build ESLs successfully. It contains mostly technical information for the more advanced builder/designer. If this doesn't interest you now, just skim it so you will know where to go if you have a question later. The next chapter on electronics construction gives the critical "need-to-know" information.

If you are an electrical engineer with a test bench full of instrumentation, you should already have this knowledge. I designed this chapter for the interested amateur who wants to move beyond the "how-to" stage, but is not an engineer. In the discussions which follow, the large, hybrid-system ESL described elsewhere will be the subject of many examples. Note its specifications.

Before examining the electronics, you need to understand the ESL's electrical properties. These involve voltage, current, capacitance, and reactance. I discuss capacitance and reactance here, and voltage and current later under power amplifiers.

CAPACITANCE. The ESL's capacitance is the electrical load the power amplifier must drive. Sometimes it is useful to know the capacitance, although it usually isn't necessary if you follow the guidelines in this book.

Below is a mathematical equation commonly used to find the ESL capacitance based on D/S spacing and area.

$$C = \frac{8.85 \times 10^{-6} \times A}{D}$$

Where:

C = Capacitance in μF
A = Stator area in square meters
D = D/S spacing in meters

Unfortunately, these formulae are usually inaccurate, not because the math is wrong but because you don't know all the values.

For example, what is the exact D/S spacing? You can't simply say that the spacer thickness is the D/S spacing. Other factors are involved which can change the spacing.

SPECIFICATIONS:
Large hybrid ESL

D/S Spacing: 70 mil

Frequency response: 400 Hz–32kHz

Overall dimensions: 1.8′ × 6′

Radiating area: 9 ft^2

Polarizing voltage: 3.5kV

Audio drive voltage: 7kV

Capacitance: 2,200pF

Spacer ratio: 80:1

For example, the spacer material manufacturer may say it is 1/16-inch thick. But production tolerances may allow it to be anywhere from 55–70 mil. Also, there may be variations in the spacers. They could be 60 mil at one end and 65 at the other.

What about the adhesive bond thickness? How thick are the glue bonds holding things together? Is the thickness uniform throughout? Not likely.

Distortion and warpage in the stator material is another problem. Perforated-metal sheets are not perfectly flat, wire is not perfectly straight, and construction techniques are not flawless.

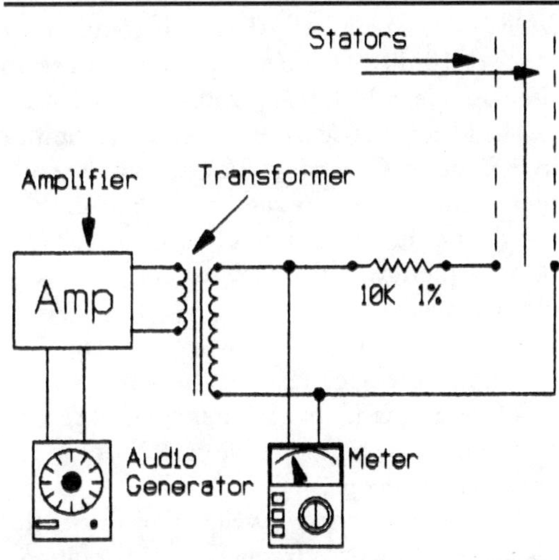

FIGURE 7-1: Capacitance measuring set-up.

GUIDELINE #11:

ESL Capacity

200–250pF per square foot

In short, you have no way of knowing the exact D/S spacing. Since a small change in D/S spacing makes a large change in capacitance, these variants are important.

Standard capacitor formulae are based on the capacitance found between solid plates. ESL stators are full of holes or slots. Depending on their size, the stator's field density may not be as high as that of a solid plate.

Insulation (if used) is another variable, because its dielectric characteristics change the capacitance. Also, it increases the D/S spacing for a given diaphragm excursion.

These and other unknowns make it impractical to compute the capacitance and it is usually more practical to simply measure it. The easiest way to do this is with a capacitance meter—if you can get one.

Since you probably don't have access to one, I'll show you a way to do it with common instruments. You'll need an audio generator and a high-impedance audio voltmeter with an input impedance at least 10 times the value of the reference resistor you use in *Fig. 7-1*.

An audio voltmeter is nothing more than an AC voltmeter (usually analog) with linear frequency response in the audio band, such as a **V**acuum **T**ube **V**olt**m**eter (VTVM) or **F**ield **E**ffect **T**ransistor **V**olt-**O**hm**m**eter (FET VOM). These typically have an input resistance of 10–11MΩ. It's also possible to use an oscilloscope, which usually loads with 1MΩ.

The idea is to use the speaker's capacitance in a **R**esistor/**C**apacitor (RC) **network** while you measure the frequency response. You use the point where the RC network rolls off the frequency response to calculate the capacitance.

Set up a basic RC network per *Fig. 7-1*. Choose a reference resistor that will cause a rolloff at a measurable frequency—10kΩ usually works. Use a 1% precision resistor for this important reference.

Run a frequency sweep and plot the response. The corner frequency is the point on the frequency-response graph where the response falls by 3dB. Use this frequency, with the resistance, in the following formula to calculate the speaker's capacitance.

Here, I introduce a formula that will appear a number of times in this chapter. The more familiar formula for the reactance (a resistance to alternating current) of a capacitor is:

$$X_c = \frac{1}{2\pi f C}$$

with C in *farads*.

Since C will usually be in microfarads (μF), a millionth of a Farad (10^{-6}), and 2π is a constant, we can simplify to:

$$\frac{10^6}{2\pi f C} = \frac{159{,}155}{f C}$$

resulting in a much less unwieldy bit of arithmetic.

Once you find the 3dB corner, calculating the speaker's capacitance is easy:

$$C = \frac{159{,}155}{f R}$$

Where:

C = Capacitance in microfarads
R = Resistance in ohms
f = Frequency in hertz

If this seems cumbersome, that's because it is. But it works accurately, and you can do it with common instruments.

If you can't measure the capacitance, perhaps you'll find some guidelines helpful. I'll caution you that these are only estimates, because there are so many unknown vari-

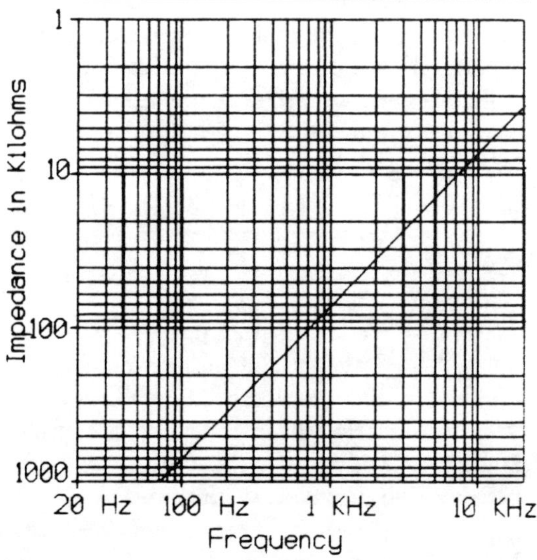

FIGURE 7-2: ESL impedance versus frequency.

FACTORS THAT DEFINE AMPLIFIER CURRENT REQUIREMENTS

- Frequency
- Speaker Capacitance
- Drive Voltage

ables. Still, some idea of the speaker's capacitance is better than none.

These are for cells with 70–80-mil D/S spacing, moderately uniform construction, no insulation, and high-field-density stator construction. The capacitance is inversely proportional to the D/S spacing.

CAPACITIVE REACTANCE. An ESL's reactance often gives power amplifiers fits, but most manage anyway and more important things take precedence in ESL design. The frustrating thing about ESL reactance is that it varies widely with frequency and is always very high compared to magnetic drivers. A large impedance mismatch exists between most amplifiers and ESLs.

The term **impedance** is normally used to describe the behavior of magnetic speakers which contain all three impedance elements. Therefore, I'll use the term as well, although a simple electrostatic speaker is a pure capacitor and only exhibits capacitive reactance.

A conventional amplifier will drive an ESL through a step-up transformer which adds inductive reactance and resistance to the circuit. Such a complex load is properly described as having impedance. If these terms are not clear, review them in *Chapter 2* on technical terminology.

You can use the related formula to find the impedance for an ESL at *one* frequency, if you know the speaker's capacitance:

$$Z = \frac{159,155}{fC}$$

Where:
- Z = Impedance in ohms
- C = Capacitance in microfarads
- f = Frequency in hertz

(Note: This assumes that the impedance Z is predominately capacitive—which, of course, it is.)

The reference ESL has an impedance at 1kHz of 72.3k. At 10kHz the impedance is 7.23k; at 20kHz it is 3.6k. The impedance falls linearly with increasing frequency (*Fig. 7-2*).

At best (a large, high-capacitance ESL operated at high frequencies), the impedance still is going to be much higher than in a magnetic speaker. The high impedance of an ESL causes problems with power transfer from the power amplifier to the speaker.

We usually use a step-up transformer to generate the required high drive voltages. Conveniently, it also helps match the impedance of the power amplifier to the ESL.

A transformer with a step-up ratio of 50:1 has a secondary impedance of 10k, when the primary is connected through its 4Ω taps to the power amplifier. Note that 10k matches the impedance of the reference speaker at 7.2kHz. At lower frequencies the power transfer isn't as good. But it works far better than the power amplifier's basic impedance of only a few ohms. The impedance matching of the step-up transformer is so important it is often called an **impedance-matching transformer**.

From a practical standpoint, the main problem with ESL impedance is that it can get so low at high frequencies some power amplifiers can't handle it through high step-up ratio transformers. They may oscillate, blow fuses, or damage internal circuitry.

ELECTRONICS FOR ESLs. ESLs require more electronics than magnetic speakers. There may be five electronic devices associated with ESLs: the power amplifier(s), step-up transformers, crossovers, equalizers, and high-voltage polarizing supply.

Although many ESL systems use all these parts, some specialized ones may use only a few of them. Let's look closely at each.

GUIDELINE #12:

Stator Current

$$I = \frac{V \times C}{165}$$

Where:
- I = Current as mA
- V = Drive voltage as kV
- C = Capacitance as pF

POWER AMPLIFIERS. The major problem when driving ESLs is generating high audio voltages with a wide-frequency bandwidth. What is the maximum voltage you can get from conventional, commercial power amplifiers?

If you want to know precisely, you need to measure the amplifier's output voltage just below clipping. If you measure it with an AC voltmeter while measuring sine waves, multiply the value by 1.414 to convert RMS into peak volts. Remember that in the discussion following, I use peak voltages.

A voltmeter works well on steady-state waveforms like sine waves, but it doesn't respond fast enough to read musical peaks. To measure music and read the peaks you need to use a very fast responding instrument like an oscilloscope.

Calibrate your oscilloscope with an audio voltmeter. Then you can use the 'scope to measure the peak voltages.

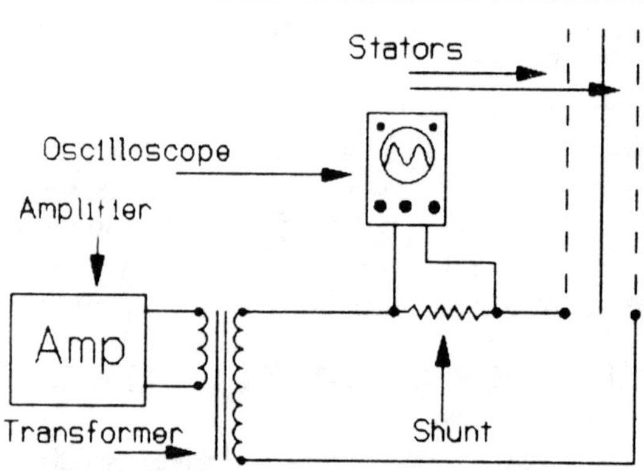

FIGURE 7-3: Current measuring set-up.

If you don't want to go through the hassle of measurements, a reasonably close value is 90% of the power amplifier's power supply voltage. Most solid-state 100W power amplifiers have ±50V power supplies, so they'll only have about 90V of peak output.

Since 90V is far short of the several thousand required, you need a step-up transformer to increase the voltage. For reasons I'll discuss later, a step-up ratio of about 1:50 is about the best that can be obtained in a wide-bandwidth transformer. Therefore, using a transformer between the power amplifier and the ESL, you can expect to get around 4.5kV with a 100W power amplifier.

This will do a respectable job, but it is a long way from the 7kV desired for a midrange/ tweeter not to mention the 10+kV we would

like for a woofer. To obtain higher voltages, you need power amplifiers with higher power-supply voltages. As a rule, the higher the power rating of an amplifier, the higher its power-supply voltage will be.

In short, you need a very large power amplifier for high drive voltages. A 250W power amplifier will usually swing about 160V, and when stepped-up with a 1:50 transformer, you can get 8kV. This represents a substantial improvement over a 100W power amplifier and is well worth doing.

So far, I have only discussed voltage. ESLs need current, as well.

Because of the many variables, calculations become complicated. We are not interested in the current required at only one frequency; rather, we want to know the current required during loud, complex music containing thousands of frequencies. Also, the capacitive reactance determines the current requirement—and it varies with frequency. As if all that isn't enough, recall that the output requirements of music decrease with increasing frequency— while the impedance of the speaker falls.

Calculations that accurately model such complex interactions are difficult. A much simpler method is to measure the current, as we did with capacitance.

Again, you need an oscilloscope. Meters don't respond quickly enough to read peak current, which is what counts. Calibrate the 'scope with a shunt so you can read peak milliamps. A full-scale reading of 100mA should suffice for most ESLs. *Figure 7-3* gives the general setup.

A **shunt** is nothing more than a low-value resistor in series with the load. An oscilloscope or meter can read the voltage drop across this resistor. The purpose is to obtain a small percentage of the total voltage or current in the circuit, when the total would otherwise overload the measuring instrument. By selecting an appropriate resistance for the shunt and adjusting the gain of the instrument, you can get any percentage of the total you wish.

Measure a sine wave with a meter while observing it on the oscilloscope. Reference the top of the waveform against a grid line. Convert the RMS meter reading to peak current by multiplying by 1.414. This will tell you the value of the top of the waveform.

Place the shunt in series with the ESL and the transformer. Then, while playing music at an appropriate volume, measure the stator current. To find the peak current in the amplifier's output stage, multiply the reading by the step-up ratio of the transformer.

Examine the amplifier's output waveform to be sure it is not clipping. You can't reliably detect mild to moderate clipping by listening, and it will render the test inaccurate. Go to the trouble to do it right. Look for clipping on the 'scope (*Fig. 7-4*). Better yet, use a distortion analyzer.

A simpler method to determine peak current is to use stator current values already gathered by others. Here is another guideline. Note the values in the equation are not farads, volts, and amps. Rather they are the more usable picofarad, kilovolt, and milliampere.

This formula is subject to error, because it involves capacitive reactance and is therefore frequency dependent. Different types of music have different energy spectra. Still, a ballpark approximation should meet most needs.

Using the above technique, and ignoring any small losses in the system, the reference ESL has peak currents of 25mA at 2kV, which is close to the 26.6mA predicted by the formula. A 255W power amplifier with a 44.7:1 step-up ratio transformer produces 7.4kV and has a peak current of 99mA. Multiply that by the transformer's step-up ratio, and the current required in the power amplifier's output stage is about 4.5A.

Many builders wish to solve the drive-voltage problem by increasing the step-up ratio of the transformer so they can use smaller power amplifiers. Unfortunately, this is not usually successful because of TINSTAAFL (there is no such thing as a free lunch). Increasing the step-up ratio also increases the current demands.

Small power amplifiers have more limited current capabilities than larger ones. A high step-up ratio develops higher voltages, but the increased current demands now cause the power amplifier to current clip instead of voltage clip.

An exception is when driving ESL woofers: the large D/S spacing reduces their capacitance, and this reduces the current required. Simultaneously, they need the higher voltages supplied by a high step-up ratio.

Murphy's Law has a short attention lapse here. This works in our favor, as it allows us to use higher step-up ratio transformers without current clipping. The trade-off between high frequency response and the step-up ratio is not a problem, if we only use the transformer at low frequencies.

A **watt** is a measurement of power: the product of volts times amps. ESLs are "wattless" speakers. They do not dissipate power by turning it into heat, as a magnetic speaker does. Instead, they store the current momentarily, then as the polarity of the signal reverses they return the current to the power amplifier. They act like a spring. It takes energy to compress the spring (the amplifier drives and "compresses" the electrons on the stator)—but the energy comes back when you release it (when the signal polarity reverses).

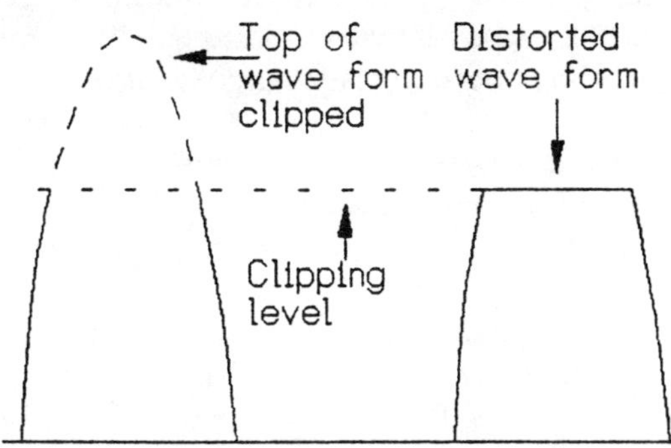

FIGURE 7-4: Amplifier clipping distortion.

Since insignificant amounts of power are being lost in this process, an ESL technically is an extremely efficient speaker. But because of the large currents moving through the amplifier's output stage and high voltages required, they seem inefficient. Despite their high efficiency, they require large power amplifiers.

DIRECT-DRIVE, HIGH-VOLTAGE AMPLIFIERS (D/D AMPS). To avoid the problems of step-up transformers, some builders and manufacturers use high-voltage, high-impedance amplifiers which drive ESLs directly without transformers. Such amplifiers work well, but they also have serious problems.

Consider for a moment the voltage and current requirements previously outlined. To produce the output of the conventional amp with transformers, a D/D power amplifier must deliver about 8kV and 100mA to the speaker.

D/D amplifiers usually operate in Class A (although there are Class AB types). A Class A amplifier must continuously dissipate the full power required by the speaker.

That power is impressive: 800W per channel, 1.6kW for a stereo amplifier. These amplifiers usually use vacuum tubes, so additional power is needed to run the tube heaters and grids. We're looking at around 2kW of power.

You would have to operate it on 240V house current. It would be very large and have dangerously high voltages and currents; operating expenses would be high; the cost of building it would be prohibitive; and it would make a great room heater.

Obtaining high-voltage parts is difficult. Tubes are becoming more rare and expensive. The B+ power supply requires a transformer rated at a minimum 5.65kV @ 200mA for a power rating of over 1.1kW. When was the last time you saw one? Where do you find large capacitors rated at over 8kV? Etcetera, etcetera.

Finding the necessary parts would be an adventure. Additionally, practical experience reveals that D/D amplifiers tend to be unstable and suffer major breakdowns.

But amplifier designers are clever. As transistors become more capable and less costly, perhaps someone will design a highly efficient solid-state D/D amplifier. For now, practical considerations limit D/D amps to small, low output ESL systems.

In summary, ESLs work best with powerful, conventional amplifiers combined with step-up transformers. You can use small power amplifiers when low outputs are satisfactory, or when driving ESL tweeters.

AMPLIFIER SELECTION.

Most audiophiles have very strong opinions about what is the "best" amplifier. If you fall into this group, I want to encourage you to lay your biases aside. I can assure you that high power is by far the most important factor when selecting a power amplifier for ESLs. I strongly urge you to place it above all other sonic criteria when you select an amplifier.

If your favorite amp also happens to be extremely powerful—great. Use it and you'll be deliriously happy. But if your preferred amp is of modest power, you are likely to be disappointed with the sound. A superb power amplifier that clips always sounds worse than a mediocre one that doesn't!

Also keep in mind that ESLs present a completely different load to an amplifier than a magnetic system. Just because an amplifier sounds great on magnetic speakers doesn't mean it will sound wonderful on ESLs.

An amplifier you think sounds bad on magnetic systems probably will sound just fine on ESLs. ESLs don't seem as amplifier sensitive as complex magnetic systems. Keep an open mind. Above all, go for *POWER*.

ESL STEP-UP TRANSFORMERS.

Step-up transformers greatly increase the voltage of conventional amplifiers so they can drive ESLs. They also help match the high impedance of an ESL to the low impedance of the amplifier.

It is beyond the scope of this book to delve into transformer design and theory. I'll limit my comments to a general discussion and follow it with practical guidelines.

Step-up transformers are hard to find, because manufacturers don't identify them as ESL transformers. They usually are output transformers which are designed for small, high-quality tube power amplifiers.

Since manufacturers don't make small tubed amplifiers anymore, transformers for these devices are uncommon. Let me assure you they are available and reasonably priced—but you may have to hunt for them.

These highly-specialized transformers must meet rigid requirements. I'll define and comment on each specification.

1. The **step-up ratio** is the ratio between the transformer's input and output voltage. For example, a transformer that converts 10V on its primary winding to 100V on its secondary winding would have a step-up ratio of 1:10.

The electrical engineering term for step-up ratio is **turns ratio**. You can easily determine the turns ratio by driving the transformer with your power amplifier and measuring the input and output voltages. When making this measurement, use a mid-band pure tone—for example, a 1kHz sine wave. If you don't have an audio generator, you can use the test tone off one of many available test CDs.

$$\text{Turns Ratio} = \frac{V_2}{V_1}$$

SPECIFICATIONS:
Transformers

- Step-up ratio
- Frequency response
- Power handling
- Voltage handling

Where:

V1 = Primary Voltage
V2 = Secondary Voltage

This technique has one serious flaw. Usually you want to know the turns ratio *before* you buy the transformer, but you can't measure it unless you buy it. Since the manufacturer didn't make the transformer for ESLs, his specifications won't include the turns ratio.

Fortunately, there is a solution to this problem. The specifications always include the impedance, and calculating the turns ratio based on impedance is easy. Simply calculate the square root of the impedance ratio.

$$\textbf{Turns Ratio} = \sqrt{\frac{T_2}{T_1}}$$

Where:

T1 = Primary Impedance
T2 = Secondary Impedance

For example, if you had a transformer with a 4Ω primary and an 8kΩ, the turns ratio would be 1:44.7.

$$\textbf{Turns Ratio} = \sqrt{\frac{T_2}{T_1}} = \sqrt{\frac{8,000}{4}} = \sqrt{2,000} = \textbf{44.7}$$

A tubed power amplifier is inherently a high-impedance device, while magnetic speakers have low impedance. Therefore, the manufacturer of tube power amplifier audio output transformers calls the high-impedance winding the **primary** winding. The **secondary** winding usually has 4, 8, and 16Ω speaker taps.

When driving ESLs, the opposite is true. The speakers are high impedance and transistor amplifiers are low impedance. Therefore, you use the transformer backwards. Connect the low-impedance taps to your power amplifier and the high-impedance ones to the ESL.

Since you will be feeding a signal into the low-impedance winding, it becomes the primary winding, and the high impedance becomes the secondary. The transformer doesn't know or care which way you use it.

You will need a turns ratio of at least 1:40—and higher ratios are better, if you can find them. But high turns ratios reduce high-frequency response, so you must compromise this specification against the frequency response. Fortunately, you can get good high-frequency performance with turns ratios as high as 1:50.

2. The **frequency response** must be reasonably linear across the desired bandwidth when driving the ESL's capacitance. **High-frequency response limits** are a major problem since few transformers can drive capacitive loads at high frequencies.

Transformers behave differently when driving capacitors than when driving resistors. Manufacturers determine frequency response when driving resistive loads—not capacitive loads. You cannot trust published specifications to define the frequency response when driving ESLs, although you can use them as guidelines.

You must measure the frequency response into a capacitor to find the transformer's true specification, but you run into the same problem that you did with turns ratio. You want to know the frequency response *before* you buy the transformer.

Predicting a transformer's behavior into a capacitive load based on manufacturer's specifications is useful. Let me give you some idea of how to estimate it.

Figure 7-5 depicts the frequency response of a *good* transformer driving an 8Ω load. *Figure 7-6* shows the same transformer driving the 2kpF capacitive load typical of a large ESL.

Note the frequency-response peak at 24kHz, after which it crashes. The leakage (series) inductance of the transformer resonating with the shunt capacitance of the transformer windings and the capacitance of the ESL is the cause. This resonance is the main reason for the difference in frequency response between resistive and capacitive loads.

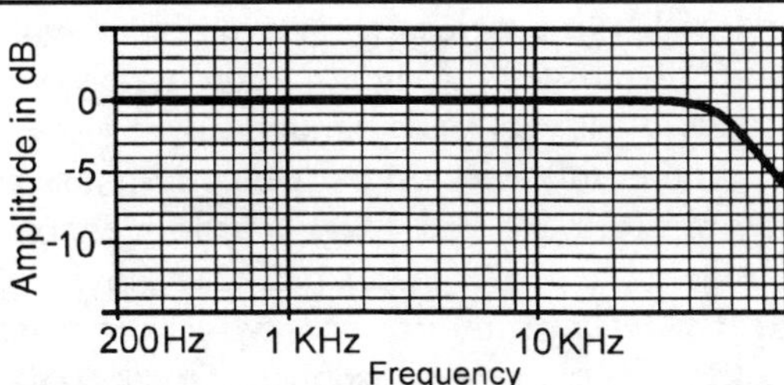

FIGURE 7-5: Transformer reponse into resistive load.

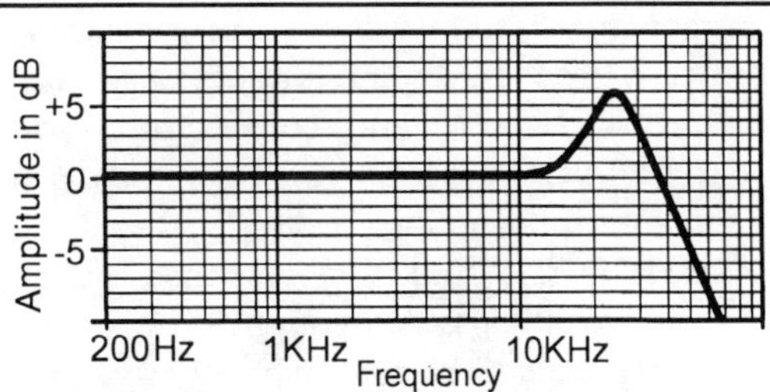

FIGURE 7-6: Transformer response into capacitive load.

TABLE 7-1

Impedance	Turns Ratio
4Ω	44.7:1
8Ω	31.6:1
16Ω	22.4:1

The linear part of the graph extends to about 10kHz, above which there is a gradual rise in the frequency response to the resonant peak. Most audiophiles expect their systems to have flat-frequency response from DC to light. They don't like the idea of less-than-perfect performance.

Realistically, the gradual rise in frequency response between 10 and 15kHz is not audible. Even if it were, most source material benefits from a slight increase in output in the top octave. Since few of us can hear anything above 15kHz, we can ignore the continued rise above there.

You can reduce the magnitude of the resonance by putting a resistor across the transformer's secondary winding. This will lower the Q of the resonant peak and flatten the response. You will have to pick the resistor value by trial and error while taking repeated frequency-response sweeps. Few builders bother to do this, because the slight rise in the top octave is not objectionable—in fact, most listeners prefer it.

Usually it is not practical to get perfectly linear response in the top octave—and fortunately it is not necessary. While you may think such performance is poor, keep in mind that the typical magnetic speaker has multitudes of resonances in the top octave which are far in excess of the transformer's flaws. Their *average* response may look better on paper, but the ESL sounds much better because it lacks resonances, and a small, smooth, gradual rise is either inaudible or preferred.

I'm trying to encourage you to keep a reasonable perspective. Perfectly linear, supersonic, frequency response from a step-up transformer is impossible and unnecessary. Save yourself frustration and compromise your ideals slightly. Trust me—you won't be disappointed with the sound.

Based on the frequency-response graphs, you will find the usable frequency response of this transformer to be somewhere between 30 and 50% of the published specifications. However, this varies greatly depending on the turns ratio you choose.

You have some control over the turns ratio, because you will almost certainly have a choice of primary and secondary taps. If you use the same secondary winding, the 0–4Ω taps will give you a higher turns ratio than the 0–8Ω or 0–16Ω. For an 8kΩ secondary winding, the turns ratios are shown in *Table 7-1*.

The high-frequency response varies inversely with the turns ratio. Higher turns ratios degrade high-frequency response so you must compromise output against frequency response.

Bass response generally is not a problem unless you want to use your ESLs into the bass. You can take low-frequency response for granted—down to a reasonable upper bass crossover frequency. When you use transformers in the bass, there is still not much of a problem with frequency response. It will be down 3dB when the value of the primary inductance equals the sum of the source resistance and the internal winding resistance.

In the small, high quality transformers necessary for good high frequency response, the above conditions are rarely a limiting factor. The problem with transformers used in the bass is not usually frequency response but power handling, which I discuss shortly.

An advantage of segmented ESLs is using each segment over a limited frequency bandwidth. This eases the problem of compromising turns ratio against high frequency response. You can select transformers that best match the frequency response and turns ratio you want.

For example, you need very high turns ratios for ESL woofers, because they demand the most excursion. Large transformers for power handling are also needed.

Both high turns ratios and large size ruin high-frequency response but, conveniently, woofers don't need high-frequency response. The transformers for ESL tweeters need excellent high-frequency response, but don't require high turns ratios. So you can see the trade-off between high frequency response and high turns ratios is not a problem with limited bandwidth designs. Unfortunately, we usually want very wide bandwidth.

Another influence on frequency response is the transformer's size. Large transformers can handle more power than small ones, but have greater leakage inductance, which impairs their high-frequency response. Small transformers offer good high-frequency response but poor bass power handling.

3. **Power handling** is not as much of a

problem as you might expect—if you don't want to reproduce bass. For ESLs used in the mid- and upper-frequencies, you don't select the transformer size based on the power amplifier as you would with a magnetic speaker. Instead, you base it on a combination of the amplifier power and the lowest frequency you want the ESL to reproduce.

Magnetic saturation of the transformer's core limits power handling, which is frequency dependent. Low frequencies demand large transformer cores.

The relationship between power and frequency is linear. For example, if a transformer is rated at 10W @ 20Hz, it can handle 100W @ 200Hz.

To determine the power handling requirements of your ESLs, you can use the following proportion:

$$\frac{Power_R}{Frequency_R \times 2} = \frac{Power_M}{Frequency_L}$$

Where:

$Power_R$ = Manufacturer's power rating in watts (or VA)

$Frequency_R$ = Manufacturer's low frequency specification

$Power_M$ = Maximum permitted power at $frequency_L$

$Frequency_L$ = ESL's lowest linear frequency

The reason I doubled the manufacturer's rated frequency in the formula is because his frequency is the -3dB corner frequency. Since 3dB down is only half power, we can't use that for our calculations. If we go up an octave (double the frequency), we'll be close to full power.

Let's look at a specific example. Triad rates their S-142A transformer at 15W @ 7Hz. Can I use a 250W amplifier with this transformer with a crossover frequency of 400Hz?

Plugging in the known values and cross multiplying shows that the transformer will handle about 429W @ 400Hz, comfortably in excess of the 250W required.

$$\frac{15W}{(7 \times 2)Hz} = \frac{X}{400Hz}$$

$$\frac{15 \times 400}{14} = \frac{6000}{14} = 429Hz$$

Or, put another way, it would handle a 250W power amplifier down to about 233Hz.

$$\frac{15W}{(7 \times 2)Hz} = \frac{250W}{X}$$

PUBLISHED TRANSFORMER SPECIFICATIONS

Frequency response:
7Hz–50kHz, +0, -3dB into 8Ω

Impedance:
Primary—4, 8, 16Ω
Secondary—4, 8kΩ, with center tap

Power handling: 15W

Insulation rating: 1.5kV RMS

Size: 3″ × 2.5″ × 2.5″

Weight: 4 pounds

$$\frac{14 \times 250}{15} = 233Hz$$

If you only use ESLs in the middle- and upper-frequency ranges, you might be surprised at how small a transformer will do the job. But what if you need to produce bass with your ESL? What size transformer would you need to handle a 100W power amplifier at 50Hz? In the calculations below, assume the manufacturer rates the power at 10Hz. Solving the proportion shows you would need a 40W transformer.

$$\frac{X}{(10 \times 2)Hz} = \frac{100W}{50Hz}$$

$$\frac{20 \times 100}{50} = 40W$$

4. **Voltage handling** means transformer insulation. Newer insulation materials have produced large improvements in this area.

Manufacturers continue to rate their transformers based on the older insulating materials. The result is extremely conservative ratings. For example, most of the audio output transformers I've seen and used are only rated for 1.5kV RMS. Yet I use them to over 7kV without trouble.

Recall that what counts is not the RMS rating, but the peak rating. The 7kV mentioned above is P-P voltage. Multiply RMS volts by 2.828 to convert to peak volts. A 1.5k RMS volt transformer is a 4.2kV P-P transformer.

Because manufacturers rate their insulation very conservatively, let's double the rating. Now you have an 8.4kV transformer. This should explain why a 1.5kV transformer can handle 7kV. You will find that transformer insulation is not a problem in the typical ESL.

To help you get an idea of published specifications, here they are for a suitable wide-bandwidth ESL transformer (Triad's S-142A):

If you translate the manufacturer's specifications to electrostatic specifications, the above transformer would read:

TRANSLATED TO ESL SPECIFICATIONS

Frequency response:
300Hz–15kHz, +2, -0dB into 2,000pF

Turns ratio: 44.7:1

Power handling: 250W above 250Hz

Voltage Handling: 8kV

TRANSFORMER TESTING. You must test transformers for adequate performance, unless you know they operate ESLs satisfactorily based on reliable testing by others. Testing is not difficult, but requires a sine wave audio generator and high-impedance audio meter such as a VTVM or the more modern FET VOM. There are also DVMs (**D**igital **V**oltmeter), but I prefer analog readings for frequency response measurements.

These need not be exotic meters. The one you already own may work. It just has to have high-input impedance (10MΩ is typical) and linear frequency response in the audio range. Connect these components per *Fig. 7-7*.

You may use your speakers for the load while testing, but it's more convenient to use a small capacitor of similar value. Do not exceed 300V at the output of the transformer when testing, or your meter may arc internally.

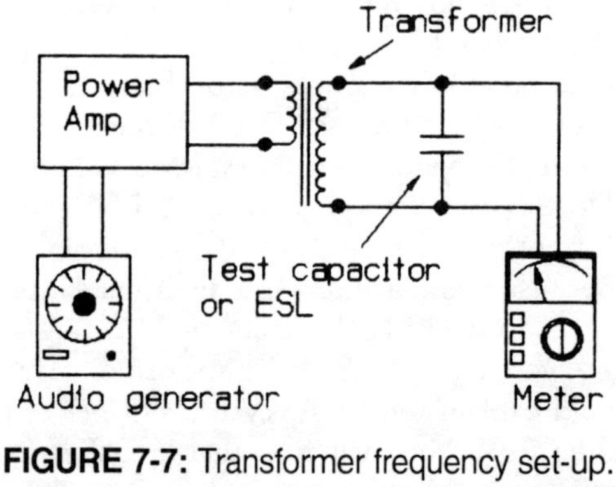

Connect the 0–4Ω taps to your amplifier, but ultimately you should test all the taps to see which gives you the best compromise between high-frequency response and output (turns ratio). Also, test by connecting between the 4- and 8Ω taps. This will produce a very high turns ratio with an 8dB increase in output over the 0–4Ω taps. However, the frequency response of most transformers will not be satisfactory at this higher turns ratio. Additionally, this will represent a very low impedance to the power amplifier, which it may not tolerate.

You probably will have a choice of secondary taps. The highest impedance ones will produce the highest turns ratio.

Sweep the frequency range of interest looking for a resonant peak. You will find one somewhere between 4kHz for a poor transformer to over 20kHz for a superb one.

You may encounter transformers which have no resonant peak, but they always have poor high-frequency response in my experience. Expect the peak to be between 3 and 6dB. There will be a rapid roll-off following the peak.

Testing with low voltages is fine when measuring high-frequency response, but if you want the transformer to produce bass, you must drive it at the full voltage expected when in actual use. The greatest magnetic demands on the core occur at low frequencies and high voltages.

How can you tell when the core saturates? The best way is with a distortion analyzer, but a fall in frequency response as an indication of core saturation can also be used. Observing the waveform on an oscilloscope will reveal gross distortion.

Since the frequency doesn't start to fall until the distortion is severe, you should assume the

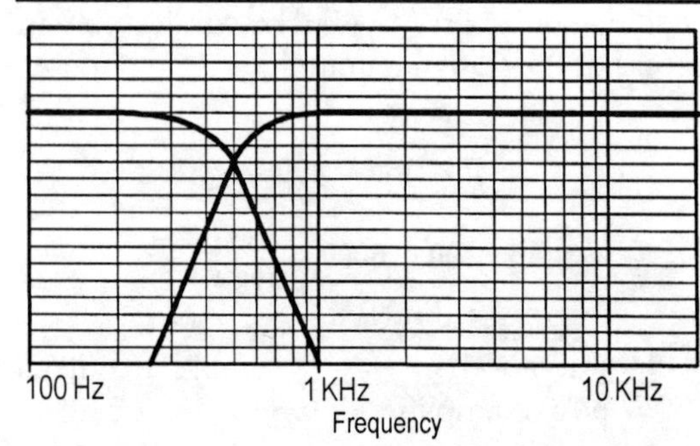

FIGURE 7-7: Transformer frequency set-up.

FIGURE 7-8: Typical response of a two-way crossover.

GUIDELINE #13:

Crossovers

- Use as few as possible.
- Use them only below 600Hz.

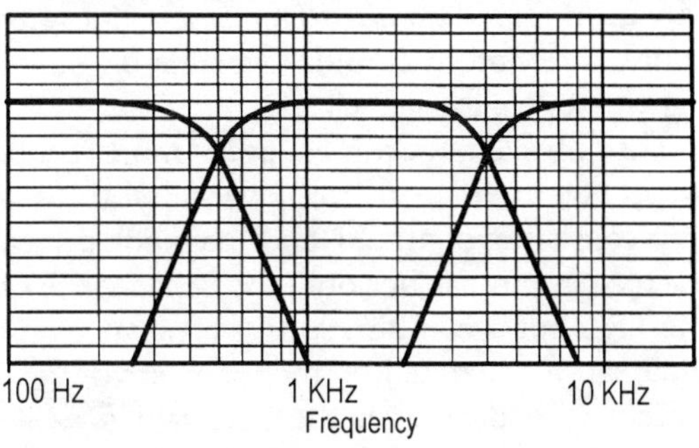

FIGURE 7-9: Typical response of a three-way crossover.

transformer is only satisfactory at an octave above the point where you *start* to see the frequency response fall. Any fall in frequency response usually means core saturation—don't wait until it's 3dB down before making a judgment.

Use a high-voltage probe with the meter to measure high voltages. I explain this measurement technique later under high-voltage polarizing power supplies.

You probably will have to use the ESL for the load when testing bass. Garden-variety capacitors will not handle the voltages generated during a high-power, high-voltage test.

CROSSOVERS. Crossovers, also known as **crossover networks**, split the audio drive signal into limited frequency bands. These bands drive the various speakers (woofers, tweeters) that best reproduce that particular range of music.

Figure 7-8 shows the frequency response of a two-way crossover with a **crossover frequency** of 500Hz and 12dB/octave slopes. *Figure 7-9* shows the response of a three-way crossover with 500Hz and 4kHz crossover frequencies. The **passband** of a crossover is the flat part of its frequency-response curve.

Crossovers come in several different configurations: **passive** and **active**, and **high**- and **low**-level. A passive crossover does not require an external power source to operate. An active one has amplifiers in its circuitry.

Magnetic speaker systems usually use high-level crossovers. They are passive arrangements of inductors, resistors, and capacitors that split the output from the power amplifier into the required frequency bands.

"High-level" crossovers are so-called because they come after the power amplifier. They intercept the high power from the power amplifier and split it into bands before it reaches the speakers.

"Low-level" crossovers come before the power amplifier and operate at low voltage and power levels. They may be a collection of passive components like capacitors and resistors, but more commonly they include small amplifiers. Low-level active crossovers are widely known as **electronic crossovers**.

High-level crossovers are more popular, because you only need one power amplifier. Their disadvantage is they isolate the drivers from the power amplifier so the amplifier can't control them directly. This loss of command degrades performance. Also, the power amplifier sees a very complex load consisting of multiple inductors, capacitors, resistors, and several drivers. This load is more difficult to handle than a single driver.

High-level crossover circuitry is simple, although here it prevents us from getting ideal frequency slopes and filter types. When compared to low-level active crossovers, high-level passive crossovers have poor detail, compromised phase behavior, and flawed frequency and power response around the crossover point.

Because low-level crossovers do not deal with high power, they can be—and often are—very sophisticated. By using small amplifiers (integrated circuits are ideal) and complex precision components, you can make many different filter types and slopes. You can easily and accurately optimize the crossover to the needs of the speaker system.

Electronic crossovers enable you to connect a separate power amplifier to each driver for direct control. Although this requires multiple amplifiers, the improvement in sound is well worth it.

Besides superior performance, there are several other reasons for using low level crossovers. Engineers design conventional, high-level, passive crossovers to drive resistive, not capacitive, loads. Since ESLs are capacitors, standard crossover design does not apply. Little information is available on building crossovers for capacitors.

This problem becomes even more acute when you attempt to cross over a hybrid system consisting of an ESL and magnetic driver. Such a system has not only the capacitance of the ESL, but a combination of resistance and inductance generated by the magnetic driver.

Trying to deal with these variables while also producing ideal slopes, phase behavior, power, and frequency response is difficult. Optimizing all the above parameters with passive, high-level crossover networks is impossible, but easy to do with active low level ones.

SLOPES, FILTERS, and CROSSOVER POINTS.

The subject of crossover type, slope, and frequency is complex and controversial. Entire books have been written on the subject, and it is beyond the scope of this book to explore the subject in depth. So, as usual, I'll present a general overview with suggestions and guidelines.

Several filter types are available. These include Bessel, Chebychev, and Butterworth. Strong opinions abound regarding the different types, and you can find proponents and opponents for each. No one type of filter is perfect. I'll summarize some important points.

Bessel filters have excellent phase and transient response, but do not cut off rapidly enough for speaker crossover use. Chebychev filters cut off rapidly, but there are problems with ripple in the passband and poor transient response. Butterworth circuits are maximally flat in the passband, and in odd-order designs have constant voltage and power levels through the crossover regions.

Most designers use Butterworth filters. However, there are Butterworth filters and there are Butterworth filters. Let's give them a closer look.

First-order Butterworth filters are low in phase shift and have constant output through the crossover point. They roll off at only 6dB/octave (an **order** is 6dB/octave). Therefore, there must be at *least* a two-octave spread beyond the crossover point for the drivers in a first-order system.

In a 500Hz crossover, the woofers should be linear up to at least 2kHz, and the tweeters down to at least 125Hz. Rarely is this practical.

Second-order filters rolloff at 12dB/octave. These slopes are steep enough and are widely used in passive high-level crossovers.

Unfortunately, there is a definite discontinuity, a "hole," at the crossover point. You can partially correct the problem by reversing one driver, but there is still an audible imperfection.

Third-order, 18dB/octave filters correct these problems. In addition to having a linear pass band, they have a constant level through the crossover point. Driver cutoff is very sharp, which minimizes out-of-bandwidth frequency response requirements.

Although careful crossover choice and design can minimize the errors and distortions associated with crossovers, there is no such thing as a perfect crossover. For example, phase shift is present in all of them. Although not a severe problem, it is undesirable and should be avoided, if possible.

Unhappily, the ideal of a full-range, crossoverless, electrostatic speaker that can produce high output and deep bass remains unrealized. High-output ESL systems continue to require crossovers. Since crossovers are imperfect, two guidelines will help you avoid most problems:

1. Minimize the number of crossovers. For example, a two-driver system is preferable to a three-driver system assuming all else is equal.

An advantage of ESLs is they can cover a much wider bandwidth than magnetic speakers. While two-driver magnetic speakers are usually seriously compromised, this is not so with ESLs. You can get splendid results with a two-driver ESL system.

2. Keep the crossover point under 600Hz. The ear is sensitive to phase and frequency response anomalies at midrange frequencies, but can't detect these problems very well below 600Hz.

A big problem with crossovers is audiophile "overkill." It has to do with certain perfectionist attitudes common to the type of person who is looking for the best in sound reproduction—like someone trying to build a better ESL. (Sound like anybody you know?)

This well-intentioned individual logically thinks that since crossovers are bad, and 600Hz the highest acceptable, then lower is better. So he picks an excessively low crossover point; 100Hz is a common choice.

This unreasonably low crossover point compromises both the frequency response and output in the system with a suck-out between 500Hz and 100Hz, because phase cancellation is not adequately compensated. Output is diminished, because of the greater excursion required by the ESL and magnetic core saturation of the step-up transformers.

The problem is reminiscent of the story about Mark Twain's cat: One day Mark Twain's cat sat on a hot stove lid. He never sat on a hot stove lid again . . . but he would not sit on a *cold* one, either!

This issue is one of perspective. You can't hear the minor problems of high-quality crossover networks below 600Hz. ESLs have a very difficult time with bass, and magnetic woofers work fine that high. So make your crossover as high as possible consistent with flawless sound. A reasonable compromise is a crossover point around 500Hz.

EQUALIZERS. An equalizer modifies the frequency response of the speakers by changing the frequency response of the electronic signal being fed to the speakers. This manipulation can be done at any point in the signal chain, but is most commonly done before the power amplifier(s).

Equalizers are very similar to crossover networks, which are just a special type of equalizer. Equalizers can be **passive** or **active**, and operate at either **low** or **high levels**. Also like crossovers, and for most of the same reasons, the best systems use active low-level equalizers.

Passive equalizers have **insertion loss**, which is a reduction in electrical signal voltage when you put the equalizer in the circuit. The effect is as though you turned down the volume. Insertion loss is an important reason why we seldom use passive equalizers.

Active equalizers solve the problem of insertion loss by incorporating small amplifiers in their circuitry. The amplifier compensates for the insertion loss so the level is either unchanged or is moderately increased.

Most electrical engineers design their equalizers to operate at **unity gain**. This means the signal voltage comes out at the same level it went in.

An important function of the amplifier in an active equalizer is that it acts as a buffer between the various components at its input and output. The component impedance an equalizer sees at its input and output can affect its frequency response. This may cause the response to deviate from what you want. Active equalizers usually are immune to impedance mismatch.

Most active equalizers are built using **integrated circuits (ICs)** because of their small size, high performance, and flexibility. The IC

manufacturer supplies "application notes" which make it easy for you to design custom equalizers.

Many audiophiles still avoid ICs, believing they lack quality performance. When ICs were first introduced, there probably was some truth to this. Modern ICs are much better. They perform beautifully in equalizers.

As discussed in *Chapter 4* on frequency response, you must do something about midrange and bass phase cancellation. Equalizers are ideal for correcting the midrange, where the ESL is just starting to encounter phase cancellation.

The top curve in *Fig. 7-10* is the familiar general frequency response of a dipole ESL. Look only at the frequency response between 300Hz and 3kHz. Compare it with the frequency response of the equalizer below it. Note the mirror-image relationship. The result is flat frequency response from the system in that frequency range.

DESIGNING AN EQUALIZER. Simple passive equalizers are easy to design and build. Let me run through the steps for designing one, so you can custom build just what you need.

These concepts also apply to active equalizers, but their design is complicated by the presence of an amplifier. If you use ICs, you can combine the following information with the IC's application notes to design active equalizers.

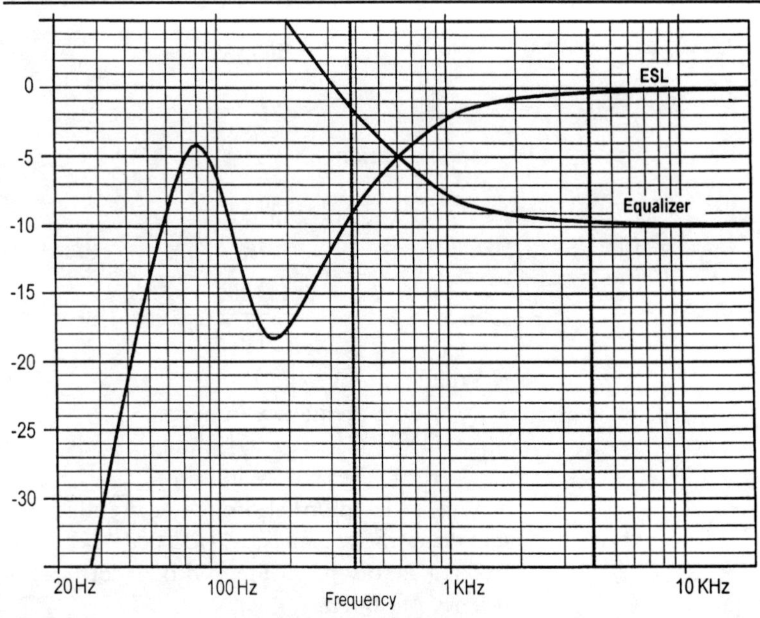

FIGURE 7-10: Equalization can correct ESL frequency response errors.

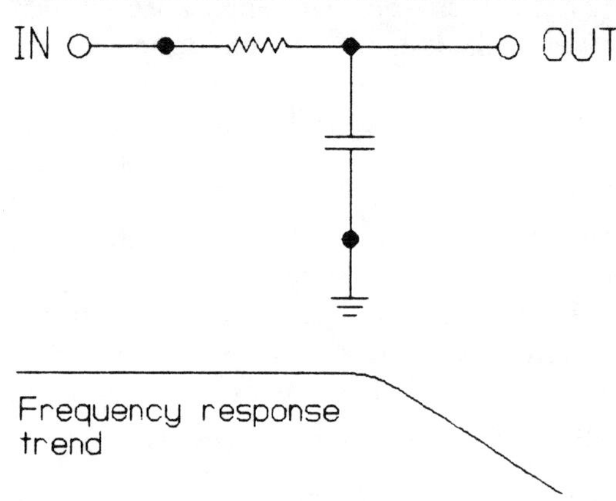

FIGURE 7-11: Low-pass equalizer.

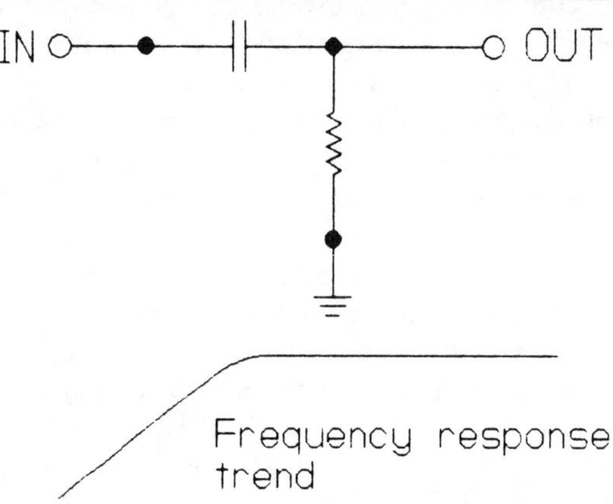

FIGURE 7-12: High-pass equalizer.

The cornerstone of equalizer design is the RC network. *Figure 7-11* shows the schematic diagram and frequency-response graph for a simple, 6dB/octave, low-pass filter. *Figure 7-12* shows a high-pass filter.

As their names imply, the low-pass filter passes low frequencies, while the high-pass filter passes high frequencies. When making a crossover, you would direct the output from the high-pass filter to the tweeter and the low-pass filter to the woofer.

The point where the output falls by 3dB is called the **corner frequency**, or **3dB corner**, or **knee** of the frequency response curve. You can easily calculate this frequency and the values of the resistor/capacitor combination that defines it with the formula introduced at the start of this chapter:

$$f = \frac{159,155}{RC}$$

and its variants:

$$C = \frac{159,155}{fR}$$

$$R = \frac{159,155}{fC}$$

Where:

f = Frequency in hertz
C = Capacitance in microfarads
R = Resistance in ohms

With this information, we can build the equalizer shown in *Fig. 7-13*. Let's begin by making a low-pass filter with a corner frequency of 400Hz.

We'll pick 10kΩ as a reasonable value for the input resistor. By inserting the values of

10kΩ and 400Hz in the capacitor solving equation above, we get a value of 0.04µF.

$$C = \frac{159,155}{Rf} = \frac{159.155}{10,000 \times 400} = 0.039\,\mu F \ (0.04\,\mu F)$$

These components produce the frequency response trend shown in *Fig. 7-11*. The next step is to stop the high-frequency roll-off to produce a **shelving equalizer**.

By adding a resistor between the capacitor and ground, we can stop the high-frequency roll-off at a frequency determined by the resistor solving equation above (*Fig. 7-14*). Let's pick a corner frequency of 2kHz.

$$R = \frac{159,155}{fC} = \frac{159,155}{0.04 \times 2,000} =$$

$$\frac{159,155}{80} = 1,989.4\,\Omega$$

The combination of these two circuits produces the frequency response shown in *Fig. 7-15*. If you want to "tweak" the frequency-response curve, you can substitute a variable

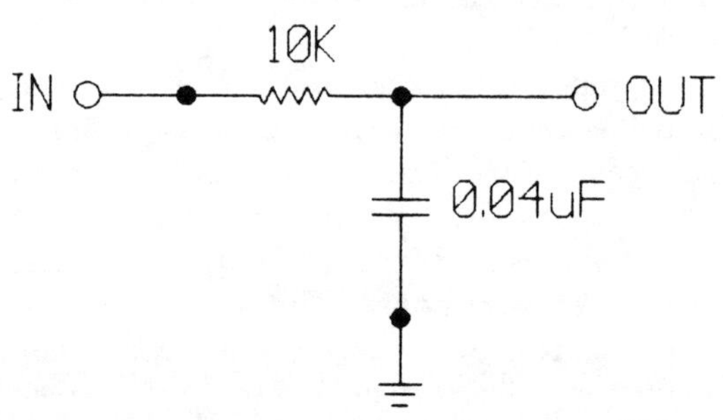

FIGURE 7-13: 400Hz low-pass equalizer schematic.

resistor (a potentiometer or pot) between the capacitor and ground.

With this you can select the high-frequency corner by just twiddling a knob—much like adjusting the treble control on a control center. It does not work exactly like a treble control, because it does not change the slope around a fixed point. Instead, it shifts the point along a fixed slope. *Figure 7-16* shows the difference between the two as you change settings on the pot.

Using a potentiometer instead of a fixed resistor is a particularly useful technique in this case, because it is hard to find precision capacitors. Garden variety capacitors vary from their specified values by ±20%. By using a potentiometer, you can easily trim the circuit to your exact needs.

This type of equalizer has 6dB/octave slopes. To get steeper slopes you can put two or more equalizers in series.

Note that these circuits don't instantly go from flat to a 6dB/octave slope. Rather, they ease into it gradually, particularly at the high-frequency end.

Conveniently, this is just how ESLs begin their frequency fall-off due to phase cancellation. The ESL and equalizer compliment each other beautifully, so we get a very close mirror image match of the frequency-response curves.

You can predict the phase cancellation behavior of an ESL by the minimum dimension of the speaker, as previously described, and the frequency response of an equalizer by the above formulas. Therefore, you can design and build an equalizer to flatten the response of the ESL in the midrange by completely mathematical methods. You do not need to go through the difficult task of measuring the frequency response of the speaker to build a midrange equalizer.

The equalizer we just designed is passive and suffers from all the problems of such devices—including an insertion loss of around 18dB. Most sound-reproduction systems do not have enough extra gain to compensate for this loss, particularly when driving powerful amplifiers at high levels.

If the impedance of the components at the input or output is near or below the input resistor (10kΩ), their impedance will lower the effective value of the input resistor. This will change the frequency response of the equalizer. You cannot rely on the formulas to give you accurate results under these circumstances.

For all the above reasons, you will almost

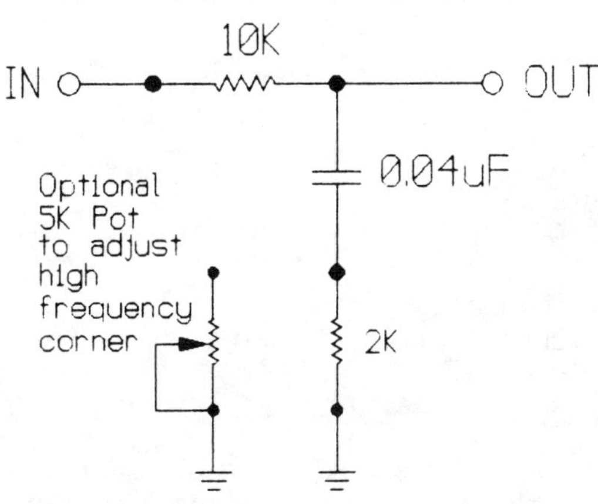

FIGURE 7-14: Shevling equalizer schematic.

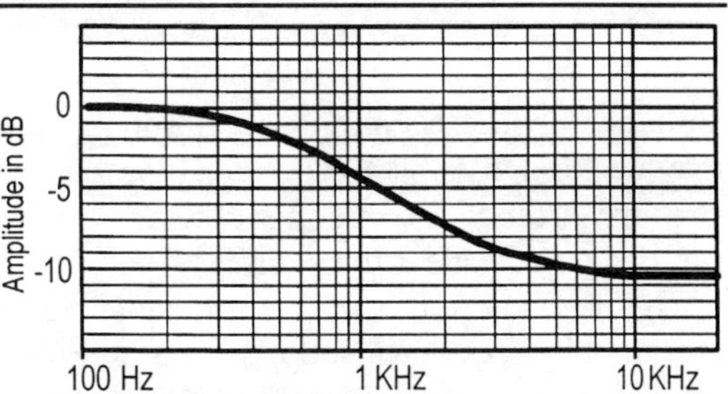

FIGURE 7-15: Shelving equalizer response.

surely need to make an active equalizer. One way to do this is simply to add a small amplifier to the output. The amplifier can add the gain you need and buffer the output against a low-impedance load. *Figure 7-17* shows a generic schematic using an IC.

While this works well and is easy to do, it fails to buffer the input from low-impedance loads. One way to solve this problem is to put an amplifier at both the input and the output of the equalizer. However, this puts two

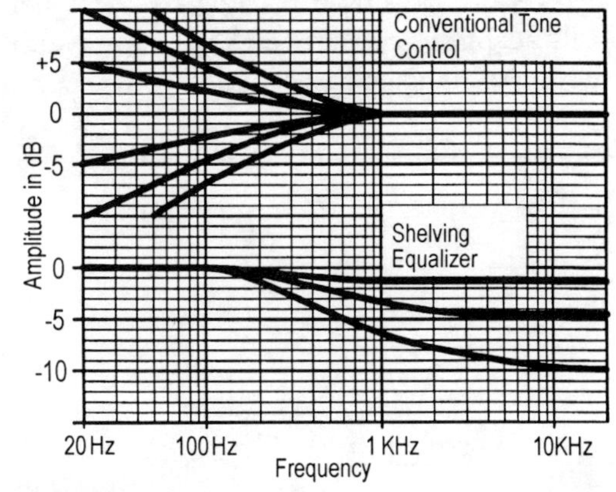

FIGURE 7-16: Compare "bass" control and shelving equalizer.

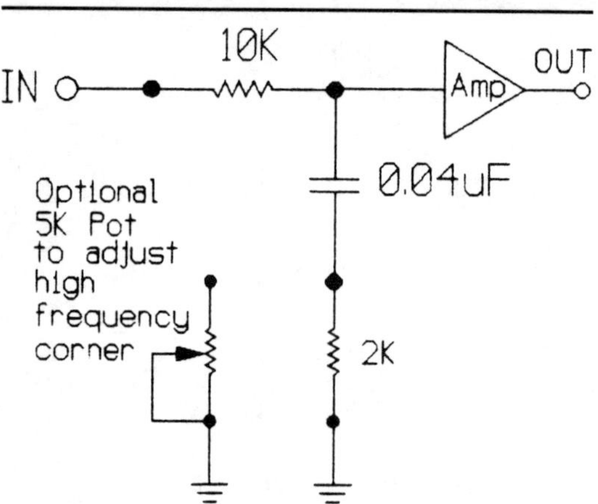

FIGURE 7-17: Equalizer with amplifier.

amplifiers in the circuitry, and although they may be good, it's still a good idea to minimize circuitry. After all, a component that isn't there can't adversely affect the sound.

The best solution is to put the equalizer in the amplifier feedback loop, so the amplifier can act as a buffer for both the input and output. *Figure 7-18* shows a generic schematic of this idea.

The amplifier's feedback loop also establishes the circuit gain. The application notes for your specific IC show how to select various ratios of resistors to put in the feedback loop to achieve the gain value you want.

Figure 7-19 shows a complete active equalizer that will correct the midrange phase cancellation problems in ESLs. It has 7dB of midrange boost and a frequency-response curve similar to the passive equalizer we designed above.

Changing the value of capacitor C4 in the feedback loop shifts the entire frequency curve up or down as shown in *Fig. 7-20*. You can easily adapt this circuit to an ESL of any width by simply changing this capacitor.

This active circuit increases the signal gain passing through it by about 10dB. It can drive your power amplifier to its maximum level without the additional gain normally provided by a preamplifier or control center. Therefore, you can operate your system without a preamplifier.

Of course, you'll quickly point out that preamplifier/control centers do more than simply add gain to the signal. They usually have some type of tone controls, allow you to adjust balance and output levels, select various music sources, and contain special equalizers for playing analog disks (LPs). How can you get by without them? *Why* would you want to eliminate them?

Many reasons exist for rejecting them, but the best two are they are expensive and they introduce distortion. Admittedly, the better preamplifiers have very little distortion, but you must agree that a component that isn't there can't color the sound.

Let's look at each function of a control center and how we might get by without it. Gain can be obtained elsewhere, such as in the ESL equalizer or crossovers. You can easily make completely passive balance and output controls. Selecting sources requires nothing more than a switch or two. Tone controls of the basic bass and treble type are inadequate and obsolete.

Most ESL builders find their new speakers so good they no longer need tone controls. You probably will discover the same thing. If

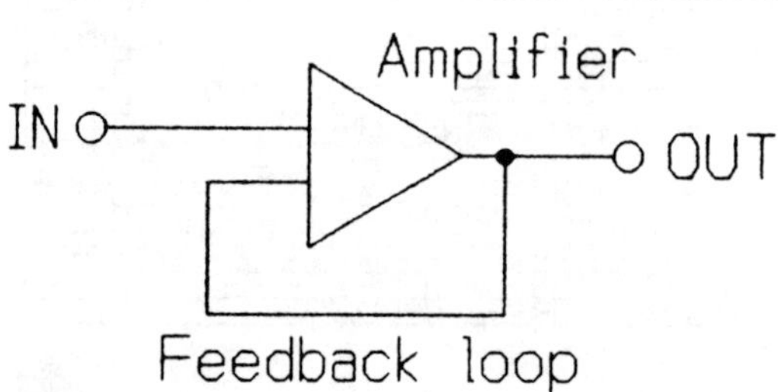

FIGURE 7-18: Generic schematic of IC equalizer.

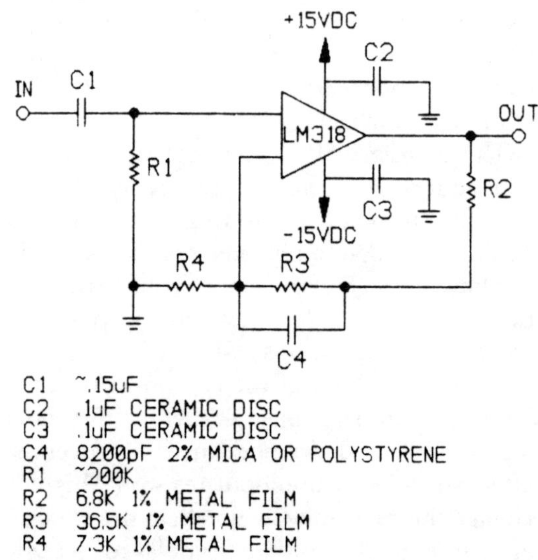


```
C1   ~.15uF
C2   .1uF CERAMIC DISC
C3   .1uF CERAMIC DISC
C4   8200pF 2% MICA OR POLYSTYRENE
R1   ~200K
R2   6.8K 1% METAL FILM
R3   36.5K 1% METAL FILM
R4   7.3K 1% METAL FILM
```

FIGURE 7-19: ESL equalizer/gain schematic.

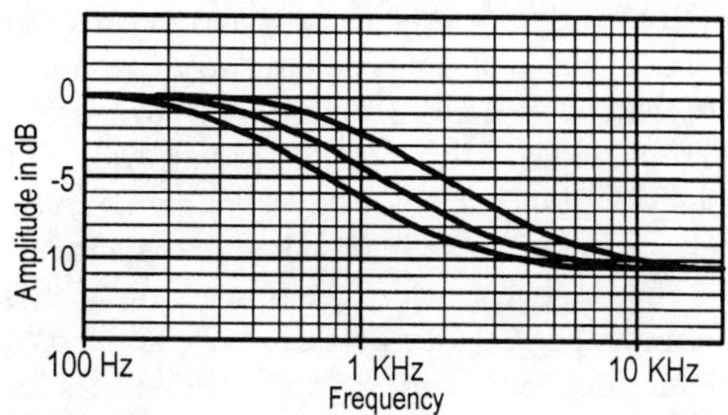

FIGURE 7-20: Effect of capacitor C4 on frequency response.

CONTROL CENTER FUNCTIONS

- Gain
- Balance control
- Tone controls
- LP RIAA equalization
- Loudness control
- Output level control
- Source selector

you must use tone controls, a 10-band octave equalizer is far better.

If you still listen to analog disks, you will need an RIAA preamplifier. But LPs deteriorate each time you play them, are very inconvenient to use, and must be cleaned constantly. You cannot automatically edit them for play.

Modern tape systems are so good that it makes more sense to copy any LP music of interest to tape; then dispose of the LPs and turntable. In this way, the music will last indefinitely and you can edit the LP when you record it so you have just the music you want. You can have long uninterrupted playing times, and you won't need an RIAA equalizer/preamplifier.

At this point, you probably realize you can build a "control center" that will perform better than commercial units, cost very little, and can be customized to your exact needs. If you use equalizers and crossovers, you can put them in one compact chassis with a power supply and the pots, switches, and jacks needed to make a "crossover/gain/equalizer/control center." The upcoming chapter on electronics construction will present a suitable design, and information you can use to customize it to your needs.

TURN ON/OFF VOLTAGE SURGES.

Active equalizers and crossovers usually have voltage surges when switched on or off. These surges can damage power amplifiers, particularly those that are DC coupled. Worse, high-power amplifiers can damage your speakers when they amplify these voltage surges.

Most power amplifiers are AC coupled. They have a capacitor at the input that blocks the passage of direct current. This type of power amplifier will tolerate turn-on and turn-off surges, but it makes a heavy "thwump" that can damage speakers.

Powerful DC-coupled amplifiers can rip woofer cones apart, if presented with a large DC signal. It is best to avoid voltage surges for all types of power amplifiers and speakers.

The simplest solution to this problem is always to leave the low-level electronics energized. Things like active crossovers and equalizers draw just this side of zero electricity, so power use is not a concern.

Electronics don't like to be switched on and off. You will find they last longer if you just leave them on. If you must switch them, a delay relay between the input of the power amplifier and low-level electronics will stop surges.

DC OFFSET VOLTAGES.

Most low-level electronics have a trace of DC in their output, called a DC-offset voltage. Most power amplifiers are AC-coupled, which blocks this DC voltage. A few modern power amplifiers are DC-coupled, and will amplify DC-input voltage.

Woofer voice coils and step-up transformers have very low resistance. Large electrical currents will flow through them, if there is even a trace of DC at the input of a DC-coupled power amplifier.

You may be surprised by how little DC-offset voltage is required to cause the power amplifier, woofers, or ESL-matching transformers to become overheated or have their operating parameters seriously degraded. Even a few millivolts is too much. If you use DC-coupled power amplifiers, you must be certain there is no DC component to the input signal.

You can solve this problem by putting a **DC-blocking capacitor** between the crossover and the power amplifier. This AC couples the power amplifier and prevents the flow of DC.

Some audiophiles hate having capacitors in the signal path, and it may be true that cheap electrolytic capacitors degrade sound. But sci-

CAUSES OF AMPLIFIER OVERHEATING

- Input DC
- Supersonic oscillation

entifically valid double-blind A-B comparison listening tests prove you can't hear a *quality* DC blocking capacitor. Nonpolarized capacitors such as polystyrene and silvered mica work well. This cure is simple, reliable, and inexpensive.

Line voltages are around 1V, so low-voltage capacitors are adequate. The capacitor's value needs to be high enough, however, to prevent low-frequency roll-off in the bandwidth of interest. For example, assume your power amplifier is driving ESLs above 400Hz. To be conservative, you can start the roll-off an octave lower—200Hz.

On the other hand, if the power amplifier is driving the woofers, the roll-off should be at 20Hz or lower. You can calculate the value with the by-now-familiar formula:

$$C = \frac{159{,}155}{fR}$$

Where:

f = -3dB corner frequency in hertz
R = Input impedance of the amplifier in ohms
C = Blocking-capacitor value in microfarads

For a power amplifier driving ESLs above 400Hz, the capacitor would need to be only 0.08μF. The calculation assumes a corner frequency of 200Hz and a power amplifier input resistance of 10kΩ:

$$C = \frac{159{,}155}{fR} = \frac{159{,}155}{200 \times 10{,}000} =$$

$$\frac{159{,}155}{2{,}000{,}000} = 0.0796 \ (0.08\,\mu F)$$

The same procedure for the bass power amplifier with a corner frequency of 16Hz and a 10kΩ input resistance yields a capacitor value of 1μF:

$$C\frac{159{,}155}{fR} = \frac{159{,}155}{16 \times 10{,}000} =$$

$$\frac{159{,}155}{160{,}000} = 0.995 \ (1\,\mu F)$$

To avoid capacitors, there is another way to solve the DC offset problem. You can put a **DC-offset null circuit** in your electronics.

You only need to put this circuit in one of your electronics stages, as the correction can be anywhere in the signal path before the amplifier. The stage just before the power amplifier is the most popular. To do this you will need to obtain the manufacturer's application notes for the IC used and follow the guidelines.

All types of electronics have DC offset problems, not just ICs. To deal with the problem, tube circuits usually use blocking capacitors, and discrete transistor circuits often have null circuits.

When using DC null circuits, you need high-stability power supply voltages. Regulated power supplies are a must. Leave the electronics on always, and check the output for DC occasionally, particularly when the unit is new.

When first testing electronics for offset, you may test the component itself. However, for final "tweaking" you should measure the output from the power amplifier. The power amplifier amplifies any DC, so it is easier to see the effects of your adjustments—and it's DC in the power amplifier output that ultimately is the problem.

The ESL power amplifier should run at idle temperature or slightly above—even when playing the ESLs at high output levels. The "wattless" characteristic of ESLs, as previously discussed, is the reason.

The ESL step-up transformers should always run stone cold. If either the power amplifier or transformers get hot, the problem should be corrected immediately.

The problem usually is DC at the input of the power amplifier, but occasionally some part of the electronics chain will be unstable and cause supersonic oscillation. Power amplifiers will sometimes be unstable with capacitive loads, but this is very rare in commercial units. Unstable amplifiers usually blow fuses at turn-on and never get the opportunity to go into supersonic oscillation.

The problem is more likely to be in the low-level electronics built by home constructors. The culprit may be ground loops, poor layout, or bad shielding. I discuss construction techniques to prevent these problems in the next chapter. Buying commercially built electronics is one way to avoid problems.

TESTING DC & ELECTRONICS INSTABILITY. Detecting DC in the output of the

> ## CAUTION
> **Although supersonic oscillation is inaudible, the amplifier probably is at full output power! Dangerous voltages and currents may be present.**

power amplifier is simple: connect a DC voltmeter across the output terminals when the power amplifier is idling. Recognizing supersonic oscillation is more difficult. Usually such oscillation gives itself away subjectively by generating a raspy quality to the sound. Often it occurs only at moderately high output levels.

Positively identifying the problem objectively may be difficult. To do so, connect an oscilloscope across the power amplifier output terminals. Don't connect it to the speaker terminals, as maximum drive voltages are likely to be present. Oscillation may only be present under conditions of stress. In other words, the oscillation may not be present when the power amplifier is idling. You may have to use music or a test waveform to excite the electronics into oscillation.

Use an audio generator for this test, since it is easy to differentiate between the test wave and an overriding "fuzz" on the waveform that represents oscillation. If the problem isn't apparent on sine waves, try square waves, then use music. It will require some experience to detect oscillations on top of music, however.

You can easily develop this skill. Observe the source output to get a good idea of what the music should look like. Be sure the power amplifier and low-level electronics are off while observing the source waveforms. If you leave them on, they may oscillate and distort the image, and you'll think the distorted image is normal.

You can then detect the tell-tale "fuzz" that supersonic oscillation superimposes on the waveform. By selectively switching out one section of electronics at a time, you can isolate the offending electronic stage and fix it.

These tips are "just in case" a problem arises. Problems are not common with commercial electronics. However, when using high-voltage D/D electrostatic amplifiers, instability is common.

High-voltage D/D amplifiers seem to drive associated electronics crazy. Troubleshooting this type of problem is difficult and beyond the scope of this text, but I'll give you one important tip: keep any D/D power amplifier several feet away from any other component.

TURN-ON INSTABILITY. Many power amplifiers have difficulty at turn-on when driving capacitive loads. They often blow fuses or circuit breakers, but are stable once they have had a couple of seconds to stabilize their internal voltages.

Fuse-blowing behavior at turn-on is NOT a sign that a power amplifier is unsuitable for driving ESLs. Some very high quality power amplifiers have this problem, and yet make superb ESL amplifiers.

The solution is to put a delay relay in the circuit between the power amplifier output and the primary winding of the step-up transformers. Many power amplifiers have such relays when they come from the manufacturer, as this is a very desirable feature.

LEVEL MATCHING CONTROLS. Your various amplifiers and drivers will not be matched in loudness. In a hybrid system, the woofer usually has more output than the ESL. You need some way to adjust the output from the various drivers. I think all amplifiers should have input-level controls specifically for this purpose, but not all do.

If yours doesn't, you can put an attenuator in it, or you can put the attenuator in the low-level circuitry just prior to the power amplifier. **Attenuators** are nothing more than a variable resistor—a potentiometer—not simply series resistance. See *Fig. 7-21* for the correct schematic diagram.

The potentiometer value needs to be lower than the amplifier input impedance. If excessively low, it may load down the low-level electronics feeding the amplifier. This is rarely a problem with ICs, since they can feed a 600Ω load without difficulty. Tube electronics are inherently high-impedance devices, and may have trouble with low-impedance loads.

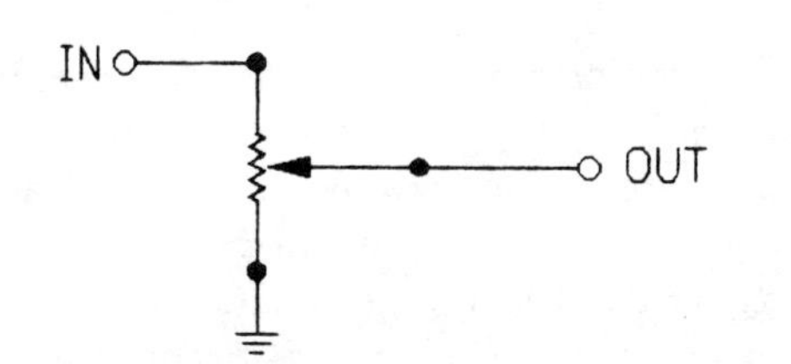

FIGURE 7-21: Generic attenuator schematic.

A 10kΩ potentiometer usually works. It is low enough to feed most low-impedance power amplifiers, but high enough not to tax most low-level electronics.

When in doubt, you can test the system by running a frequency response sweep. If the value of the pot is too low for the low-level electronics to drive, their low-frequency response will roll off at a higher frequency than expected. If the value of the pot is too high for the power amplifier, its low-frequency response will roll off.

HIGH VOLTAGE POLARIZING POWER SUPPLY.

Recall from *Chapter 5* on Output that many variables determine the polarizing voltage. For maximum output, you need to maximize the polarizing voltage, but the diaphragm instability threshold is touchy.

The guideline of 50V/mil will get you in the ballpark, but this voltage is so important for high output, and there are so many variables, that having some way to adjust the polarizing supply voltage is very desirable.

The power supply needs to deliver only a tiny amount of current—0.1mA is plenty. Power-supply ripple will not cause hum in this application, so quality standards are very low. You don't need filtration, storage, regulation, or full-wave diode bridges.

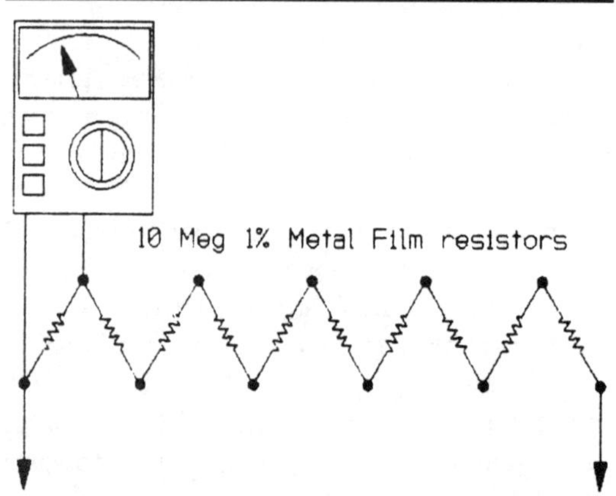

FIGURE 7-22: High-voltage probe schematic.

The diaphragm's voltage may be either positive or negative; it doesn't matter if you are using a step-up transformer. Some D/D amps require a specific diaphragm polarity (usually negative), but since the polarizing supply is usually built into them, you needn't be concerned. If it isn't, follow the manufacturer's recommendations.

Hopefully, you won't have to measure the voltage, since it can be hard to do. Standard voltmeters are limited to 1kV. You must use a series resistor with them to measure higher voltages. This is called a **probe resistor**.

If you must measure the voltage, you can buy commercial high-voltage probes that have a built-in resistor. You also can make your own.

Find the input impedance of your voltmeter. The manufacturer states its value. For most DVMs and FET VOMs it's 10MΩ. For VTVMs, it's usually 11MΩ. Don't use a meter with a lower impedance.

Multiply the input impedance by nine. For 10MΩ meters that will be 90MΩ. Use that value for a probe resistor.

This procedure decreases the load presented by the voltmeter by a factor of ten. Thus, a 10MΩ meter becomes a 100MΩ meter (10MΩ meter + 90MΩ probe resistor).

This also shifts the decimal point of the meter scale one place. For example, with the added probe resistor, if the voltmeter reads 350V, it is actually seeing 3.5kV.

The probe resistor must have a voltage rating higher than the voltage you will connect to it. Since most resistors are only rated at a kilovolt, this can be a problem. The best way around this is to use nine equal resistors in series. For 90MΩ, use nine 10MΩ resistors.

These resistors should be metal-oxide film types. Carbon resistors have a poor **voltage coefficient** and are not precise enough. A poor voltage coefficient means they change value as you increase the voltage. *Figure 7-22* shows the schematic of a probe resistor.

Measuring the output voltage from these high-voltage power supplies can be tricky because they have such limited current capacity. The total resistance of your meter and series resistor must be very high so it doesn't load the circuit excessively, which will drag down the voltage and give you an inaccurate reading.

Ohm's Law will find the lowest resistance without overloading the circuit. For example, let's say you are trying to measure a 7kV power supply that has a current rating of 0.1mA. To find the lowest resistance, divide the voltage by the current.

$$\text{Resistance} = \frac{\text{Voltage}}{\text{Current}} = \frac{7,000}{0.0001} = 70M\Omega$$

You can see that the minimum value for the meter/probe resistor will be 70MΩ. This represents the *minimum* permissible load resistance, but it's better to be conservative and use well in excess of 100MΩ.

To avoid measurements, get close to the 50V/mil range with calculations. Make your power supply adjustable within that general range, so you can "tweak" it empirically. With the supply connected to your ESL, turn up the voltage until the diaphragm becomes unstable and "caves in." Back off the voltage just a tad, and you've got it.

Although there is no commercial source for ESL polarizing supplies, they are easy to build. Most ESL designers use surplus or junked copy machine power supplies.

You can get these from surplus electronics outlets. They are usually inexpensive, often less than $10, and typically produce about 7kV.

ESLs with 70 mil D/S spacing can only use about half that. Those with 135 mil spacing can use the full 7kV—if you use the techniques that produce high-stability diaphragms.

There are many ways to reduce the output for 70 mil D/S spacing. One way is to buy a power supply designed to operate on 240V. When you use this on the usual 120V house line, the output voltage will be half the 7kV—just about right.

Adding series resistance at the input often works and is cheap, simple, and easy to do. Try a 1kΩ 20W wire-wound ceramic variable resistor. These have a clamp that you can move to "tweak" the output. If the resistor gets too hot, use a larger wattage resistor. If the full 1kΩ doesn't drop the voltage enough, use a higher value resistor (*Fig. 7-23*).

If you need more than 1kW of series resistance, the output voltage may be unstable. You can improve stability by shunting the input with a 10W wire-wound potentiometer of around 1kΩ (*Fig. 7-24*).

If the potentiometer gets too warm,

increase the value of the series resistor to reduce the current in the circuit. These resistors are cheap enough that you can experiment freely.

Don't shunt the high-voltage output with a high-value resistor. This requires very high-value resistors (several MΩ) and heavily taxes the power supply.

No potentiometer exists that you can use at this location, because they will arc internally. Therefore, you can't just "dial in" the voltage you want.

The most elegant (and expensive) way to make your power supply adjustable is to use a **variable autotransformer**—commonly known as a **VARIAC®**—at the input of your power supply. *Figure 7-25* shows a schematic diagram.

These physically appear to be a very large wire-wound potentiometer, but are actually transformers with a single winding. The wiper attached to a rotating knob allows you to vary the voltage from zero to slightly above line voltage by tapping the winding at the location of your choice. An interesting feature of VARIACs is they have their highest voltage tap somewhat short of the end of the winding. This allows you to get slightly *higher* output than input voltage, if necessary (*Fig. 7-26*).

VARIACs are extremely efficient and allow essentially infinite variability. If you put one at the input to your power supply, you no longer have to worry about adjustability.

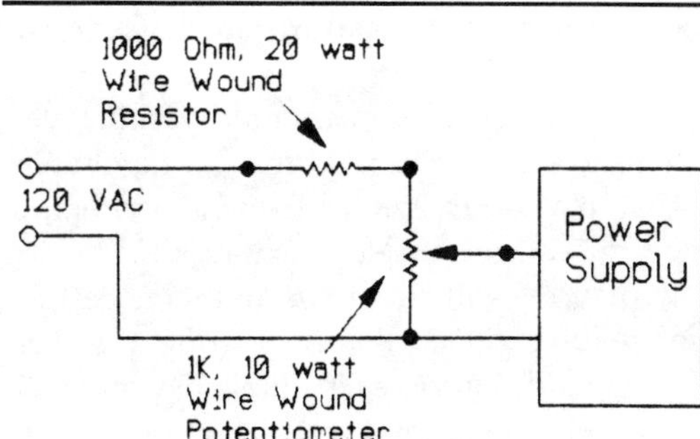

FIGURE 7-24: Voltage divider reduces voltage.

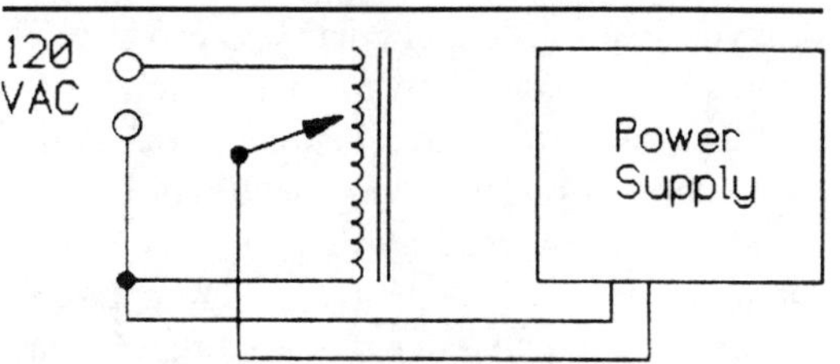

FIGURE 7-25: VARIAC voltage adjuster.

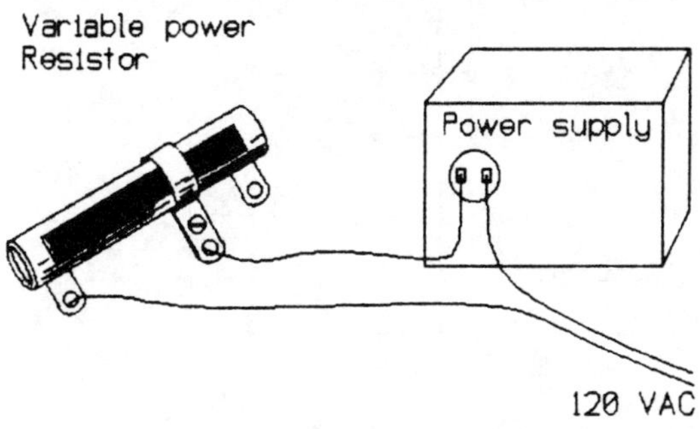

FIGURE 7-23: Series resistor reduces voltage.

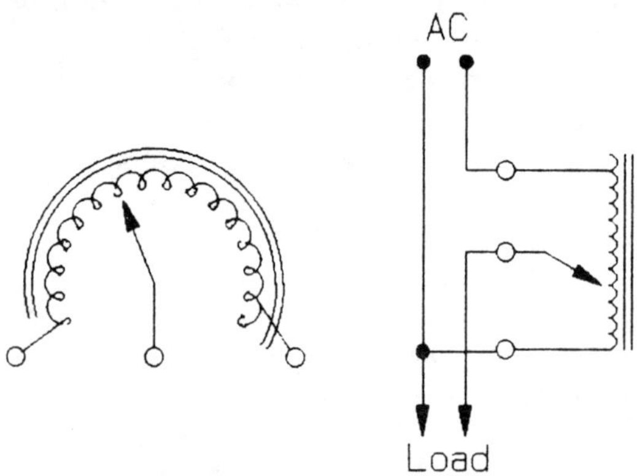

FIGURE 7-26: VARIAC schematic.

Although most builders use copy machine power supplies, there are other sources. Neon-light power supplies should work, although I've never used one. You can often get these from old beer signs at a local bar or grocery store.

If you can't locate a ready-made power supply, you don't need much electronics background to make a simple transformer/diode power supply. The catch is getting a small, high-voltage transformer.

Remember that transformers use RMS ratings, but when rectified (changed from AC to DC output) their DC voltage will be the peak value that is 1.414 times the RMS value. For example, to get 3kV DC, the transformer need only be rated at 2.1kV AC.

If you can't find a suitable transformer, you have several options. You can use lower-voltage transformers ganged in parallel/series as shown in *Fig. 7-27*.

By using multiple diodes and capacitors, you can make voltage doubler, tripler, or quadrupler circuits that do exactly as their name implies. See *Fig. 7-28* for suitable schematics.

It may be difficult to find the necessary high-voltage diodes. Putting a string of lower-voltage diodes in series is unreliable; it is best to get high-voltage diodes. Here's a helpful tip: a good source for diodes rated to 5kV is from microwave oven repair shops. If all else fails, or if you want the experience, you can custom wind a high-voltage transformer, which is easier than you might think. I explain how in the next chapter on electronics construction.

Note that you need only one power supply for all your speakers, not a separate power supply for each channel or speaker. You may use more than one, if you wish to install a power supply within each speaker enclosure, but it isn't necessary.

POWER SUPPLY LOCATION. Where should you put the polarizing supply? You can mount it with the step-up transformers in a separate chassis kept near the power amplifiers, or you may put them in the speakers. There are pros and cons for both. Mounting them in the speakers has three disadvantages:

- You need two high-voltage power supplies unless you run a separate polarizing voltage wire from one speaker to the other.
- You must run an AC power cord to a wall socket from each speaker to energize the power supplies.
- When builders put the high-voltage power supply with the speaker, they usually also put the step-up transformers there. This requires that large wire be used to connect the speakers to the amplifiers.

The only advantage of putting the electronics in the speakers is there will be one less electronics chassis to park near your amplifier(s). On the other hand, the advantages of having a separate chassis are many:

- Only one power supply is needed.
- The wires running from the transformer to the ESL may be very small.
- The speakers won't have AC power cords.
- The chassis may be specifically designed for high voltage. The best insulation when building high-voltage electronics is plenty of room around the parts.

If you need assistance designing and building a power supply, just look for a large "ham" antenna. An amateur radio operator lives in the house under it. They not only have infor-

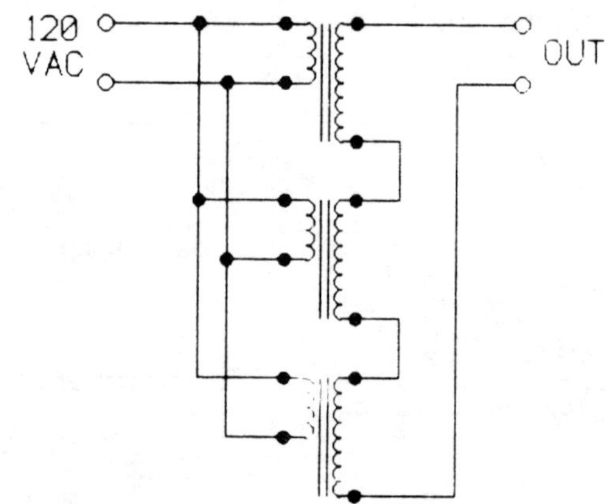

FIGURE 7-27: Using many transformers to increase voltage.

mation on power supplies, but are usually friendly folks who enjoy helping with interesting projects. If they can't help, they will know someone who can.

ELECTRICAL SAFETY.

ELECTRICAL SAFETY. ESLs have several sources of high voltages. Despite this, they are not very hazardous when you analyze the situation. Still, caution is appropriate.

It is *current* that kills, not voltage. Voltage itself is not dangerous. High voltages only make it possible to drive current through a person.

The way to kill someone, with the least amount of electricity, is to drive current through his heart. About 50mA will disrupt the heart's electrical system. Any other electrocution method requires large amounts of current to literally burn and coagulate human tissues.

The maximum current a large ESL system can deliver is only a fraction of an amp. Therefore, high-power electrocution doesn't apply to ESLs.

Polarizing voltage is the most common source of shocks. Most builders get zapped by the diaphragm voltage. While unpleasant, it isn't hazardous. It's like touching a door knob after having walked across a new carpet on a dry day.

The charging resistor isolates you from the polarizing power supply, and prevents significant transfer of current. Assume, for example, that the polarizing voltage is 3.5kV and there is a 22MΩ resistor in series with the diaphragm. Using Ohm's Law, the maximum current that can flow through this resistor would be only slightly more than one tenth of an mA. Even ten times the voltage would deliver less than 2mA, which is far below the 50mA required to stop a heart.

The diaphragm coating stores electricity even when the polarizing supply is disconnected or switched off. In fact, it is when working with a disconnected cell that most shocks occur. Again, the diaphragm has a high resistance that prevents significant current flow.

While the diaphragm of an ESL is not an electrocution danger, this is not true of the polarizing power supply. Although these usually have a rating of only a few milliamperes, they may have a power-supply capacitor that can deliver enough current momentarily to be fatal. Treat the power supply with caution and mount it so that children, pets, and you can't touch it.

A practice used by technicians who work with high voltages is to keep one hand in a

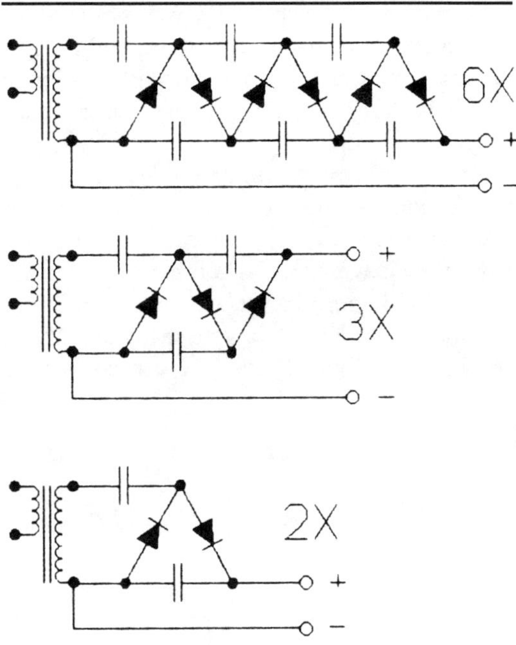

FIGURE 7-28: Voltage multipliers.

pocket when working around high voltages. For current to pass through the heart, it has to travel from one side of the chest to the other. The obvious path is from arm to arm. By using only one hand, the technician makes it nearly impossible for current to pass through their heart.

If you don't do this, you tend to forget where your other hand is while probing the innards of a high voltage source. Most likely it will end up resting on the chassis, where it can complete the electrical circuit if the other hand touches something "hot."

The main shock risk is the audio-drive voltage. It can pass several hundred milliamperes because there is no resistance in series with the voltage source.

Fortunately, it is not easy to get shocked by the stators. Since they are "floating" and aren't referenced to ground, you can't readily get shocked by touching just one. You have to touch one stator with one hand and the other stator with the other hand while loud music is playing.

Even this does not guarantee a fatal shock, but I wouldn't want to try it. How likely is it that somebody will touch both stators while you're playing loud music? Not likely, but then people do weird things. At low output levels or when the music is off, there is no hazard.

The connections between the cell and amplifier are the greatest risk. These are close together, easily reached, and we often work with them while testing. I recommend the following precautions. First, use plastic grille cloth. It offers significant protection from touching the stators.

Second, be thoughtful when designing your connections between the ESLs and the step-up transformers. Do *not* use screw terminals, since they leave exposed contacts and wiring.

Safer and more convenient connectors are banana plugs. These are well insulated, give a solid plug-in contact, and you can color code them. If you mount them on a piece of insulating plastic, and the connections between the plugs and the stators are protected from human contact, they are quite safe (*Fig. 10-49*).

One final caution: direct-drive high-voltage amplifiers are extremely dangerous. Not only are high voltages present with high current capacity in the amplifier, but the amplifier will have large storage capacitors in the high-voltage power supplies. This setup can truly fry someone. A conventional amplifier and step-up transformer is much safer.

CHAPTER 8:
ELECTRONICS CONSTRUCTION

Many speaker builders have good, basic wood-working and general construction skills, but feel insecure when it comes to building electronics. If you find yourself in this state, this chapter should prove very helpful. If you're familiar with electronics construction and have previously scratch-built projects, flip to *Chapter 9*.

As with ESLs, there is no magic to building electronics. You just need some guidance in the basics to get around the common pitfalls.

I'm going to take you through the construction of an integrated circuit gain/equalization/crossover unit, and a polarizing voltage/step-up transformer assembly. Along the way, I'll explain the basics of electronic parts, construction details, and how to customize components.

SCHEMATIC DIAGRAMS. Schematics are simply maps showing how electrical parts connect to each other. Symbols represent the components. *Figure 8-1* shows some common symbols. I've shown a simple schematic diagram in *Fig. 8-2*. In *Fig. 8-3*, I've drawn components and connected them together as shown in the schematic.

Figure 8-4 is a schematic of a basic gain/equalization/crossover unit. *Figure 8-5* shows it turned into a full-blown control center. I designed these to work with the hybrid ESL/TL systems shown in this book. You may use them as shown, or customize them to your special needs. Don't be intimidated by the complexities I've just outlined. By the time you finish this chapter, you will be able to understand and build electronics.

Building electronics is less a technical skill than a mechanical task—it's just putting pieces together. Most of the work is in the chassis. This involves no more than drilling holes, bolting things together, and dealing with cosmetic details.

To make the chassis, you should have all the parts on hand, so you can do a physical layout.

CIRCUIT BOARDS. Before turning to chassis construction, I'm going to describe how to make **printed circuit (PC) boards**. My intent is to discourage you from making circuit boards stuffed with high-precision parts because it's better to use commercial units.

Oh sure, you *can* fabricate your own, but it requires an inordinate amount of work and expense if you only need a few. For mass production this is not an issue, but for a small project I don't think it's worth the trouble.

Making PC boards is a multi-step project that begins with designing the circuit board on paper, or computer, or by laying it out using precut "donuts" and tape on transparent film. You can purchase circuit-board layout elements made of tape. You must then turn your drawing into a photographic negative. Take a picture of your work with a large-format camera and develop the film.

Attach the negative to a piece of blank circuit board which has an etch-resistant, photo-sensitive coating. Following the manufacturer's recommendations, expose the board to a high intensity light or the sun for the specified time.

Develop the etch-resistant coating on the board by soaking it in the specified chemical baths. This is similar to developing photo-

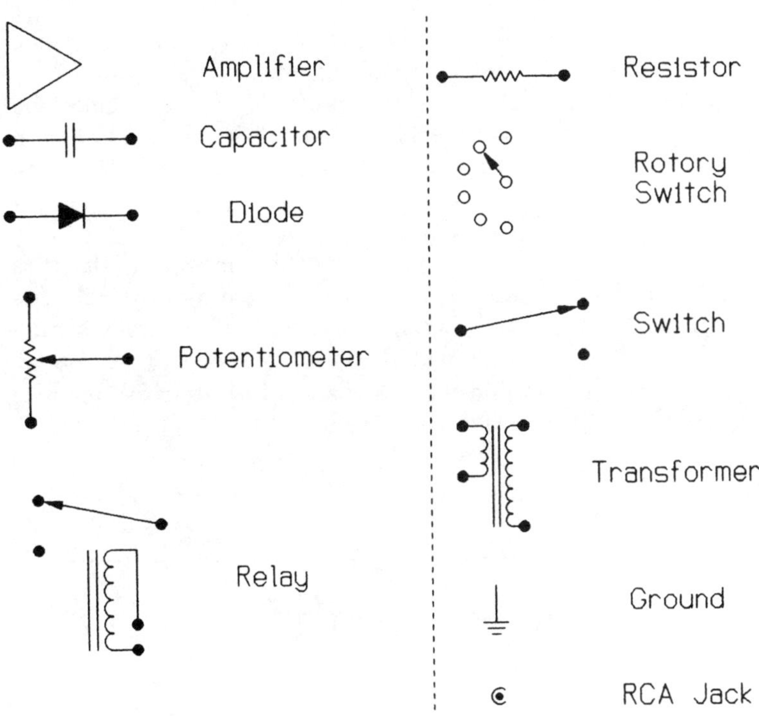

FIGURE 8-1: Symbols for electronic parts.

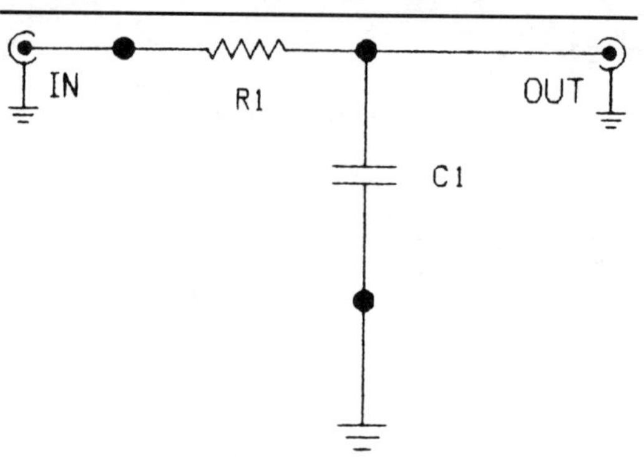

FIGURE 8-2: Sample of a simple schematic diagram.

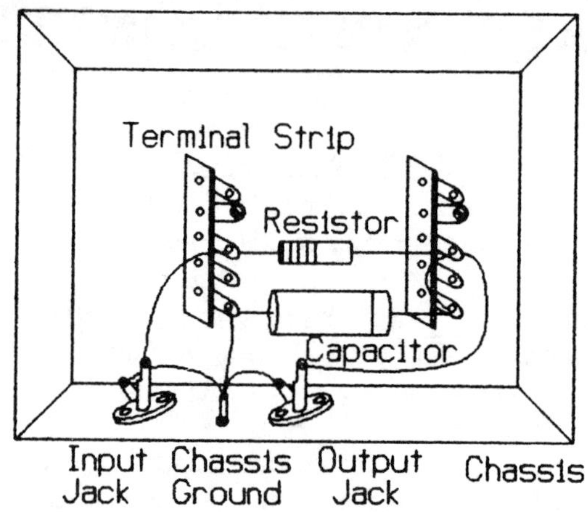

FIGURE 8-3: Components in postion and connected based on schematic in *Fig. 8-2.*

graphic prints, but instead of a picture you have etch-resistant tracks where light struck the board.

Place the board in an etching solution. This dissolves the copper from the unexposed areas, while leaving the photographically-reproduced, etch-resistant coated areas unaffected. Wash the board thoroughly.

Drill holes where the various component leads will stick through. You probably will be surprised by how many holes you must drill, and by how fast fiberglass circuit boards dull standard drill bits. A drill press is extremely helpful. You will also be frustrated by the fact that the long, tiny drill bits like to wander instead of drilling in the desired location.

This experience will help you understand why circuit board manufacturers prefer not to use fiberglass for their circuit boards. It also explains why they use very short, solid, carbide drill bits.

Now find high-quality, precision-value resistors and capacitors. The run-of-the-mill resistor is a carbon device, and is typically rated to only within 10% of its stated value. You will find that the standard values may not meet your requirements.

You will need 1% tolerance, metal-oxide film resistors and similar precision, tight-toler-

ance capacitors for steep-slope crossovers. You won't find these at your local Radio Shack or TV repair shop. When you find a source, you'll be surprised by the price and high minimum-order requirements.

Ways around some of these problems exist, but have their own difficulties. In the old days we used to **breadboard circuits** instead of building photographic circuit boards.

The breadboard was circuit-board material the manufacturer drilled with jillions of small holes—usually 10 holes/inch—100 holes/square inch. You could stick component leads through the holes and solder them together on the other side.

Although messy, unsightly, and error prone, breadboards usually worked for simple circuits with discrete devices. In this age of miniature integrated circuits, however, they frequently *don't* work.

The maze of wires running haphazardly on the back side of the board often destabilizes the circuitry and causes oscillation. This doesn't mean you can't successfully breadboard, but it's not as easy as it used to be.

You can sometimes get around the problem of expensive precision resistors and capacitors by hand-selecting the values you need from a stock of wide-tolerance ones. This is time-consuming and the parts are not of high quality, but you might find that your local TV repair shop will let you rummage around in their stock bins.

On balance, I think it makes more sense to buy commercial modules. They usually are of very high quality, you don't have to test them

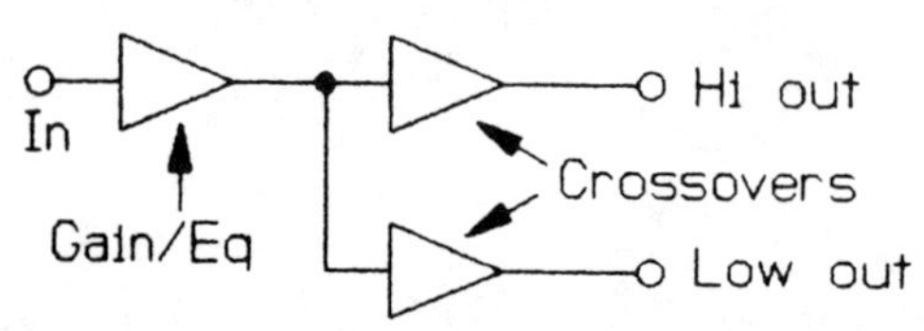

FIGURE 8-4: Gain/equalizer/crossover unit schematic.

or worry that they work correctly, and you will have a guarantee in the event of problems.

Also, they are cheaper than a home project. You can usually buy commercially for less than the cost of your do-it-yourself project.

I don't mean to suggest you shouldn't build circuit board electronics. Do so if you want the experience, or if you work where supplies and equipment are available. Most speaker builders only need the electronics, and are put off by having to build them. If you're one of these, I recommend you buy commercially.

The chassis and the power supply are expensive for manufacturers. You can save money in these areas and make things exactly the way you want.

PARTS AND RATINGS. Before getting into the nitty-gritty of building chassis and power supplies, you need to understand the basics of the various parts you'll be using. I'll discuss four parts: resistors, capacitors, semiconductors, and transformers.

RESISTORS. There are three general types available : **carbon**, **wire-wound**, and **metal-oxide** film. Others exist, but these are the ones you are likely to use. They in turn have three ratings: resistance, wattage, and tolerance.

Resistance refers to the resistance value in ohms. Printed values on small resistors are hard to read, so they usually have color-coded bands to indicate their resistance and tolerance.

Four colored bands are typical. The first two represent numbers, the third is a multiplier, and the fourth is the tolerance. Occasionally, a fifth band is present, which is a reliability rating for military use.

The band nearest the end is the first band. Bands 2, 3, and 4 are progressively nearer the center (*Fig. 8-6*).

The passage of electrical current through a resistor generates heat, which it must dissipate without failure. Its ability to do this is a function of its construction type and the surface area available for air-convection cooling. These factors define its **power** or **wattage** rating.

Tolerance refers to how closely the resistor meets its stated resistance value. For example, a 20% tolerance, 100Ω resistor may have an actual resistance value anywhere between 80Ω and 120Ω, while a 1% resistor would lie between 99Ω and 101Ω.

Voltage is not usually specified, as most resistors will handle at least a kilovolt without difficulty. However, if you have higher voltage requirements, you must either buy special types or string several stock-value resistors together.

Voltage coefficient is a resistor's tendency to change value at higher voltages. Usually unspecified, it isn't a problem for most applications.

For precision high-voltage applications, you must keep this factor in mind. A good example is the feedback circuit of a high-voltage D/D amplifier. Carbon resistors have the

---

TABLE 8-1

PARTS LIST FOR FIGURE 8-5

A	Gain/Eq module, Channel 2
B	Gain/Eq module, Channel 1
C	Crossover module, high pass, Channel 2
D	Crossover module, high pass, Channel 1
E	Crossover module, low pass, Channel 2
F	Crossover module, low pass, Channel 1
1	Output jacks, ESL amps
2	Outputs to woofer amps
3	Inputs from main tape deck
4	Outputs from main tape deck
5	10-high level inputs from CD, FM, DAT, Cassette, etc.
S1	Rotary switch input selector
S2	Tape monitor switch
R1	Gain control pots, 10Ω, audio taper
R2	Balance pots, 10Ω, double log taper

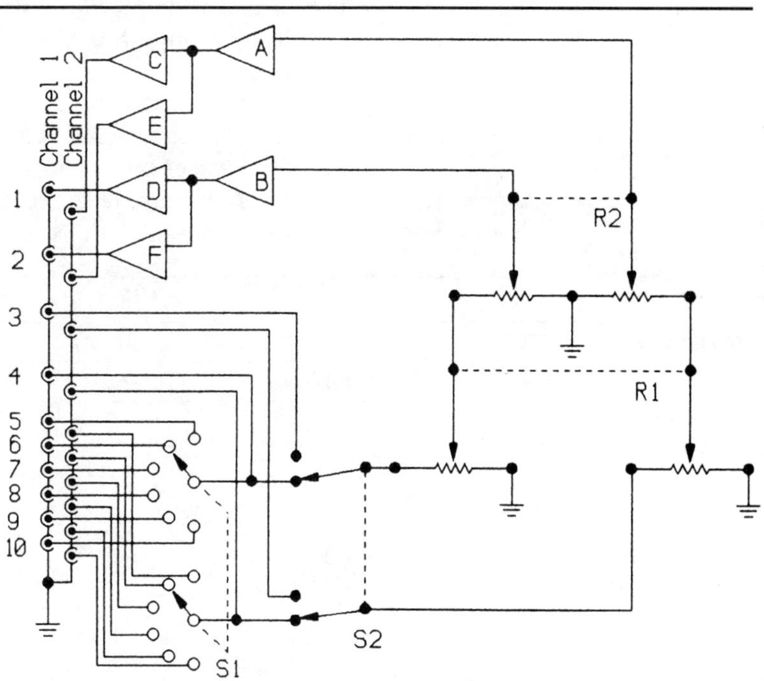

FIGURE 8-5: Schematic for a control center.

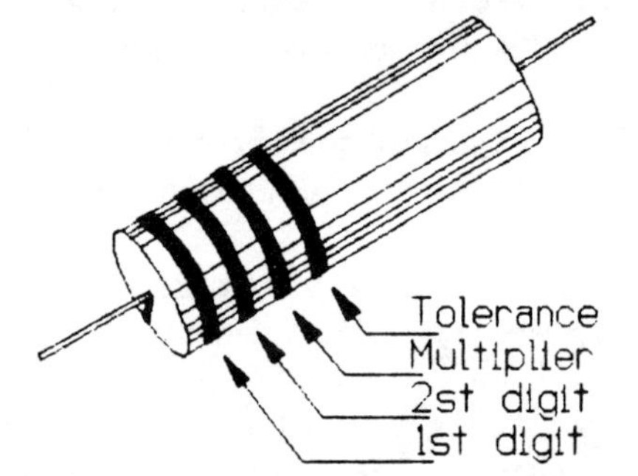

FIGURE 8-6: Color-coded resistor bands.

worst voltage coefficients, and metal-film resistors have the best.

Carbon resistors, made from compressed carbon granules, are cheap, common, and have low tolerance values—10–20%. Common wattage ratings lie in the ¼–5W range—the usual value being ½W.

They have a limited life span and will eventually fail. If overheated, they sometimes burst into flame. Because of product liability issues, even the TV repair shops use "flameproof" metal film resistors.

Carbon resistors are OK in noncritical areas, but are not precise or stable enough for steep slope equalizers. They occasionally fail. If you don't ever want to fix things, you'll be better off with a metal film resistor.

Metal-oxide film resistors are commonly called metal film resistors, or simply film resistors. They are made by coating a tiny glass cylinder with a metal oxide.

Metal oxides have much more resistance than elemental metals. The manufacturer controls the resistance and wattage by varying the type of film, its thickness, the size of the unit, and by spiral tracks cut into it.

TABLE 8-2

BAND COLOR TABLE

Color	Number	Multiplier	Tolerance
Black	0	.0	
Brown	1	0	
Red	2	00	
Orange	3	000	
Yellow	4	0,000	
Green	5	00,000	
Blue	6	000,000	
Violet	7		
Silver	8	Divide by 100	10%
White	9		
Gold		Divide by 10	5%
No 4th Band			20%

RESISTOR RATINGS

· Resistance

· Wattage

· Tolerance

Metal-film resistors have tight tolerances— 1–2% is typical. They are high quality and rarely fail. Like carbon resistors, they have low wattage ratings. Most high-quality precision electronics use metal-film resistors of the type you will need for steep-slope equalizers.

Wire-wound resistors are used in high-power applications. They usually consist of a ceramic core wound with resistance wire, and are often hollow to increase surface area and heat dissipation capabilities. Some have an open section and a movable clamp to adjust the resistance. They come in a wide range of physical sizes. The largest may be several feet long and handle hundreds of watts. While carbon and metal-film resistors rarely even get warm, wire-wound are designed to operate at very high temperatures.

Tolerances are low (10% is typical), but since they are not used in precision applications, this is not a problem. If you need greater tolerances, the adjustable types are usually satisfactory.

Potentiometers (**pots**) are a special form of adjustable resistor with a moveable wiper

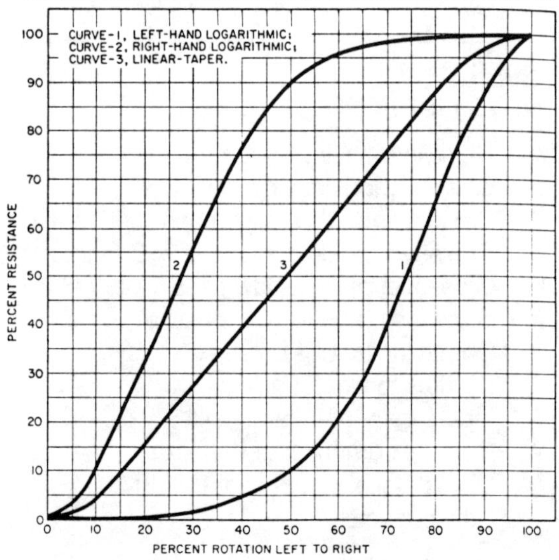

FIGURE 8-7: Resistance taper curves for potentiometers.

which varies the resistance. Most are of rotary construction where the wiper shaft will rotate about 270°, but high-precision, multi-turn types exist. Pots usually have three contacts—at each end of the resistor and at the wiper.

Like resistors, cheap potentiometers have a carbon track and are low tolerance, low power, and low reliability. The metal-film and plastic-film types are usually more precise and of higher quality—and also more expensive. Some carbon-track resistors are very precise and expensive. The wire-wound types can handle large amounts of power.

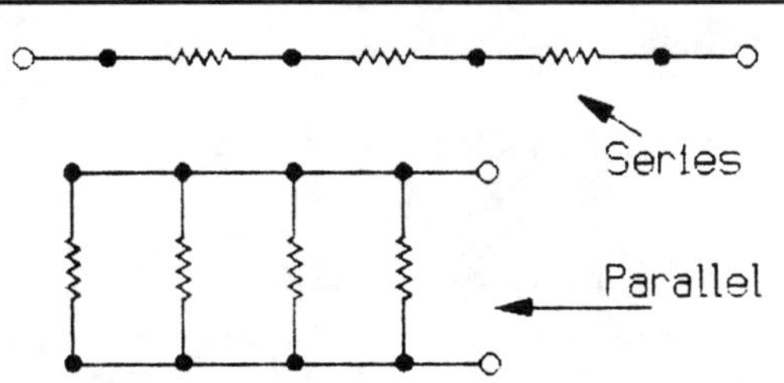

FIGURE 8-8: Types of resistor connections.

Pots have an additional characteristic: **taper**. This refers to the change in resistance as the wiper swings from one end to the other. Typical tapers are **linear**, **logarithmic**, and **reverse logarithmic**. With linear tapers, the resistance changes uniformly. Log or audio tapers have a progressively changing taper, where the resistance at one end changes much faster than at the other (*Fig. 8-7*).

An audio pot changes resistance only slightly at low-gain settings, so the loudness doesn't change dramatically when you adjust the volume. As you turn the music louder, the change becomes more rapid.

The reason is because humans do not hear loudness changes linearly—we hear geometrically. For this reason, the decibel (dB) scale is not linear. Recall that if we double the loudness, we only perceive a small change (3dB). Therefore, if we want the volume to increase smoothly, the pot must have a geometric taper.

If you wonder about this, just use a linear pot as a volume control and see what happens! The first 10% of the rotation makes the sound subjectively go from nothing to about 90% of full output. The last 90% of the rotation makes almost no difference.

The issue of tolerance, or more specifically **tracking tolerance**, becomes very important

here. In a stereo system, the pots for each channel should change resistance identically as you rotate them. If they don't, the balance will shift. Because of the audio taper, manufacturers find it very difficult to get good tracking—20% is about the best you'll find in a *good* American-made pot. Japanese and French manufacturers supply dual log pots that match to within 2–3%.

Some audiophiles use fixed-resistor switching networks to get precise tracking. Another excellent alternative is an electronic remote control. These units use ICs to control the gain, and are much more precise than a mechanical pot. I highly recommend them both for their convenience and for precision tracking.

CONNECTIONS & CALCULATIONS. You can connect resistors either in series or in parallel (*Fig. 8-8*). The total resistance of a string of series-connected resistors is simply the sum of their resistances. The string's power rating will be approximately the sum of the ratings.

Resistors connected in parallel will have resistance lower than any of the individual resistors. They add according to the formula:

$$\frac{1}{R_T} = \frac{1}{R_1} + \frac{1}{R_2} + \frac{1}{R_3} + \text{etc.}$$

The power rating of the resistor group is *not* the sum of all the resistors. You must consider each resistor's power-dissipation capacity separately using Ohm's law in order to find how much current that resistor will draw. Multiply the current by the voltage to determine the wattage. Always be conservative—you don't want hot resistors in your electronics if you can avoid it.

CAPACITORS. Like resistors, they come in several types and tolerances, but with other characteristics which are different. These include voltage and capacitance ratings.

Voltage refers to capacitor tolerance before it fails (arcs internally). Like an ESL, there is a voltage limit. Because higher voltages require thicker insulators, higher voltage capacitors will generally be larger for a given capacitance and design.

Capacitance is the amount of electricity (number of electrons) the capacitor can store. Think of a capacitor as a battery. A large battery will hold more electrons than a small one.

Tolerance is how closely the capacitor matches its capacitance rating. Electrolytics may vary by 50%. Nonpolarized capacitors

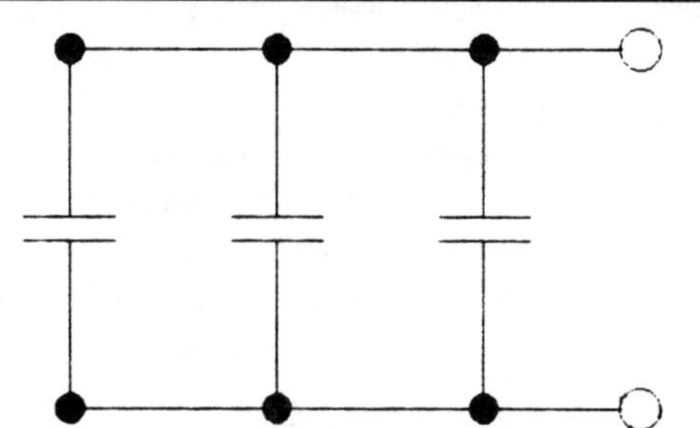

FIGURE 8-9: Capacitors connected in parallel.

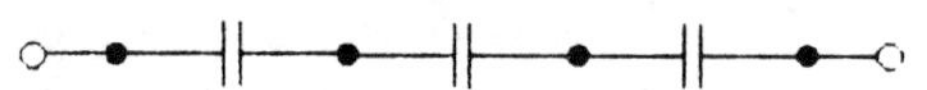

FIGURE 8-10: Capacitors in series.

are usually 20%. Tight-tolerance (2%) capacitors are expensive and hard to find.

Capacitor **type** mostly has to do with the insulator separating the plates inside the capacitor. These may be made of mica, polystyrene, tantalum, polyester, Mylar, glass, or other materials. Each of these insulators has various electrical characteristics which have significance depending on the application.

Nonpolarized capacitors don't care which lead is positive or negative. They work either way. They are high quality, and are the type of capacitor you should use in your audio-signal path.

Polarized capacitors are touchy about polarity—much like a battery. They work correctly only when connected correctly. The manufacturer marks the capacitor's polarity.

Polarized capacitors are commonly called **electrolytic** capacitors, because this is the most common type. Compared with nonpolarized types, they are of lower quality, cannot handle as much voltage, are unreliable, tend to leak their charge, and some leak oil when they fail.

If electrolytic capacitors are so bad, why do we use them? They have one redeeming quality: they hold more energy for a given size than nonpolarized types. An electrolytic may have more than 100 times the capacitance of a nonpolarized cap. When you need a large "electron reservoir," you have little choice but to use electrolytics.

CONNECTIONS & CALCULATIONS. If space is a problem, you may put several capacitors in parallel where their total capacitance is the sum of the group, and the voltage rating is unchanged. This assumes that all the capacitors have the same voltage rating (*Fig. 8-9*).

CAPACITOR RATINGS

· Capacitance

· Voltage

· Tolerance

To handle higher voltages, you can put capacitors in series as per *Fig. 8-10*. The voltage capability will be additive, but the capacitance will *not* remain the same. Putting them in series reduces the capacitance. They add according to the formula:

$$\frac{1}{C_T} = \frac{1}{C_1} + \frac{1}{C_2} + \frac{1}{C_3} + \text{etc.}$$

POWER TRANSFORMERS. Power transformers have three ratings: **voltage** (both primary and secondary windings), **current**, and **frequency**. The primary winding's voltage is your house current; the secondary's is the voltage you feed to your electronics. These voltages are RMS (multiply them by 1.414 to find the peak voltage).

The transformer should deliver its current rating continuously. It will get fairly hot if you run it at its limit, so you should be conservative. Choose a transformer that will deliver at least 20% more current than you'll need. You can draw up to twice a transformer's current rating momentarily, but a continuous overload will cause overheating and failure.

Frequency refers to the input alternating current. Like audio transformers, higher frequency transformers are more costly, but they offer higher performance.

House current is 60Hz in the US and 50Hz

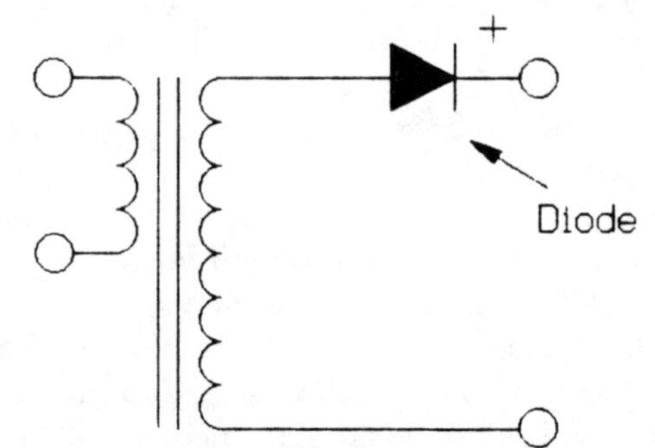

FIGURE 8-11: Diode circuit.

in Europe. Most power transformers are built to operate at these relatively low frequencies.

Some power transformers operate at higher frequencies. The advantage is that a transformer can transfer more power at higher frequencies than at lower ones.

A good example of this is in military aircraft, where weight is important. Large transformers are heavy. Aircraft types work at 400Hz so that small, lightweight transformers can handle high power. For example, a transformer rated at 50W at 40Hz can handle 500W at 400Hz.

SEMICONDUCTORS. These include diodes, transistors, and integrated circuits (ICs). This topic is too large to discuss in detail, so I'll only touch on diodes.

Diodes are one-way check valves for electricity: current will flow through them in only one direction. As with all electrical circuits, that direction is from the negative to the positive.

The diode manufacturer will have marked the positive end with a ring, a plus sign, a color dot or band. *Figure 8-11* shows a typical connection and schematic symbol.

Since we often use diodes as full-wave bridge rectifiers (to be discussed shortly), they also come in integrated packages with the necessary diodes connected internally. *Figure 8-12* shows a schematic diagram.

A **full-wave bridge rectifier** will have four leads, and looks like one of the packages in *Fig. 8-13*. Two of the four leads will be marked "AC," and the others will be marked positive and negative. Sometimes only the AC and positive leads are marked, and the negative is blank.

Diodes are rated by their voltage and cur-

rent capacities. The voltage is rated as **PIV**, which means **P**eak-**I**nverse **V**oltage. Notice the word *peak*. You must never exceed this voltage or the diode will instantaneously fail.

What trips up many amateurs is that AC voltage is RMS voltage, unless otherwise specified. Recall that peak voltage is 1.414 times the RMS voltage.

Well-intentioned but uninitiated amateurs will connect a diode rated at 15V with a 12V transformer, and are surprised when it fails.

They don't realize that a 12V RMS transformer develops almost 17 peak volts. They also overlook the occasional power-line voltage spikes that can easily drive a diode over its limit.

You must be conservative. If you are using a 24V transformer, the peak voltage will be almost 34V. The diodes should be rated for at least 50V—and 100V is an even better choice.

Because diodes have some resistance, they get hot when they pass current. Like resistors, they must dissipate this heat, or they will fail. Also, their internal junctions are limited in the amount of current they can handle. Therefore, manufacturers give them current ratings.

You should not run diodes at their current limits, because they will run hot. Be conservative and use them at no more than half their capacity.

Also, to keep them cool, manufacturers design high-power diodes to be used with heatsinks. Typically, the diode is in a metal

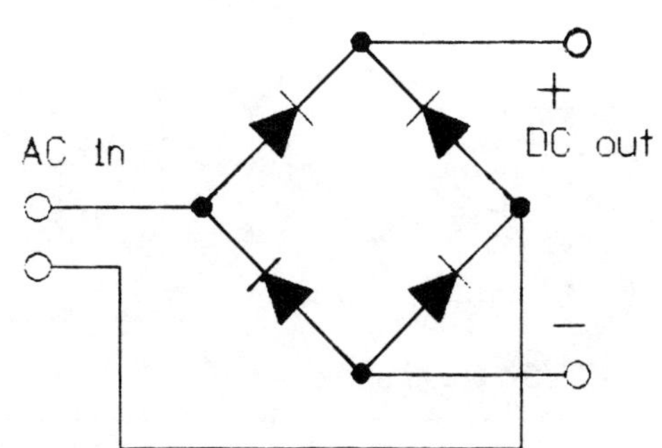

FIGURE 8-12: Full-wave diode bridge.

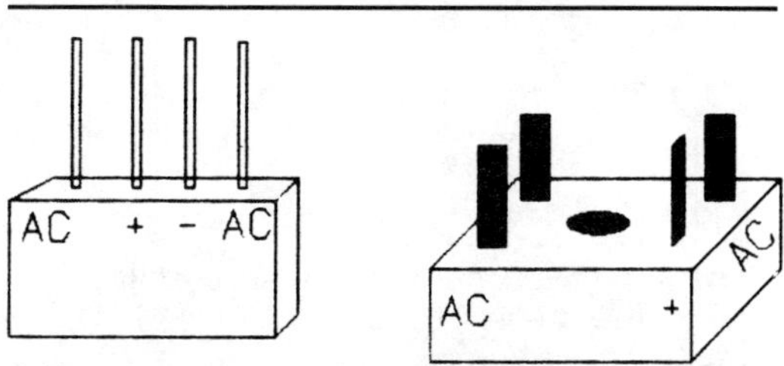

FIGURE 8-13: Full-wave diode bridge packages.

case that you can bolt to the chassis. This not only makes it possible to mount the diode conveniently, but allows the heat to be drawn away from the diode into the chassis.

If you use this type of diode, be sure you mount it on a metal chassis so it can rid itself of heat. It must be mounted against bare metal, not paint.

You should put **heatsink grease** between the diode and the chassis. This transfers heat more efficiently. When mounting power transistors, heatsink grease is essential.

Zener diodes act as voltage regulators. They behave like ordinary diodes until you exceed their voltage rating. Then they conduct current backward—from positive to negative—and are said to be **reverse biased**.

Think of a zener as a spring-loaded, one-way gate that normally lets electrons through in only one direction. If there are enough electrons pushing against the gate hard enough, the spring gives and lets the electrons go through the other way.

Unlike a normal diode that permanently fails when you exceed its PIV, zener breakdown is reversible. When the voltage drops below the reverse-bias threshold, the diode stops conducting backward and acts like a normal one-way diode again.

You can regulate circuit voltage by shunting the power supply with the zener (*Fig. 8-14*). If the power-supply voltage creeps up too high, the zener conducts and drags it down to its reverse-bias threshold. The zener can switch on and off very rapidly to control the circuit voltage.

This rapid switching generates electrical noise. Standard practice is to shunt each zener with a small value capacitor to quiet it.

To use a zener as a voltage regulator, you need to make your power-supply voltage slightly higher than the zener's voltage rating so it has something to regulate down to. To be sure you don't exceed the zener's current rating, the power supply must have some series resistance.

Select the resistance value that will pass enough current for the needs of your electronics, but not so much that it overloads the zener. *Figure 8-15* shows a generic schematic diagram.

BUILDING A LOW-VOLTAGE POWER SUPPLY.

Most ICs operate from 15V power supplies. Their current requirements are only a few milliamperes, so a small power supply is all you need. However, you may make it very large if you wish. It will not harm your electronics to have more current available than necessary. Your electronics are like a 100W light bulb that draws 0.83A from a 120V AC house circuit. The circuit can probably deliver 20A, but the bulb doesn't care—it only uses what it needs.

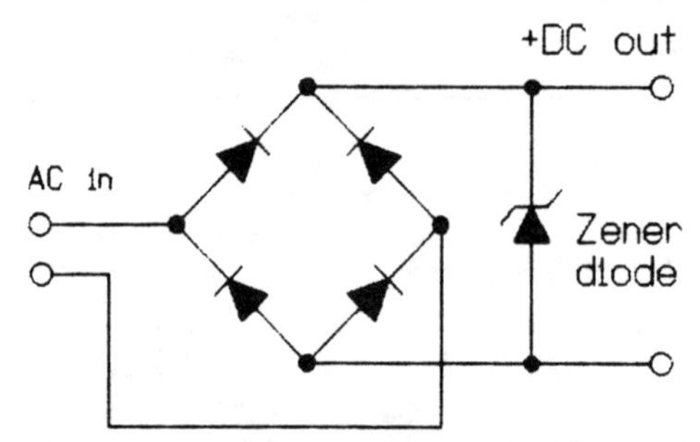

FIGURE 8-14: Zener diode regulator.

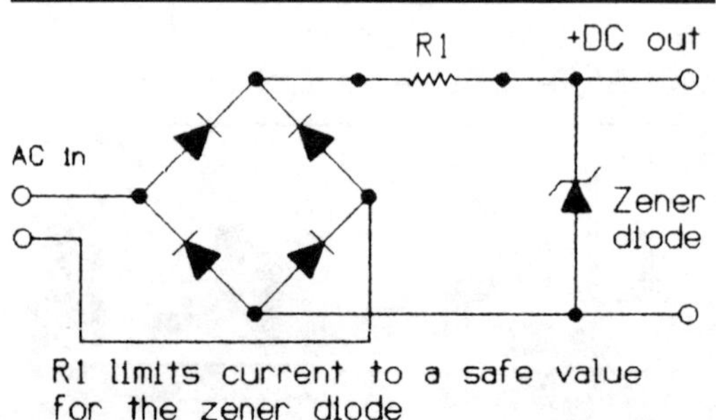

FIGURE 8-15: R1 limits current to a safe value for the zener diode.

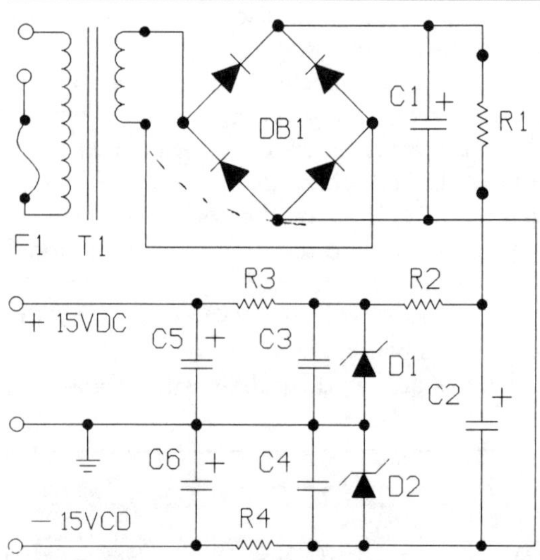

FIGURE 8-16: Regulated ±15V DC power supply.

You can buy power-supply modules from many sources, but it is not difficult to build one and save money in the process. *Figure 8-16* shows a simple, regulated power supply which will deliver 100mA at ±15V. Although not a fancy power supply, this has the advantage of low cost, reasonably good regulation, simplicity, and it's easy to find the needed parts.

Many ways to build one are possible. For example, you could replace the zener-diode voltage regulators with IC voltage regulators having better performance, but you may have a problem finding the hookup information for them. Sources like Radio Shack package them with connection diagrams. Still, zeners work well and are readily available.

I like to know what the various parts do, and I think you'll find it useful to know, too. I'll discuss what each part of this circuit does.

The transformer's 24V secondary winding (T1) produces about 34V when rectified (converted) to DC by the full-wave diode bridge (DB1). A **full-wave diode bridge** is simply a group of diodes that rectifies *both* the positive and negative waves from the transformer to DC.

Figure 8-17 shows the output from a simple, single diode rectifier. *Figure 8-18* shows the output from a full-wave diode bridge.

The pulsating nature of the DC, produced by the transformer/diode, causes hum in the electronics. We must **filter** out this **ripple** to stop hum. The place to start is to have many small ripples instead of a few large ones. The full-wave diode bridge has twice the number of pulses as the single diode rectifier. Additionally, a full-wave bridge delivers twice the power of a single diode.

The relatively large **filter capacitors** (C1 and C2) act like reservoirs to store the pulses of electrons from the transformer/diode bridge. This greatly smooths the flow and reduces ripple. These capacitors should be as large as practical: the bigger they are the more they smooth the ripple. For this reason, many audiophiles greatly value large power-supply capacitors in their electronics.

The resistor R1 limits the current which further smooths the flow. These two components form an **RC filter network**. C2 and R2 are another stage of RC filtration. The values of R1 + R2 + R3 are designed to limit the total current flow to about 100mA.

The 15V zener diodes (D1 and D2) regulate the voltage down to 30V. By connecting to ground between them, I've split the +30V into ±15V.

C3 and C4 shunt the zeners to suppress their switching noise. The RC networks of R3/C5 and R4/C6 form a last stage of filters. This power supply has no audible hum.

The values for the capacitors and diodes are minimums. You may use higher values, if you wish. However, larger ones occupy more room in the chassis and cost more. The diodes may have higher values with no penalty, except cost.

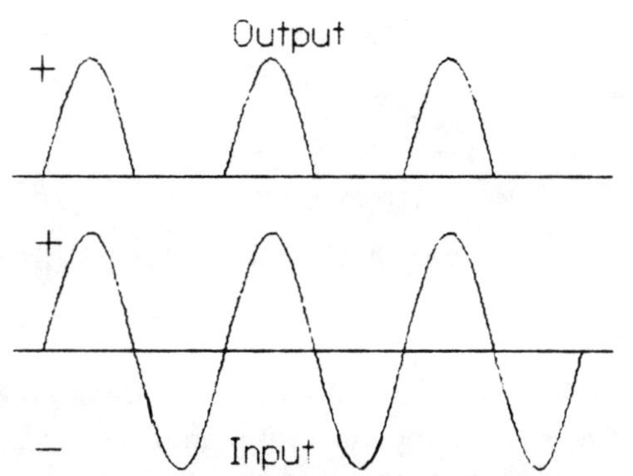

FIGURE 8-17: Output from single diode rectifier.

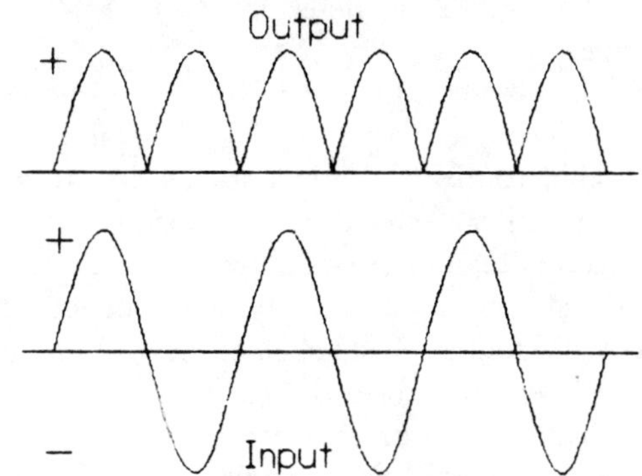

FIGURE 8-18: Output from full-wave diode bridge.

TABLE 8-3

PARTS LIST FOR FIGURE 8-16

F1	Fuse, ½A, slow blow
T1	Transformer, 120V AC primary, 24V AC secondary @ 200mA
DB1	Diode bridge, 50 PIV, 3A
D1, 2	Zener diode, 15V, 1W
C1, 2	Electrolytic capacitor, 50V, 250µF
C3, 4	Ceramic disc capacitor, 0.1µF
C5, 6	Electrolytic capacitor, 25V, 100µF
R1–4	Resistor, 100Ω, ½W, 10%

PARTS. Before you begin constructing electronics, you need to understand some things about hookup wire, circuit board stand-offs, pots, switches, jacks, and solder. Let's look at these in greater detail.

Wire comes in several types. The best hook-up wire is **silver-coated copper**, **multi-stranded**, and **Teflon-coated**. It conducts better than ordinary copper wire, the insulation doesn't melt from the soldering heat, and the insulation has the highest dielectric rating.

Teflon-coated wire is expensive, not readily available, and unnecessary. Ordinary copper wire with vinyl insulation is acceptable.

Stranded or solid wire: which should you use? Stranded wire is more flexible and therefore less prone to failure, but it's difficult to handle. The ends tend to fray, and it can be hard to wrap around terminals or push through holes. Since it's springy, it doesn't like to lie neatly in the chassis or turn neat corners.

FACTORS THAT HARM SWITCHES

- Direct current
- Capacitive loads
- Inductive loads
- High voltage
- High current

Solid (nonstranded) wire solves these problems. Most component kit manufacturers use it—so can you. Its greatest limitation is that it can't tolerate high vibration environments.

Solder comes in several sizes and types. You'll want a diameter in the 1/16–3/32″ range.

Solder is a mixture of lead and tin. Although the percentages vary (40/60%, for example), I haven't found the percentages to matter much—if it's electrical solder, it will work satisfactorily regardless of the percentages.

What *does* matter is the type of flux you use with the solder. Most solder comes with a flux core of either a rosin or acid type. You must use only **rosin-core** solder.

Never use **acid-flux** or **acid-core** solder on electronics. The acid will eat away the metal, and the joint will eventually fail.

While I'm talking about solder, I should mention soldering irons. The controlled-heat types are worth every cent. Do yourself a big favor and spend the money to get one.

Jacks are highly varied. Many different types can be used to connect your stereo components to your custom electronics.

Unless the jacks are used to transfer high power, they should be electrically isolated from the chassis, which is necessary to avoid ground loops. Manufacturers are aware of the problem, and make both individually insulated jacks and gangs of jacks mounted on a common insulator. If you can't find insulated mounting types, you can use grounded jacks and mount them on an insulator to keep them from touching the chassis.

Switches come in many different types. I recommend you avoid the slide type—they are unreliable and difficult to mount. Small toggle switches are common and of better quality. They cost more, but are worth it.

Switches have current ratings. When you break an electrical current with a switch, an arc forms between the contacts. This arc is hot (like an arc welder), and it transfers metal from one contact to the other. This action pits and damages the switch contacts. Higher current generates more heat and causes more damage.

Larger switch contacts will tolerate more heat and have more material to transfer. Therefore, they last longer, but they also take up more space and cost more.

The switch on your home's electrical box is huge (it will switch 200A) and expensive compared with the little switches you use in your music system. The essential difference between them is their current capacity.

When you turn a switch on, a small arc forms across the contacts just before they touch. However, this is very short lived.

The big problem is the arc which forms when you turn the switch off, beginning the moment the contacts separate, and continuing until they are far enough apart that the voltage is insufficient to push electrons across the gap.

Current type has a large effect on the duration of, and the damage caused by, the arc. DC is tough on switches, because it produces a continuous arc until the contacts are separated widely enough to extinguish it. AC is less harmful.

US house current is 60Hz, so it pulses on and off and changes polarity 120 times each second. Switches generally open much more slowly than 1/120/second. Under these conditions, the arc will stop as the current revers-

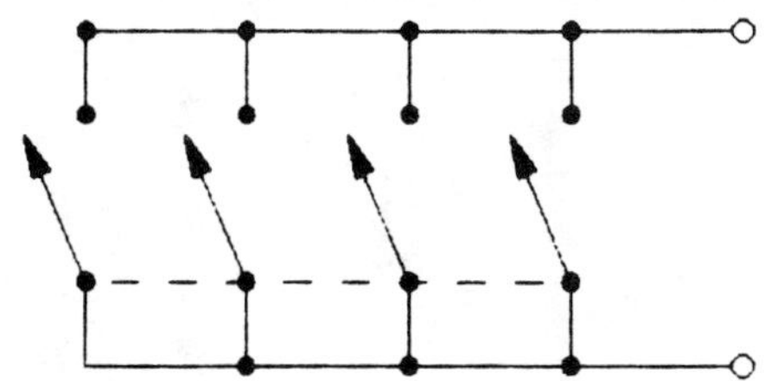

FIGURE 8-19: Parallel switch contacts.

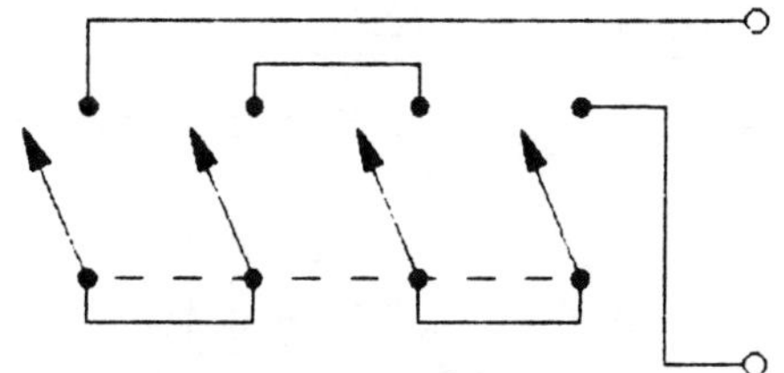

FIGURE 8-20: Series switch contacts.

es—well before the contacts open widely enough to extinguish the arc if it is DC. Therefore, manufacturers rate switches to handle more AC than DC current.

Though voltage does not damage the switch contacts, higher voltage prolongs the arc's duration by driving it across a wider gap. Therefore, higher voltages increase switch-contact damage.

Capacitive or inductive loads tend to store electrons, which they feed into an arc when you turn the switch off. The result is more damage to the contacts.

Now you understand why manufacturers rate their switches differently for different conditions. For example, they may rate a switch for 6A @ 120V AC with a resistive load. If the load is inductive or capacitive, they may reduce the rating to only 4A.

The manufacturer will rate the same switch at 240V at perhaps only 3A resistive and 2A inductive. He may rate it at 6A at 120V AC resistive, and only half that for DC. You can see there is considerable interaction among these factors.

These ratings, however, aren't absolute. They don't mean the switch won't fail if used within the ratings. The contacts are damaged every time you use the switch. Manufacturers' ratings just give a *reasonable* life span. As usual, it's best to be conservative.

You will discover that it's nearly impossible to find a small switch to do the job when you need high-current capacity. A good example is the power switch on your control center which you use to switch your system.

The huge power-supply capacitors in large power amplifiers appear as a dead short at the moment you switch on the system. Momentary switch current can be over 100A, but most component switches are only rated at a few amps. Is it any wonder they often fail by melting or welding their contacts?

Unfortunately, it is difficult to find a switch with enough capacity to fit in the cramped space of audio components. Even if you have the room, it's challenging to find *any* switch that will handle that much current.

You can solve the problem in several ways. First, consider using multiple-switch contacts. You might logically think that putting four switch contacts rated at 6A in parallel would produce a 24A switch, since each contact would only see 25% of the load (*Fig. 8-19*).

Well, you'd be wrong. Current doesn't damage closed-switch contacts. It's the arc that forms at turn-on and turn-off that does the damage. Specifically, it's the arc's *duration* while the contacts are moving apart that causes most of the damage.

FACTORS THAT ARE EASY ON SWITCHES

- Alternating current
- Resistive loads
- Low voltage
- Low current

Therefore, the contacts need to be in *series* to increase their rating (*Fig. 8-20*). Putting four switch contacts in series increases the total air gap four times faster than a single contact. The arc only lasts 25% as long so arc damage is greatly reduced. Furthermore, there is four times as much metal in four contacts as in one.

A way to deal with turn-on surges is to use a delay relay with series resistance (*Fig. 8-21*). This limits the current flow in the arc to minimize damage. Once the power-supply capacitors are charged, the switch shorts the resistor so full power can flow.

Some amplifiers have a **thermistor**. This device has a moderate amount of resistance when cold, so it limits turn-on current. The

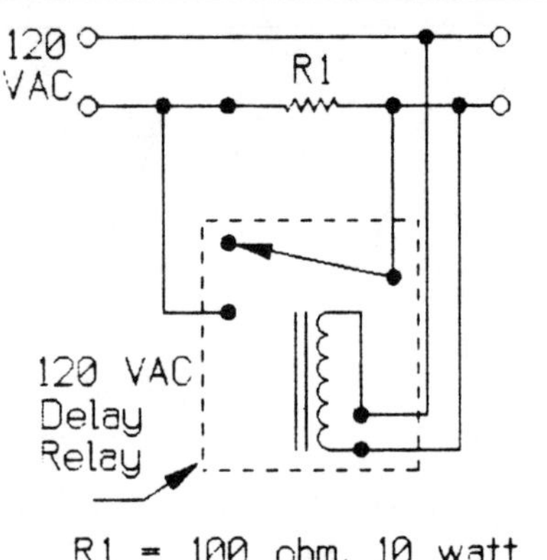

FIGURE 8-21: Reduce switch damage with delay relay.

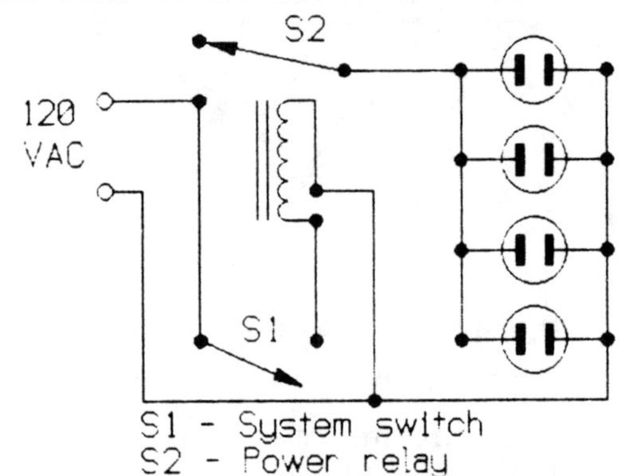

FIGURE 8-22: Power relay and AC plugs.

current heats the thermistor and lowers its resistance after a few seconds.

Another technique is to use the normal power switch to energize a massive power relay, which is mounted in a small chassis with many AC outlets. If you have several components, you've already discovered that there are never enough plugs available. You can solve the plug problem as well as turn-on surges with this gem. *Figure 8-22* is a schematic diagram of such a device.

When buying switches, keep in mind that you must make some type of hole in your chassis to mount them. It's easy to make a round hole with an electric drill, but impossible to make a rectangular one.

If you buy parts that require rectangular holes, such as **slide** switches, you must laboriously file each round hole to a rectangular shape, or buy a special punch which makes square holes. Besides being time-consuming and a bother, it's difficult to make a hand-filed hole cosmetically attractive. You can

buy special tools (nibblers) to make this job easier, but they can be expensive and may not do a professional job. Therefore, whenever possible, you should select parts that mount in round holes.

Each switch has a configuration code that designates the number of **poles** and **throws** it has. Poles refer to the number of circuits in the switch. Throws (or positions) refer to how many terminals/switch positions each circuit can contact. *Figure 8-23* shows various pole/throw configurations. These combinations are referred to as SPST (single-pole, single-throw), DPST (double-pole, single-throw), and so on.

If you are building a control center, you can connect the device to the sources you have available in several ways. The most common way is with a **rotary switch**.

This must be in double-pole (for stereo), and have as many throws as necessary to accommodate all your sources. Additionally, I suggest you add one extra throw and put a set of jacks on the *front* panel, which you can occasionally use to hook up other external sources. This convenience is rare on commercial components.

Rotary switches invariably have heavy

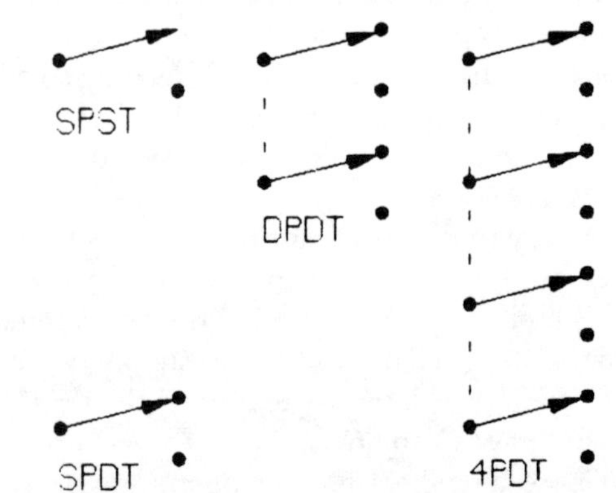

FIGURE 8-23: Various switch configurations.

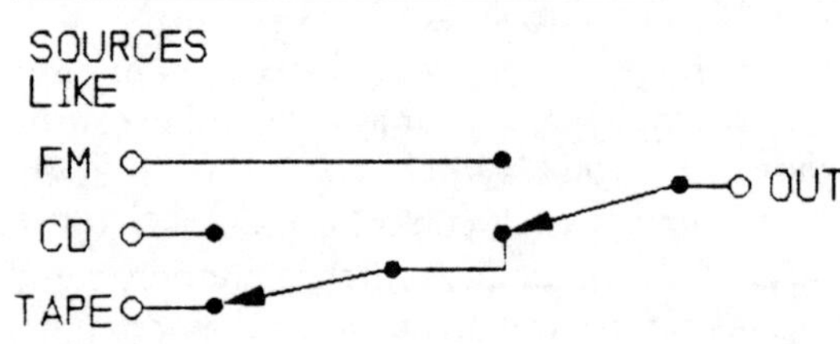

FIGURE 8-24: Switching three sources with two switches.

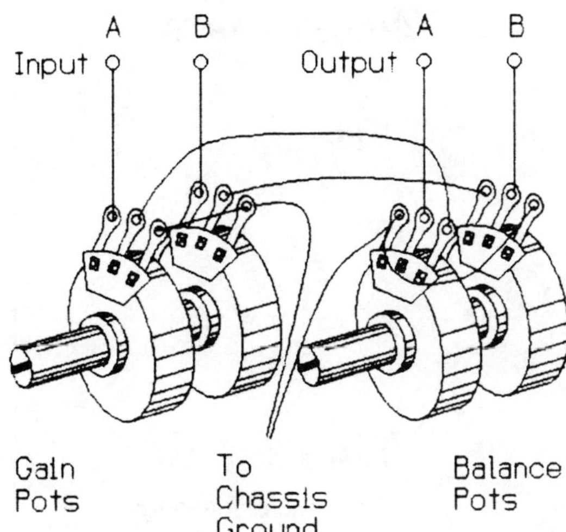

FIGURE 8-26: Mechanical connections of gain and balance controls.

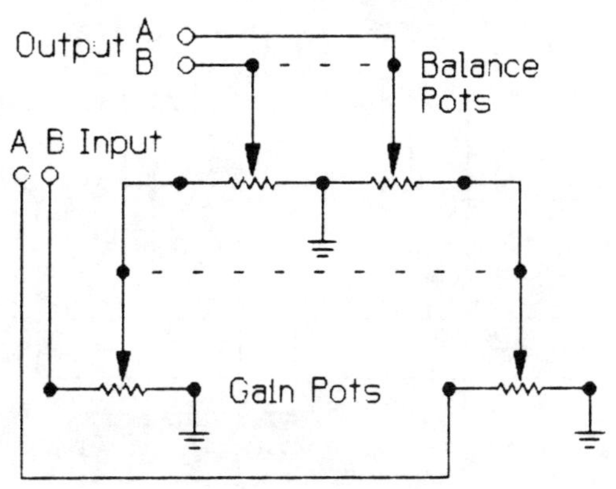

FIGURE 8-25: Gain and balance control schematic.

detent action. You can improve their feel by reducing the tension on the leaf spring that holds the ball in the detents. Just take a pair of pliers and bend it a little. You can enhance it even more by replacing the ball with one about 50% larger.

If you only have three sources, a switching arrangement using a pair of miniature, DPDT toggle switches works well (*Fig. 8-24*).

Pots are a problem: few of them are good enough. Most have poor tracking, and they eventually get dirty and make crackling noises when you move them. Still, unless you use an electronic remote-control unit or stepper switch, you are stuck with them.

The typical carbon pots stocked by your local electronics parts house just won't track well enough for use in a stereo volume control. In particular, you must avoid the types where you snap sections together to make a dual pot—they track horribly.

The best source for volume controls is a manufacturer of high-quality stereo components. He has the same problems as you, but he has the resources to demand higher tolerance parts for this critical function. If you can, find a source for Alps or Noble potentiometers, as these are very high quality and track quite well.

I've shown the usual arrangement of pots for a stereo volume- and balance-control system in *Fig. 8-25*. *Figure 8-26* shows it mechanically.

While this is the most popular way of controlling volume, it's not the only way. You can use ganged volume controls (concentric shafts with concentric knobs) with a light friction lock so they turn together. If they track well, this method is excellent. You often find this used in tape decks.

You can use completely independent pots for each channel, but volume adjustment is inconvenient. You must turn each knob individually, and then tweak the balance every time you adjust the volume. This gets to be a real nuisance.

STEPPER SWITCHES. You can make a volume control that tracks perfectly by using a two-pole rotary switch with many positions and precision resistors. As you string the resistors from terminal to terminal, you can tweak

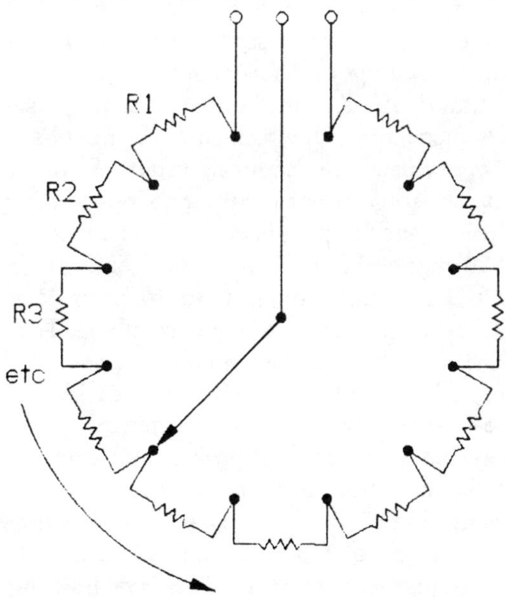

FIGURE 8-27: Schematic of stepper switch.

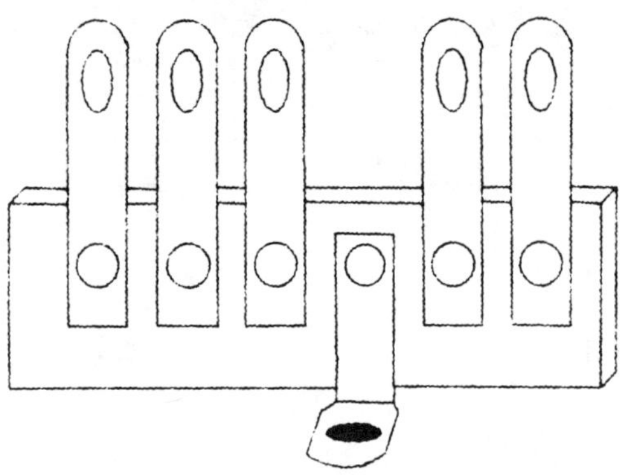

FIGURE 8-28: Terminal strip.

the values with your ohmmeter to be certain the sections match closely. Use **shorting** switches (sometimes referred to as "make before break") to avoid signal pulses.

Figure 8-27 is a schematic of one channel. For clarity it only has 12 positions. You really need at least 23 positions for adequate resolution. Remember that it must *not* be linear. Each step should be about 3dB. Calculations for these switches are complicated because of the audio taper, and construction is time-consuming. The cost of the switch, including the 50 or so precision resistors, will surprise you.

Incidentally, component manufacturers understand the value of stepper switches. Some put a detent mechanism on a conventional pot, so the unsuspecting think they are getting a stepper switch when they are not. Some, however, put detents in a regular pot which enables the user to establish reproducible settings for each recording.

Stand-offs are small pieces of tubing, usually metal, through which you can put bolts to mount circuit boards a small distance from the chassis. You can also use them to mount other devices, such as plug-in edge connectors.

It may be hard to find good stand-offs. If you have difficulty, just go to your local hobby shop and buy some brass tubing. They have a multitude of diameters between 1/8 and ½". The tubing usually comes in 12- and 36-inch lengths. They have little tubing cutters for it. It's not expensive, and you can make any size and length stand-off you want. Some switch manufacturers, such as Switchcraft, offer switch wafers and switching frames as separate kits. The wafers come in a variety of configurations and base materials. The frames accommodate two, four, six, or eight wafers.

CHASSIS PARTS

- Pots
- Knobs
- Switches
- Jacks
- AC cord
- AC cord strain relief
- Stand-offs
- Fuse holder
- Fuse
- Dry-transfer lettering
- Nuts and bolts
- Terminal strips

BUILDING A CHASSIS. To build a chassis you must first get a stock, blank, undrilled chassis box. But first, you need to know what size box to get.

To figure out the size, lay out the various circuit boards, switches, transformers, and power supplies, and carefully decide what dimensions the chassis must be. Here is where you are likely to make a serious mistake. Most first-time builders use a chassis that is too small.

More is involved with building a chassis than just stuffing the circuit boards in a box. You need room to work, and wires have to go somewhere; you need room around transformers to prevent hum. Give yourself lots of space.

You can get the raw chassis from any electronics parts house. Most builders prefer aluminum boxes because they are lightweight and don't rust.

Fancy chassis come with feet. If yours doesn't, be sure to get some. The stick-on types are tolerable, but bolt-ons are more rugged.

Consider the cosmetics of your chassis. A few extra dollars spent on an attractive one will save you work and money later, when you try to "dress up" a drab gray box.

You will need a multitude of detail parts to build your chassis. Get them all at once and save yourself frustration and trips to the store.

Please don't omit the fuse. It doesn't cost much or harm performance in any way, and by preventing a fire it may save your house and the lives of your family.

Dry transfers are letters you rub off a carrier sheet onto your chassis to make professional-looking labels. With a little planning, care, and effort you can add a quality look to your equipment. You may have to get the dry-transfer lettering from a stationery store. It comes in many sizes and styles.

Terminal strips are just strips of insulating material with terminals attached. The terminals can take many forms.

Figure 8-28 shows a typical configuration you bolt to the chassis. You can attach wires, capacitors, resistors, and other electrical parts to it. You also can get screw-type terminal strips, which are used for connecting loud-speaker or transformer wires.

You will need the first type for mounting supply parts like diodes, capacitors, resistors, and transformer leads. You can use a screw-type terminal strip to connect your amplifier output to the step-up transformer chassis, since this is low voltage. However, do not use screw terminal strips for the high-voltage connections to the ESLs.

Banana plugs are excellent for high-voltage and/or high-power, plug-in connections like speaker wires, because they are well insulated, cheap, and reliable. You can mount female banana plug jacks directly into a chassis hole without additional insulation.

When locating parts in the chassis, watch out for magnetic fields produced by the power transformer. They induce tiny currents in wires, circuit board traces, or any other conductor within range, and cause *hum* in your electronics.

To avoid this, you need *space* between the power transformer and any low level circuitry. How much space? It varies depending on the transformer's power and design, and the electronics. Usually 3″ is enough.

Normally, you put the power transformer in a corner, and the jacks, circuit boards, and wiring at the other end of the chassis. If you don't have enough room, you can *shield* the electronics from magnetic fields in various ways.

The most common way of doing this with wires is to use the shielded type. Such wire has a braided or metal-foil jacket to bleed the induced current to the chassis. Just connect *one* end of the shield to ground. Don't connect both ends, as this could cause a ground loop. The exception to this is when you are using the shield as an electrical conductor.

Metal cages around the circuit boards, or a metal wall between the power transformer and the electronics, also helps. Shielding is a

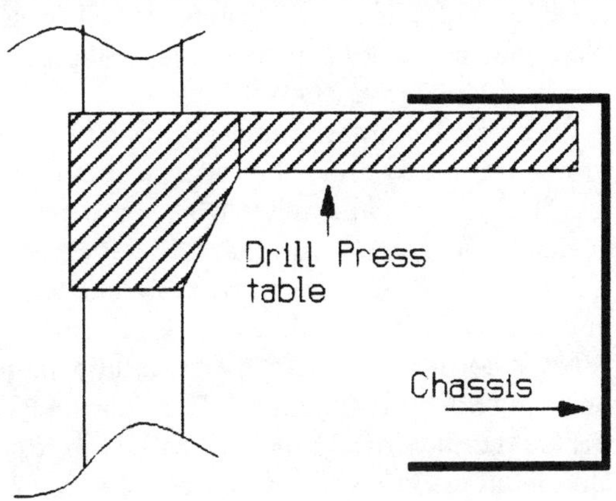

FIGURE 8-29: Supporting chassis with drill press table.

lot of trouble and isn't necessary, if you give yourself enough room.

The next step is to do all the chassis drilling. If you want it to look professional, it is important to do your layout work very carefully so the holes are even and in-line.

Precise drawings on a slick aluminum chassis are difficult to do. It's better to draw the dimensions of your chassis' front and rear panels on quadrille paper first. (Quadrille paper is printed with lines in both directions in a variety of widths.) With a good ruler, you can place the centers of the holes. Tape or glue the paper to the chassis. Use a center punch to mark each hole. The aluminum is very soft and can be easily dented and damaged with a center punch. Prevent this by backing up the aluminum with something massive.

Since the chassis usually is "U" shaped, you can't put it on a heavy, flat surface. Putting the panel over the edge of a drill press table works

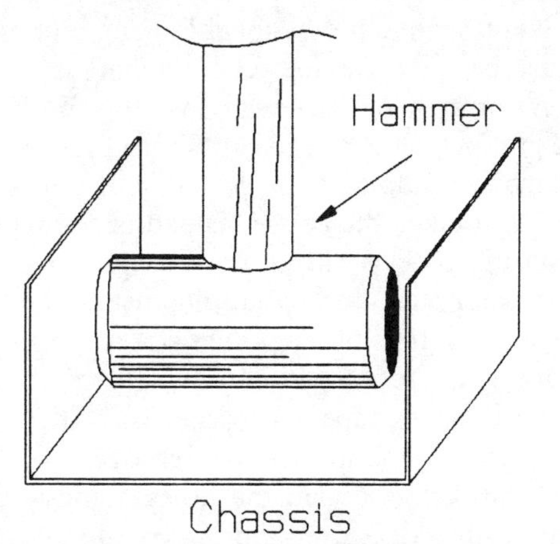

FIGURE 8-30: Supporting chassis with hammer.

well (*Fig. 8-29*). If you don't have a drill press, then you can back it up with a large hammer or wood block (*Fig. 8-30*).

With all the holes center-punched, it is a simple matter to drill them with a small drill bit. The bit should be under 1/8″. Some newly available bits coated with titanium have a special split tip which does not wander. These outlast regular bits six times.

Now comes the tricky part: drilling large holes in soft thin aluminum. The typical electric hand drill with a ½-inch bit will "grab" thin aluminum and twist it into a pretzel—usually cutting you in the process.

When machining metals, rigidity is king. Drilling is no problem, if you can hold the chassis absolutely rigid in a milling vise on a drill press while you *slowly* downfeed the quill. Unfortunately, not everybody has a drill press and milling vise.

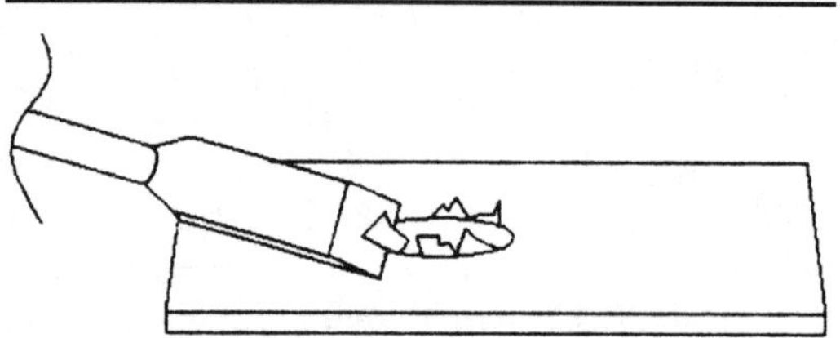

FIGURE 8-31: Deburring with a chisel.

If you must do it by hand, the trick is to drill the holes larger in small increments. You can use 1/8-inch increments when the hole is small, and reduce to 1/16-inch increments once you get above ¼″. Also, once you get a hole larger than about ¼″, it is necessary to use slower drill speeds.

If you want a professional-looking unit, you must be very careful when drilling so you don't scratch or otherwise mar the finish. A helpful precaution is to put about four layers of masking tape on the panel, so if you slip, the tape takes the abuse instead of the paint. Masking tape is surprisingly tough—but it's not armor plate—and you still must be careful. Peeling the tape off later does not damage the paint, if you do it within a few hours. If you leave masking tape on for weeks in the hot sun, it's nearly impossible to remove.

After you've drilled the holes, remove the burr with a deburring tool. If you don't have one, a large countersink (commonly called a rosebud) spun slowly in the hole works. If you

have neither, you can use a large drill bit. The bit will tend to "chatter" and give a ragged cut. It works better just to twist it by hand.

Deburring the holes outside the chassis and on the chassis floor is no problem. But without a deburring tool, it's difficult to reach the side of the holes inside the chassis on the front and rear panels. Often there isn't enough room to get a rosebud or drill bit on the inside.

Although not a very good solution, you can lay a wood chisel flat against the inside of the panel. Carefully tap it with a hammer to peel off most of the burr (*Fig. 8-31*).

PAINTING. If you've damaged your chassis during the drilling operations, or otherwise don't like the paint, feel free at this point to repaint it. If you've not painted much before, let me offer a few pointers.

Remember that the finished paint job will only look as good as the surface below it. In other words, paint won't fill-in and cover scratches and dents. Only when you have the surface flawless are you ready to paint.

Prepare the surface by using 180-grit, silicon carbide, "wet or dry" sandpaper. Wrap a small piece around a 1″ × 1″ × 2″ gum eraser. Use it wet: the water keeps the sandpaper from getting filled with paint and aluminum. Just wet the chassis and sandpaper and start sanding. Sand for perhaps 20 seconds, wash off the resulting "sludge" and check the surface.

You are striving for a uniform matte finish on the aluminum or old paint. Shiny spots are bad. Keep sanding until they are gone and the surface is uniform. As you get it near perfection, you'll need to wipe off the water with an old towel to see the surface better.

If you have deep scratches that you can't sand out, you can fill them with Spackle™. Apply it with a putty knife. After it dries, you can easily sand it. It sands better *dry*, so use the Spackle after you have finished wet-sanding the old paint/aluminum.

If you are painting over old paint, sand the 180-grit scratches out with 400-grit and you are ready to paint. If you are going to paint bare metal, you must first spray on a light coat of primer.

If you make the primer uniform and just thick enough to cover the surface completely, you can paint directly over it. If it is uneven and full of runs, you will have to sand it with 180- and 400-grit before painting.

Good painting requires at least two coats. You can do a surprisingly good job with today's spray cans, but they don't come with really good paint.

If you want a **crackle finish**, you can get suitable paint from an electronics store. Crackle paint is applied differently than other types. While normal paint works best if you use two or three light coats, you must put crackle paint on very thickly to get the effect. Follow the directions on the can and ignore what I say about normal paints.

The best paint is **epoxy**. You can make your chassis look like it is coated with wet glass or porcelain if you use epoxy, and it is also extremely tough and hard. You can also make it into a matte finish. You can get small quantities of epoxy from a hobby shop. Get the color you want and a can of catalyst for each can of paint. If you want an extremely "wet" look, also buy a can of clear epoxy.

If you want a matte finish, you can either buy matte catalyst or regular gloss catalyst. I prefer to use the gloss catalyst, and then sand the finish with 400-grit to dull the finish.

You *must* use a spray gun for a high-quality finish. You'll get a better looking job with a cheap spray can than you will by applying the world's best paint with a brush.

Mix the epoxy 1:1 with its catalyst. Dilute the mixture an additional 1:1 with epoxy thinner or reducer. Let it sit in your spray gun for one hour to get the chemical reactions going before spraying. Epoxy catalyzes within a couple of minutes after spraying. If you "fog" on a light coat, it won't hide the undersurface, and you can spray on a wet coat a few minutes later with little fear of a run. One or two wet coats is all you'll need. You can put on a second wet coat in 15 minutes.

You need to apply the paint just heavy enough to "wet it out." If it has a sandy appearance, you haven't put on enough paint. On the other hand, you want *just enough* paint to make it flow out wet. It isn't difficult to tell. When in doubt, go light and use another coat later.

If you apply paint too heavily, the surface will either run or have an orange peel appearance. For this reason, it's necessary to use a couple of light coats instead of one heavy one, which is easy with epoxy because you only need to wait a few minutes between coats.

In an hour the paint will be dry to the touch, but still soft. Let it harden for 24 hours before handling it.

If you did a good job, you won't believe how beautiful the paint looks. It has a smooth, polished, wet look. You can stop at this point, but

you can make it look even better by putting a clear coat over it. Also, if you have runs, orange peel, bugs, or other flaws in the paint, you can get rid of them with a clear coat.

To remove flaws, wait until the epoxy is good and dry—one day is usually sufficient if the temperature is above 70°F—then wet-sand with 400-grit. Sand just enough to remove the flaw and leave a matte finish. If you sand too much, you will go through the paint and have to put on more color.

Don't fret that sanding has removed the gloss. Putting on a clear coat will restore it.

Apply two or three clear coats just as you did for the color. You won't believe the finish. It will look like glass—even when it's dry.

CLEANING A SPRAY GUN. Most amateurs clean their spray gun by pouring out the unused paint, dumping thinner in its place, and spraying the thinner through the gun.

This doesn't work well, because they are actually spraying diluted paint through the gun that later hardens and clogs it. Also, it wastes expensive thinner and pollutes the air.

Here's a technique that works much better. First, dump any remaining paint. If it's epoxy, *don't* put it back in the paint can, because the catalyst in it will cause the whole can of fresh paint to harden in a few days even with the lid closed airtight.

Take a couple of paper towels and wipe all the paint you can from the inside of the spray gun cup and the pick-up tube. Add about ½ oz. of thinner, swish it around, and wipe it out with more paper towels. The object is to get the cup perfectly clean. Note that it only takes a little thinner.

When the cup and pick-up tube are spotless, pour an ounce or two of thinner in the cup, assemble the gun, open the control valve fully, and spray clean thinner through the gun. The clean thinner will dissolve the remaining paint in the passages and leave you with a clean gun. Be sure to wipe any paint from the outside of the nozzle. That's all there is to it.

DRY-TRANSFER LETTERING. Put on dry-transfer lettering after the clear coat is dry. Although it's tricky, you may put it on before you apply the clear coat. Putting the clear coat over the lettering will protect it from dirt and abrasion.

If you want to clear-coat the lettering, you must be very careful: the epoxy thinner will dissolve and wrinkle the lettering. To avoid

> # GROUND LOOP PREVENTION
> Connect all ground wire
> to only **ONE** point
> on the chassis.

this, you must spray on a *very light* clear coat, wait two minutes, and repeat this process several times.

The thinner will evaporate from these light coats before it can damage the lettering. After about six very light coats, there will be enough pure epoxy over the letters to protect them from one wet coat. Let the wet coat dry for an hour, then put on a final wet coat. The lettering will look as though it's under a sheet of thin glass—and will be as durable.

If you put the lettering on after you've finished painting or over a factory paint job, wash the chassis with soap and hot water to remove any traces of oil, fingerprints, and tape adhesive residue first. Completely rinse off any soap residue. The lettering comes with instructions. If you follow them carefully and patiently, you can do a super job on the front-panel cosmetics.

Often there isn't room to label the rear-panel jacks using dry transfers. Also, cosmetics aren't as important on the back. You can type the information identifying the jacks on a self-adhesive, gummed label and stick it on an open spot on the back panel. You might want to use a single number from the dry transfers to key the jacks.

A good idea is to type two gummed labels. Stick one on the chassis, and the other on a 3 × 5 card that you lay behind the unit once you've installed it. The label on the back panel is hard to read when it's inside a cabinet or on a shelf that you can't get behind. It's easier to pick up the card and read it.

ASSEMBLING THE CHASSIS. Now you're ready to assemble your chassis. Be very careful not to damage the cosmetics while you bolt and screw things together. I recommend that you install all the floor-mounted parts first, then those on the back panel, and finally the ones on the front panel. While working on the back panel, you can set the chassis on its front panel on a clean, soft towel.

When installing the parts, use the proper

tools. If you try to tighten the large nut on a pot or switch with a pair of pliers, you are just asking for disaster. The pliers are going to slip and scratch the finish. Even if they don't slip (a miracle), they will scratch. Use a socket wrench to tighten these parts even if that means going out and buying one.

Build and install the power supply and *test* it before doing any electronics installation. Be very sure you have the correct polarity on the electrolytic capacitors and diodes.

Again, be absolutely sure it works correctly, and that the voltages are correct before proceeding. You must have a voltmeter to do this work. Your electronics are too expensive to test a power supply.

Wire the electronics, following the manufacturer's directions exactly. If you're not sure, call them. They would rather answer the phone than replace your damaged electronics. Yes, I know that you're working late in the evening and want to get them up and running. You're too impatient to wait until the next day to call the manufacturer. Well, you'll soon learn the meaning of the old saying "haste makes waste."

Ground loops cause buzz, hum, instability, and other problems in electronics. They are the result of having more than one ground point on your chassis.

Each circuit board, jack, pot, and the power supply must be individually grounded to the chassis. The pitfall waiting for most beginning builders is they ground the various parts at different chassis locations wherever it is convenient.

Connect the ground wire from each part to only *one* point on the chassis. The only exception is the jacks. If they are all close together (they usually are), you can run one large ground wire through all their ground lugs. Take this ground wire to chassis ground. The best ground point is near the input jacks.

Seeing our mistakes is hard. Have someone double check your wiring to be sure it's right before energizing the electronics.

HIGH-VOLTAGE CHASSIS. Because of the high voltages involved, I don't like to use metal for the polarizing supply/step-up transformer chassis. I cut ¼-inch plastic (Lexan, Plexiglas, or other hard plastic sheet material) into the parts and glue them together with cyanoacrylate glue (Krazy Glue®, Hot Stuff, etc.).

Make the bottom of the box removable for access. You may use small screws to hold it in

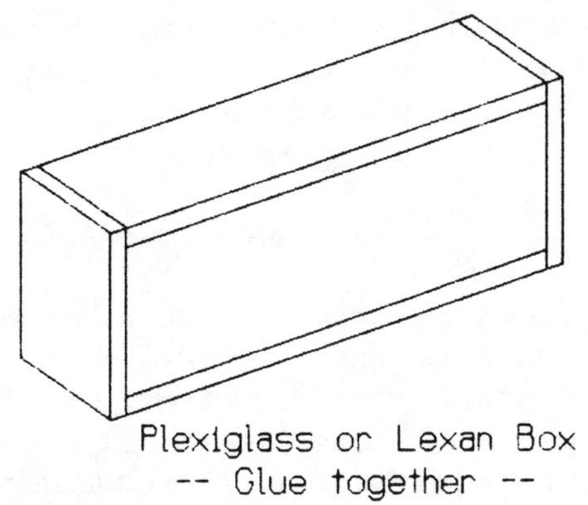

FIGURE 8-32: High-voltage parts chassis.

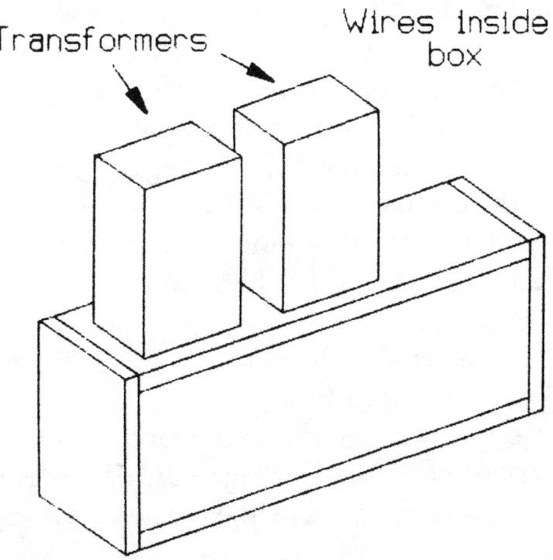

FIGURE 8-33: Mount transformers with leads inside box.

place, but it's hard to drill the holes accurately. I prefer to glue it in place with a few dabs of epoxy. The epoxy joints break easily with a putty knife or screwdriver if you ever need to get inside (*Fig. 8-32*). The transformers may be mounted outside the chassis on the top. Be sure any exposed terminals are *inside* in the interest of safety (*Fig. 8-33*).

Banana plugs make good high-voltage connectors and offer reasonable protection from the voltages involved. *Figure 8-34* shows a schematic diagram of a high-voltage, polarizing power supply and step-up transformer chassis.

WINDING A TRANSFORMER. The following are instructions for winding a high-voltage, polarizing power supply transformer. The basic concepts are applicable to all but audio transformers.

Start by buying one of Radio Shack's larger transformers. Their transformers are unique in my experience, because they wind them on flanged, plastic bobbins. This gives you a form upon which to wind a new transformer. If you use other brands, you must build a collapsible mandrel for making the windings.

Take the transformer apart by prying off the metal cover over the core laminations. Remove the laminations. They are coated with varnish which causes them to stick together, but it's not difficult to pry them apart with a knife blade.

The laminations are made of alternating "E"s and "I"s (*Fig. 8-35*). The first couple will be tight, and you probably will have to draw them out of the core with needle-nose pliers. Once there is some room, use a knife blade or sharp putty knife to separate each lamination for easy extraction.

After removing the laminations, unwrap the tape covering the windings. If the secondary

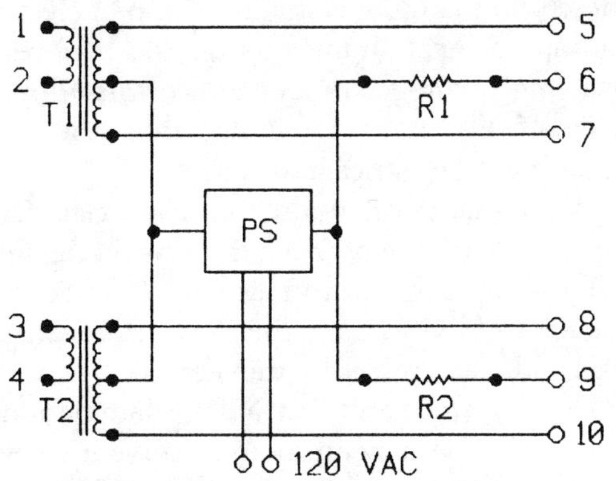

FIGURE 8-34: ESL polarizing supply and audio transformer schematic.

TABLE 8-4

PARTS LIST FOR FIGURE 8-34

T1	Channel 1 step-up transformer
T2	Channel 2 step-up transformer
PS	High voltage power supply
R1	Channel 1 charging resistor
R2	Channel 2 charging resistor (20–100Ω @ 2W, see text for details
1 & 2	From Channel 1 ESL amplifier
3 & 4	From Channel 2 ESL amplifier
5 & 7	To Channel 1 ESL stators
6	To Channel 1 ESL diaphragm
8 & 10	To Channel 2 ESL stators
9	To Channel 2 ESL diaphragm

winding is on top, count the turns while unwinding it. The number of turns is *important*. Immediately write them down so you don't forget them—you'll get no second chance. Discard the old windings.

If the primary (the 120V winding) is on top, carefully unwind it and save it for reuse. Although it should not be necessary, it is wise to count the turns as well in case there are problems reusing the wire.

Now you need to figure out the number of turns required for the secondary winding to develop the desired voltage. Divide the number of turns removed by the secondary voltage.

$$T_{NEW} = \frac{T_{OLD}}{V_{OLD}} \times V_{NEW}$$

Where:

T_{NEW} = Number of turns needed to obtain the new voltage

T_{OLD} = Number of original turns

V_{NEW} = New voltage desired

V_{OLD} = Original secondary voltage

For example, assume that the transformer's original secondary voltage rating is 24V, and you removed 78 turns of wire. The transformer needs 3.25 turns of wire to generate one volt.

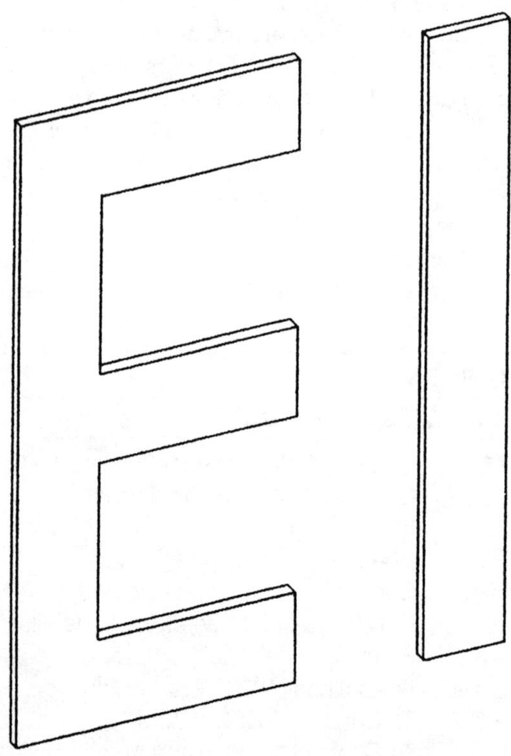

FIGURE 8-35: Transformer core laminations.

Multiply that by the new voltage to find the number of turns needed to get the new voltage. If you want a 2kV transformer, you will need 2,000 × 3.25 or 6,500 turns.

$$T_{NEW} = \frac{78}{24} = \times 2,000 = 3.25 \times 2,000 = 6,500$$

Of course, you must use much smaller wire to fit all those turns in the space left by the large secondary wire. **Magnet wire** is used for winding transformers. You can get it from electronics supply houses or electric motor repair shops.

How small a wire should you get? The wire cannot be too small in this application, since the current requirements are insignificant. Expect it to be hair sized if you have a small transformer.

Although it is possible to wind the transformer by hand while counting the turns, this is very time-consuming and hard to do without breaking the wire. It's well worth the trouble to mount the bobbin in a drill press, lathe, or milling machine and let it do the winding, while you smoothly feed the wire onto the bobbin.

Instead of counting turns, determine how much *time* it will take to wind on the required turns at a given RPM. When the time is up, the project is finished—you don't have to count.

The high voltage in such a transformer may be enough to exceed the dielectric strength of the insulation. This won't be a problem in the coil you just wound, because wires near each other have relatively low voltage potentials. This does not apply to the connecting wires that come from the primary or the start of the secondary windings. These wires pass along the side of the new coil and may see the full voltage potential.

To prevent arcing, add insulation between these wires and the coil with a strip of electrical tape or heat shrink tubing. You may also use high-voltage putty or corona dope. Separate the primary winding from the secondary with electrical tape.

Wrap the finished bobbin with electrical tape. Wind on the primary, if it was on top initially. Wrap again with electrical tape. Solder the ends of the wires to their respective terminals and reassemble the transformer.

Usually you can't put all the laminations back unless you clean them in solvent to get off the old varnish, but for this project you needn't bother. Instead, just leave out a few laminations—the transformer will work fine without them. The varnish prevents eddy currents in the core. **Eddy currents** are electrical currents generated by the magnetic fields that cut through it. They reduce transformer efficiency. Varnish insulates the laminations from each other so they don't form an electrical circuit.

While efficiency is desirable, it's not important here and may be ignored. The old varnish will suffice unless you are a perfectionist.

If the laminations are not tight, the transformer may buzz when energized. You can prevent this by coating the outside of the laminations with epoxy just before putting the band back around the core.

CHAPTER 9:
THE DESIGN PROCESS

Everything I've presented so far has been aimed at understanding the factors involved in ESL design and construction. Now it is time to discuss the design process.

Like most scientific endeavors, the design process works best if approached in a systematic and organized manner. There is no one best way to do this, but one exceptionally good method is to rank your design goals. Because ESL design parameters conflict, by deciding what is most important you automatically make most tough design decisions.

For example, if your most compelling goal is a full-range electrostatic, then you must abandon the lower priority goal of very high output. If small size and high output supercede deep bass on your list, then you will forfeit deep bass.

Goals further down the list will resolve many other design decisions. You can make any remaining choices based on the advantages and disadvantages of each, as presented in previous chapters. Your goals will fall into major and minor categories. The major issues are very important, of course, but the minor ones may prove to be more significant than you think. For example, from a design standpoint, you may believe that building your ESLs into a huge enclosure is best, but your environment may be too small.

The next step is to write your objectives. Only when you have them clearly identified can you set priorities.

I'll go through the design process for three ESL systems for examples and practice, and explain the reasoning used for design decisions.

I'll not discuss every design detail. The idea here is to show you in general terms one way to go about designing an ESL. The little details are up to you.

It happens that these are existing systems. One of these may fit your needs. If not, you can use them as the basis to design your own. Complete construction details for all three systems occupy the following chapters.

System 1 is a full-range crossoverless design. System 2 is a high-output, state-of-the-art hybrid system. System 3 is a compact/integrated hybrid system designed to be aesthetically acceptable in most fine living room environments.

The point of the following discussion is to give you an understanding of the thought processes that occur when you design an ESL system. Presenting all these thoughts in a completely organized and coherent way is difficult. I apologize if it's hard to follow.

SYSTEM 1:
FULL-RANGE CROSSOVERLESS ESL.

We will top the priority list with objectives defined by the title:

- Full-range electrostatic
- Crossoverless

If those were the only goals, we would have no problems. Unfortunately, there are other considerations. What should we do about high output, linear frequency response, size, cost, and dispersion, just to name a few?

With the first two goals defined, the real decision-making process begins. Several desir-

---

MAJOR DESIGN FACTORS

- Output
- Appearance
- Size
- Full-range electrostatic or hybrid operation
- Frequency response (what to do about the bass)
- Ease of construction
- Dispersion pattern
- Cost
- D/S spacing

MINOR DESIGN GOALS

- Crossovers
- Enclosures
- Associated electronics
- Parts availability
- Environmental concerns
- Safety
- Available tools
- Cosmetics

able goals conflict with each other. Specifically, a full-range crossoverless ESL dipole simply can't produce high output and linear frequency response simultaneously because of phase cancellation. It can produce bass, but not at high output. Conversely, it can produce high output if we sacrifice bass.

Here is where prioritizing becomes a major issue. We must choose the most important of these goals and accept the compromise this imposes. Because of phase cancellation, we can forget deep bass, but we would like good midbass performance. The drive voltage and D/S spacing seriously limit the excursion needed by the mid-bass. We need to optimize mid-bass performance, which means sacrificing output.

Output cannot be ignored. What good is bass that you can't hear? The situation isn't black and white; both must be enhanced as much as possible. You may choose differently, but in this exercise let's pick frequency response over output.

What about size and cost? Fortunately, cost is not a major concern if it is not outrageous. For example, it would be unacceptable to build a new house just to harbor a built-in ESL. In terms of the typical home speaker, however, large-dipole ESLs are not very costly and most rooms can accommodate them.

Since we want the ESLs to fit into a typical room, we can't use huge enclosures. So for sonic excellence, we'll use a dipole configuration.

Many factors influence the speaker's size. Floor-to-ceiling line sources have the highest output, image quality, and frequency response, so the maximum height could be 8'. However, it would be best if we could reduce the capacitance to improve high-frequency response.

Let's separate the speaker from the floor and ceiling by 1', and make the ESL 6-feet tall. This is very nearly a full line source, but having a 1' space between the floor and ceiling reduces the capacitance by 25%.

The remaining question is, "How wide?" Wider is better for phase cancellation, but a width of over 2' produces a speaker with so much capacitance that it is difficult to drive at high frequencies with conventional amplifiers and transformers. Also, at some point, very wide speakers become aesthetically unacceptable. To get maximum output and bass, let's select a very wide but tolerable 2'.

Again, we need to reduce the capacitance as much as possible without compromising bass.

Let's make the speaker 22- rather than 24-inches wide, and use a small baffle to limit the radiating width to 18". Let's make the speaker's frame 2-inches wide on each side of the cell. This will have the effect of adding a 4" baffle to the width of the cell and brings the total width to 22".

From the standpoint of phase cancellation, the speaker will behave as though it were essentially 2-feet wide, but by reducing the driven area to 18" we have reduced the capacitance by another 25%. The speaker will still have a relatively large capacitance, but it is practical to drive.

FIGURE 9-1: Radiating area is smaller than overall dimensions.

Dispersion, although important, is relatively easy to handle, because it doesn't involve much of a trade-off against other desirable characteristics. For example, narrow dispersion makes construction a snap. Resolution of detail and imaging is superb, and output is maximized. Wide dispersion damages imaging, output, and detail. You can make both types aesthetically attractive.

You can see that the goal priority list has resolved many of the design questions. For example, the issue of segmentation or equalization has been settled because we ruled out crossovers right at the beginning. This leaves only equalization or enclosures as a way of dealing with phase cancellation.

We've ruled out enclosures because we want

to fit the speaker into a typical room. Therefore, equalization is our only remaining option.

Important design parameters like D/S spacing are also pretty well-defined because we want full-range operation. Because there is no practical way to generate both high drive voltages and wide frequency bandwidth, we can't use large spacing. At the same time, the spacing must be as wide as possible for maximum bass excursion. Since this rules out both wide and narrow spacing, the only option left is spacing in the middle range of around 70–130 mil.

A high-power amplifier has more drive voltage than a small one. Therefore, it can drive wider spacing than a small-power amplifier. We can get high output by using a very powerful amplifier and maximizing the spacing. Because adequate output is a big problem, let's not compromise. We'll use a 250W amplifier and 130-mil spacing.

For 130-mil D/S spacing, a spacer ratio between 50:1–100:1 means the diaphragm support spacers should be between 6.5 and 13″ apart. The ESL's panel is 18″, but this includes the perimeter spacers. The actual driven area will be a couple of inches smaller— probably 16″ (*Fig. 9-1*).

Since 16″ is beyond the maximum spacer ratio, we must make either several cells of smaller size or a single cell with internal spacers to support the diaphragm. One large cell is easier to build than several small ones, so let's add spacers to a single cell.

What spacer pattern should we use? Only two patterns fall within the spacer ratio specifications. The first puts a single spacer vertically down the center of the cell that divides it into two 8″ sections. The resulting spacer ratio will be about 60:1. The other option is to run four spacers horizontally, breaking the cell into five 11-inch-wide sections. This puts the spacer ratio at 85:1 (*Fig. 9-2*). Which should we use?

The vertical configuration has a total spacer length of about 6′ while the horizontal spacer pattern has approximately 7.5′ of spacers. Spacers cause stray capacitance unless we use special construction techniques. To make construction easy, we won't use low-stray-capacitance construction. The vertical strip will have less stray capacitance.

The single vertical strip's lower ratio will have a greater stabilizing effect on the diaphragm than the wider horizontal strips. This will permit us to use higher polarizing voltages and get more output. Building it is

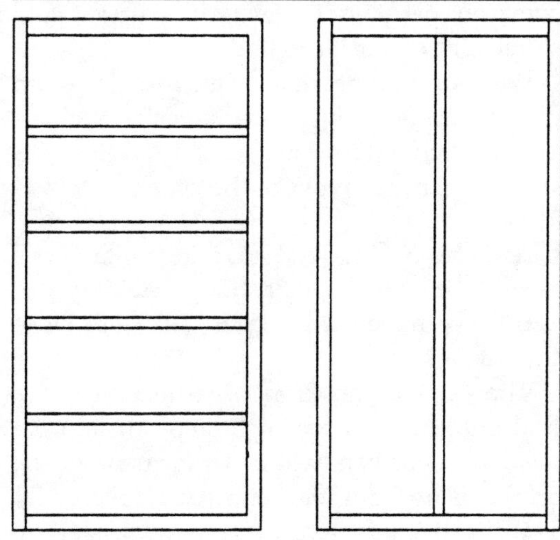

FIGURE 9-2: Different spacer patterns.

easier because we only have to make and position one strip instead of four. Clearly, the single vertical strip is better.

What type of construction should we use? Let's keep it simple and use perforated metal. There might be a small loss of output compared to the best wire cells, but it would be no more than 1 or 2dB, and the ease of construction is worth it.

Using the guideline that the perforations should be no larger than the D/S spacing, we can use 1/8″ perforations. To get an adequate percentage of open area, the perforations need to be on 3/16-inch-staggered centers. This is a

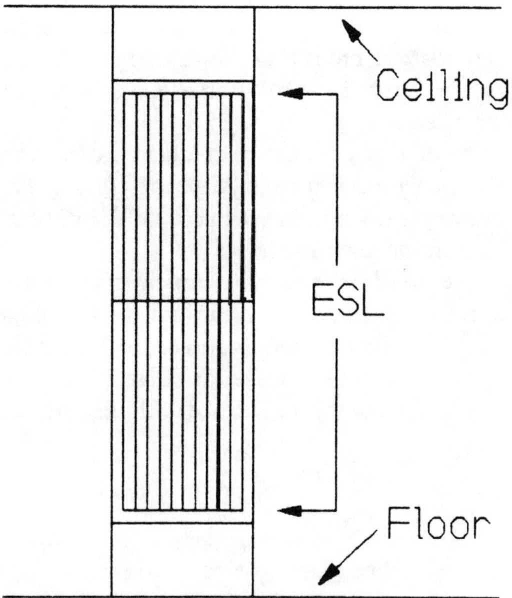

FIGURE 9-3: ESL as room divider.

common perforation size and pattern, and shouldn't be difficult to find.

We can mount the speaker as a room divider, using spring-loaded feet to hold it between the floor and ceiling. *Figure 9-3* shows a general layout of the speaker.

ELECTRONICS. The polarizing voltage will need to be as high as diaphragm stability will permit for maximum output. 7kV will be in the ballpark.

The step-up transformer is critical—can we find one that has the high step-up ratio we need *and* wide bandwidth? Unfortunately, the answer is no—but we can come close.

We can find transformers with step-up ratios of 50:1 that will drive the speaker's capacitance (around 1.2kpF) to 20kHz. Let's look closely at a specific transformer: the Triad S-142A.

Using the 4Ω primary taps and the 8kΩ secondary taps, the turns ratio is 1:44.7. A 250W power amplifier will deliver at least 160V P/P, which, when stepped up 44.7 times, gives 7,150V of drive. Although more would be better, this will work and is about the best commercial transformer technology has to offer.

This transformer is rated at 15W at 7Hz. Can it handle a 250W amplifier? Well, certainly not at 7Hz. But our dipole ESL can't produce any output down there anyway, so it's a moot point. The real question is, "What is the lowest frequency our ESL can reproduce linearly?"

Fundamental resonance determines the lowest usable frequency, since output falls swiftly below it. Although we have no accurate way of predicting the fundamental resonance, 70Hz is a reasonable guess based on experience.

At first glance, the next question is, "Can the transformer handle 250W at 70Hz?" The answer is no, but do we need 250W at 70Hz? Can the speaker use it?

We need to see how much current the speaker needs at the maximum drive voltage at 70Hz. To do that, we need to know the impedance of the speaker at 70Hz.

We can find the impedance with the formula:

$$R = \frac{1}{2\pi FC}$$

This shows the impedance to be about 1.9MΩ. Ohm's Law says to divide the voltage by the resistance to get the current. Doing so gives us 3.8mA.

We get watts by multiplying volts times

amps to get 27W. Is that within the transformer's capability?

The formula for determining the transformer's power handling at a given frequency is:

$$\frac{Power_R}{Frequency_R \times 2} = \frac{Power_M}{Frequency_L}$$

Where:

$Power_R$ = Manufacturer's power rating in watts (or VA)

$Frequency_R$ = Manufacturer's low-frequency specification

$Power_M$ = Maximum permitted power at $Frequency_L$

$Frequency_L$ = ESL's lowest linear frequency

Solving this formula for our transformer at 70Hz:

$$\frac{15}{14} = \frac{Power_M}{70}$$

$$\frac{1,050}{14\,Power_M} = 75W$$

The transformer can deliver 75W at 70Hz, while the speaker needs only 27W at 70Hz. Surprisingly, we don't have a problem.

Note that at ten times that frequency (700Hz) the speaker can use all the power the amplifier will deliver plus some—270W. But at ten times the frequency, the transformer can deliver ten times the power—750W. So a 15W transformer can do the job.

We would have a problem if the speaker's capacitance were higher. For example, if we reduced the D/S spacing so the capacitance doubled, we would need twice the power. Instead of 27W at 70Hz, we would need 54W. That's still comfortably within our transformer's capability.

But note what happens to the frequency when the speaker asks the amplifier for maximum power. Our 250W amp will no longer be maxed out at 700Hz, but rather at half that—350Hz.

Recall that the nature of music is such that as we go down in frequency, the power required rapidly increases—typically, at 6dB/octave or more. If you double the power, you get only 3dB more output. By reducing the frequency where full amplifier power is required by an octave (700–350Hz), we probably will square our amplifier power requirements. This seriously reduces the speaker system's output capability.

Why would we want to reduce the D/S spacing in the first place? Because smaller spacing *increases* output for a given drive voltage. So, to some extent, the loss of output caused by power amplifier voltage limitations is compensated by reduced drive voltage requirements. Since the usual limitation on output is voltage, it makes sense to optimize it.

But there's no free lunch. If you reduce the D/S spacing to get more output from a low-power amplifier, you'll increase the current and power requirements. In essence, you trade current for voltage. Since the small amplifier will also be limited in current, you'll find that the total output isn't increased. You need high-power amplifiers to get high output.

The next step is to correct the problem of phase cancellation and fundamental resonance with equalization. Because of the large corrections required, it is desirable to break down the problem into three sections.

Working from the high frequencies down, we can see that phase cancellation will begin to occur at around 4kHz, where the wavelength of the sound is about one-quarter of the minimum speaker dimension. The loss starts gradually and then steepens. For that reason, for about two octaves below the start of phase cancellation down to 500Hz, we'll lose only about 8dB.

This error is small enough so we can correct it with a mirror-image, 6dB/octave equalizer. Let's call this range the first section.

The second section begins at around 500Hz and extends down another two octaves, where fundamental resonance starts to push the frequency response back up. This two-octave range between 500–125Hz is the real killer because the slope starts to get steep and large losses in output occur. Depending upon where fundamental resonance is, the slope in this area will approach the 12dB/octave range.

If we lose 8dB in the first section, and perhaps 20dB in the second, the total is 28dB. It quickly becomes apparent why high output isn't possible from an ESL operated into the bass.

The third section is fundamental resonance. The frequency of this resonance is variable depending upon room size and shape, and the size and tension of the speaker's diaphragm. Its magnitude will be determined by the room resonances and whether they fall near the speaker's resonance, and whether we use any speaker damping. For these reasons, it is not practical to predict the resonant frequency or its magnitude.

What is the best way to arrange equalization for this speaker? One way is to use an equalizer that is a dedicated, low-level, active unit consisting of three sections.

The first section would be a 6dB/octave shelving equalizer starting at 100Hz and extending up to 2kHz. The second section would have another 6dB/octave shelving equalizer adding more boost to the first. It would cover the frequencies between 100–500Hz.

The equalizer's slopes do not start immediately at 6dB. You must start them at around 50Hz so the steepest part of the curve is present at 125Hz, where it is badly needed.

Finally, a notch filter could be added to suppress the fundamental resonance once it is identified in both frequency and amplitude. The advantage is nearly perfect frequency response correction at the expense of complicated equalizer design.

Another way to cope with the problem is to use a 6dB/octave shelving equalizer from 50Hz–2kHz, and supplement it with the bass tone control on a typical preamp. Such tone controls usually start to boost the bass at around 1kHz and can add 8–10dB by 100Hz. Often this is not enough, but it is a big help, and when added to the shelving equalizer, we can get reasonable results.

Most builders, when they use a preamp in this manner, won't use a notch filter to control fundamental resonance. The fundamental resonance adds to the thin bass caused by inadequate equalization, and makes the overall bass sound better. The penalty is excessive excursion at resonance which limits output. Also, the frequency response in the mid-bass is not linear.

The equalizer is simple. It can be passive, because it is used with a preamplifier which will have excess gain to compensate for its insertion loss. It cheaply and easily solves the problem to the satisfaction of many.

A third way to correct the frequency response is with a shelving equalizer from 50Hz–2kHz, but deal with the second and third sections with a graphic equalizer rather than with preamp tone controls. The advantage is more severe correction, and you can use one of the bands as a fundamental-resonance notch filter.

With a graphic equalizer, you can experiment to find the best frequency band to control fundamental resonance. The amount of suppression can be adjusted to fit the system's needs.

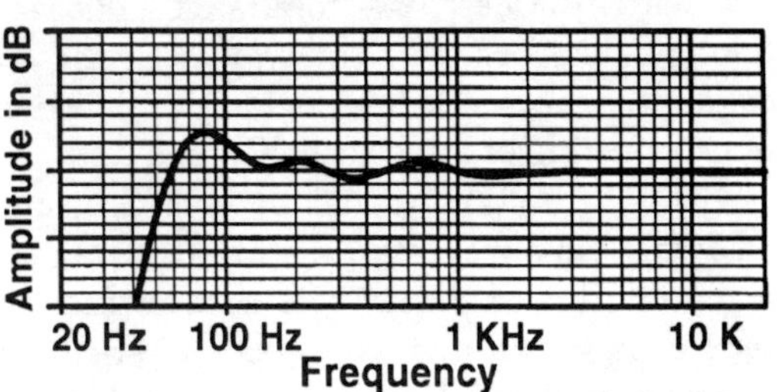

FIGURE 9-4: Non-linear response of graphic equalizer.

This flexibility is particularly worthwhile if you don't have objective measuring instruments available. You can do a good job of identifying the fundamental resonance and its amplitude empirically.

The only penalty, other than the cost of a graphic equalizer, is that frequency response in the equalized range will be wavy (*Fig. 9-4*). This becomes an insignificant problem when you consider the ragged bass response of the typical room.

Another way to deal with the fundamental resonance problem is to use mechanical damping in the form of a finely woven cloth on one or both sides of the speaker. This can effectively reduce the resonance without electrical intervention, but it has several disadvantages.

The main complaint is that the cloth, if used on the front of the speaker, will adversely affect the high-frequency clarity. You can easily test this by mounting the cloth on a frame and having an assistant alternately place it in front of the speaker, while you listen blindfolded. You should have no difficulty "hearing" the cloth.

This should come as no surprise because you are literally "listening to the sound through a blanket" (albeit a thin one). Repeat the test on the back of the speaker, and you may be surprised to hear how the cloth still affects the sound.

Another disadvantage of mechanical damping is that, in comparison to a notch filter, the amplifier load is higher. A notch filter reduces the voltage at resonance, while mechanical damping leaves it unchanged.

Before leaving the full-range electrostatic design, let's look more closely at segmented ESLs. Without a doubt, the sound quality will be inferior to the full-range crossoverless system. Using several electrostatic panels of different dimensions will not result in as smooth a frequency response as an equalized system.

Crossovers must be in the critical midrange region, which will cause audible problems.

You will need several transformers of different specifications for the different segments. However, they can be optimized to the load and frequency bandwidths required of them. This will make it possible to use a very high step-up ratio transformer for the bass, which should result in higher output than a full-range transformer.

Such systems usually use a single amplifier and multiple high-level passive crossovers. This serious compromise can be improved by using multiple amplifiers, low-level active crossovers, multiple transformers, and possibly multiple equalizers to clean up the ragged frequency response.

Such a system would be very complex and relatively expensive. The sound quality would still not be as good as the full-range crossoverless system, but the output might be slightly better.

SYSTEM 2:
NO-COMPROMISE HYBRID ESL/TL.

The two insurmountable problems of a full-range dipole ESL, low output and lack of deep bass, are both extremely important to high quality sound reproduction. System 2 resolves these issues while maintaining the outstanding sound quality available from a full-range crossoverless ESL.

Most audiophiles believe hybrid systems must compromise sound quality. In reality, you need not accept *any* audible compromise. The goal priority list can now have high output, wide bandwidth, flat frequency response, and electrostatic sound quality at the top.

These goals must be balanced against each other in an all-ESL system, but in this design we won't have to settle for less-than-excellent performance in all categories. No significant restrictions are placed on the design, other than general reasonableness. For example, the system must fit into a typical home, and the cost must not be outrageous.

The design must start with a way to obtain high output and flat frequency response. Speaker systems based on magnetic drive can easily produce high output, deep bass, and meet the goal of "general reasonableness."

The problem with magnetic drivers, of course, is they fail to meet the goal of "electrostatic sound quality"—or do they? The fact is that in the bass, the negative effects of massive drivers are not much of a problem. After all, it is not always necessary to use a dragster to travel to the end of a quarter mile. A loaded truck can do it given enough time—and in the bass, there is time.

A woofer need not have good transient response. Many audiophiles would argue this point, but the fact is that bass musical instruments do not have rapid attack or decay times. Although a magnetic driver cannot stop quickly, it will not be asked to do so in music, so ringing and overshoot are unlikely. What audiophiles perceive as "bass transient response" is the harmonics of the bass reproduced in the midrange. The problem of transient response is really a midrange problem.

Cone flexure is a troublesome area with magnetic drivers because it causes harmonic distortion. Without question, magnetic drivers have a great deal more distortion than ESLs. But this is not a problem in the bass for two reasons.

First, the distortion is primarily harmonic. While this changes the sound character to some degree, it is not unpleasant. In other words, it doesn't sound distorted unless the percentage of distortion is extremely large. Small amounts of harmonic distortion merely make the sound a bit "warmer" and "more full" than it ought to be.

The other issue is ear sensitivity to bass frequencies. The ear simply doesn't recognize distortion in the bass very readily.

Magnetic woofers have the potential to match electrostatics in the bass, but in practice they have not done so. The reasons for this are many, primarily the limitation of enclosures that we must use with magnetic drivers. This issue will be dealt with in detail in *Chapter 12* on transmission-line design. I believe transmission-line magnetic woofer systems can produce ESL sound quality in the bass, while simultaneously producing high output and low frequency.

I realize I may not have convinced you of this yet. But for now, let's assume it's true so we can continue this discussion.

By using a magnetic woofer system, we relieve the ESL of its major problems involving bass, output, and linear frequency response. The remainder of the design task now becomes relatively easy.

A large ESL operated in the midrange and high frequencies, and driven by a powerful amplifier, can produce very high output levels. We still must pay close attention to the details, but the ESL can do the job if we do ours by using good design.

The biggest remaining problem is crossover-point selection. Ideally, the crossover frequency should be where the ESL's frequency response is starting to fall due to phase cancellation. This puts the adverse effects of crossovers in the critical midrange frequencies, and so is not practical. Also, even the best magnetic woofers are inadequate in the midrange.

Remember, the woofer doesn't suddenly stop at the crossover point. It needs reasonably flat frequency response for about two octaves above it.

Several modern magnetic woofers in the 10- to 12-inch range have flat frequency response to 2kHz. You may have to search for them, but they exist.

They will work beautifully with a crossover point near 500Hz if we use steep crossover slopes. As previously discussed, the human ear is less sensitive to crossover flaws if the crossover frequency is below 600Hz, which then becomes our reasonable guideline.

The crossover frequency should not be much below 500Hz, either. Recall that phase cancellation causes large output losses between 100–500Hz in the full-range crossoverless system. These two octaves are the real "killers" of output. ESLs just don't have the output necessary to deal with such large deficits.

If we force the ESL to reproduce this range linearly, the system output will not be much better than the full-range system. The only thing a magnetic woofer will do in a system with a 100Hz crossover is produce deep bass. We won't meet our twin goals of high output and linear frequency response.

Despite this, most ESL designers continue to use unreasonably low crossover points. Such systems can only produce high outputs by failing to equalize the range between 100–500Hz. The sonic result is a "suck-out" in the upper bass that causes the speakers to lack punch and sound thin and anemic. ESLs have a poor reputation in this regard, yet it is not an inherent fault of the drive principle. Poor design decisions are the cause. This sad state of affairs has no logical reason.

What about the size of the ESL? It might seem that with a crossover point of around 500Hz a small ESL will suffice, but this is not so.

The frequency versus wavelength graph reveals that phase cancellation starts well above the crossover point in all dipoles of reasonable size. It quickly becomes apparent that we will need midrange equalization just as in

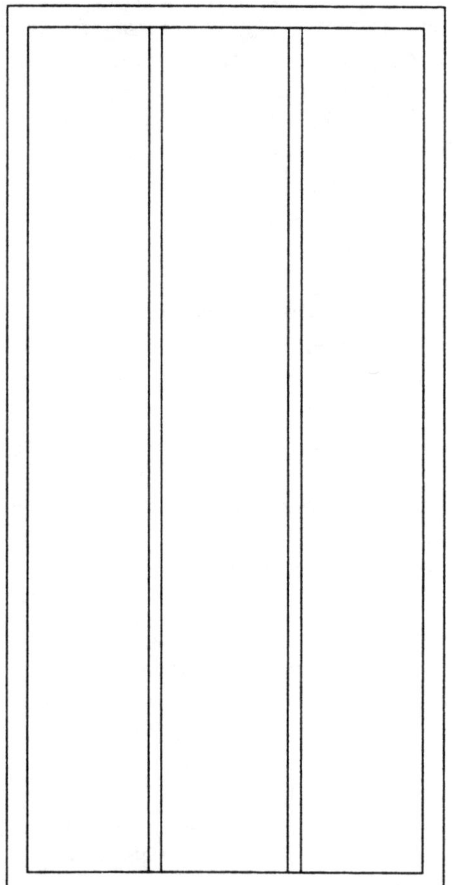

FIGURE 9-5: Spacer design for large ESL.

the large full-range design—ideally, we will need the same full-sized ESL.

Let's make the same size and design decisions as for the full-range system, but with one exception—smaller D/S spacing. By using a smaller spacing, we can dramatically improve the output from the same amplifier/transformer system used in the full-range ESL.

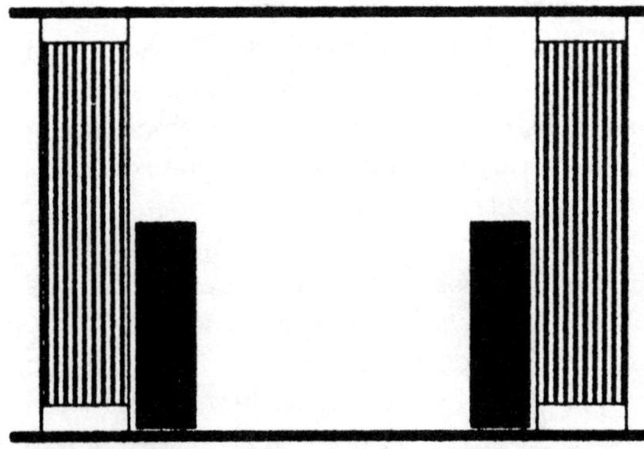

FIGURE 9-6: Large hybrid ESL/TL.

Because we're not operating the ESL below 500Hz, we can avoid the problems with excessive voltage, current, and power requirements in the lower frequencies.

What is the smallest D/S spacing we can use? There is massive acoustic coupling between a large ESL and a powerful magnetic woofer system. We can produce very high SPLs with a hybrid system, so acoustic coupling can be a serious problem. Practical experience shows that 70 mil is the minimum acceptable spacing in large hybrids. If the listening room has bad bass resonances, you should increase the spacing another 10 or 20 mil.

This change in D/S spacing forces a change in the distance between spacers to maintain a reasonable spacer ratio. Now the ideal free diaphragm length is 3.5–7″.

Again, we can use either vertical or horizontal spacers. We can use two vertical ones to make partitions slightly over 5–7″, or eleven horizontal spacers producing twelve partitions of about the same size. The spacer ratio for both is about 75:1.

The total length of spacers in the vertical orientation is about 12′, while it is 16.5′ in the horizontal design. The vertical design best reduces stray capacitance (*Fig. 9-5*).

We'll use active low-level equalization to flatten the midrange. As before, we'll need a total of 8dB of correction. This will still permit very high output if we are careful with the rest of the design.

Low-level active crossovers with Butterworth filters and 18dB/octave slopes give the best performance for this design. Refer to *Chapter 7* for a detailed discussion of crossovers.

The issue of directionality is easily decided here. We want no compromises. Since sound quality and output is better in highly directional designs, that's what we'll use.

Our system is going to be very large. Not only do we have full-range-sized ESLs, but a pair of very large magnetic systems. Aesthetics must be set aside in favor of performance (*Fig. 9-6*).

This system can produce outputs in excess of Row A concert hall levels. Deep bass is exceptional, and the system will sound totally electrostatic.

SYSTEM 3:
A COMPACT/INTEGRATED ESL/TL.

The large hybrid system offers stunning performance, but leaves much to be desired aesthetically. The large woofer enclosures and floor-to-ceiling ESLs—all of which must be free-standing—make it unacceptable for most living room environments. The challenge is to make an ESL speaker system where compactness and aesthetics are at the top of the goal priority list, while maintaining performance on a par with the large hybrid system.

Achieving these goals requires paying serious attention to finish work and general exterior design. We must also play the various design parameters against each other for best performance.

We'll have to build a smaller speaker. It will also be necessary to integrate the ESL and the transmission-line woofer system.

The biggest problem is maintaining high output and linear frequency response from a small ESL. Smaller speakers have greater phase cancellation losses, so require greater equalization which reduces output.

Let's start the design process by deciding on the ESL's dimensions. Since vertical height does not take up floor space, let's make the ESL 6-feet tall. For aesthetic reasons, let's reduce the minimum dimension to 1'.

If we make the woofer enclosure tall and thin, we can mount the ESL to it and have it act as a baffle. This will make the air follow a longer path around one side of the ESL, and will minimize phase cancellation. We also can use the transmission line cabinet as a beam splitter to improve casual dispersion qualities, so the compromises of a curved ESL will not be required. I'll discuss this TL design in detail in *Chapter 12* on transmission-line design. *Figure 9-7* is a general design layout.

To compensate for the increased phase cancellation losses, we must begin the equalization at 3kHz with the same type of shelving equalizer used on the other systems. All we are doing is moving the entire range upward.

For the same reasons, we'll have to move the crossover point as high as possible. I consider 550Hz the maximum acceptable.

We can't reduce the 70-mil D/S spacing because we need just as much (or more) excursion as in the larger designs, and acoustic coupling will be just as much of a problem. We again have a choice of vertical or horizontal

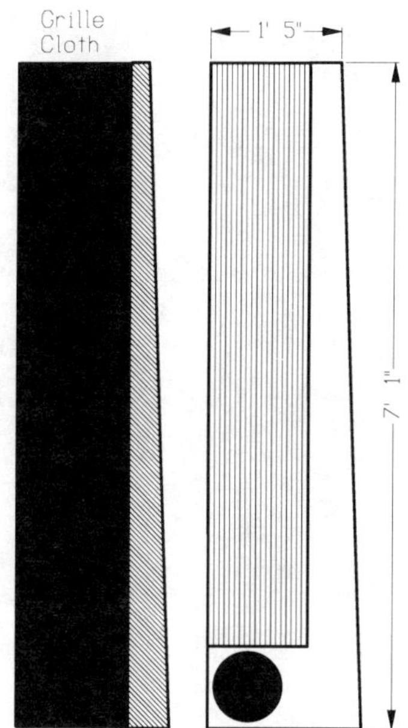

FIGURE 9-7: Compact integrated ESL/TL.

spacers, and again the vertical orientation offers the least stray capacitance and greatest simplicity.

We can continue to use the same amplifier/transformer drive system. It might seem as though a smaller system would need smaller amplifiers, but just the opposite is true.

The smaller ESL system needs all the help it can get to produce high output. The smaller woofer system also needs more power to produce the same output as a larger woofer. So the smaller system needs *larger* amplifiers to meet the goals of high output and bass response.

The laws of physics can't be circumvented. We know at the outset that the smaller system can't have as much output as the larger one. However, the difference is only about 2dB—a barely detectible difference subjectively.

Frequency response is good except in the deep bass, where the smaller magnetic woofer (9″ in the small system, 12″ in the large one) and shorter transmission line result in slightly less depth. Aesthetically, the integrated/compact system is quite acceptable.

Let's build some electrostatic panels. Although there are different ways to do so, three of the most popular cell types are:

- Perforated metal
- Rigid wire
- Tensioned wire

Let's do flat (planar) cells first. The next chapter expands on the curved-cell theme.

The various designs differ only in the way you make the stator. These designs have the same sound *quality*—their differences lie in efficiency, output capabilities, ease of construction, and cost.

Perforated-metal is the easiest, cheapest, lightest, simplest, thinnest, and fastest construction type. You can build one complete set of large ESLs in one day using this technique.

What are the disadvantages of perforated metal? The output *may* be lower than in a wire stator, and it's slightly more difficult to build perforated-metal cells with low stray capacitance than wire ones.

The output may be lower because it's difficult to get perforated metal that will produce as high a field density as a *well*-designed wire stator—not impossible, just difficult. The ideal hole pattern required for high field density is not readily available. With wire, you can build anything you wish.

Stray capacitance is often higher in perforated-metal than in wire-stator types because perforated-metal stators are usually built for the ultimate in simplicity. They need not be built that way, but they usually are.

Perforated metal requires excellent support because the perforating process produces internal stresses that warp the metal. To make the metal flat, you must hold it straight with a supporting structure.

You can do this easily by using the diaphragm support spacers as laminations to hold the metal sheet flat. This results in the metal overlaying the spacers and producing stray capacitance, particularly around the perimeter (*Fig. 10-1*).

Wire stators are often built the same way and also have some stray capacitance, but the wire only overlays the spacers at the ends. Therefore, they have less stray capacitance (*Fig. 10-2*).

You can build near-zero stray capacitance sta-tors with perforated metal, but the external framework required to hold the metal flat negates the advantages of ease, simplicity, and low cost. Wire stators with low stray capacitance also require substantial external structures.

Wire stators have the advantage of design flexibility. You can make them in almost any configuration, and with any percentage of open area and slot size.

They have no sharp edges which can produce coronas. A corona is the tendency for a sharp, high-voltage surface to ionize the air near it. The resulting ion cloud is a conductor. Usually, the cell will arc if a corona forms. The smooth, rounded surface of wire is virtually immune to coronas, while the sharp edge of a hole may produce one. This appears to be more of a theoretical problem than a real one, as I've never seen a corona in a perforated-metal stator.

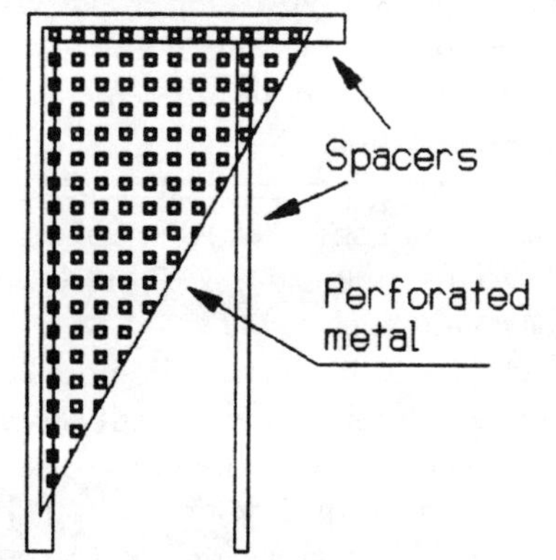

FIGURE 10-1: Stray capacitance in perforated metal stators.

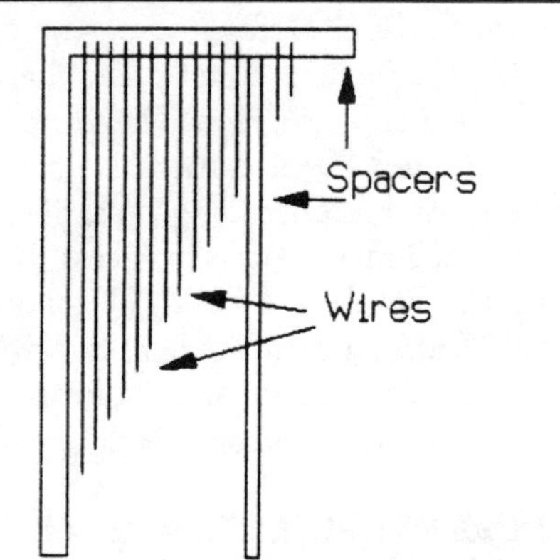

FIGURE 10-2: Stray capacitance in wire stators.

You can buy wire coated with many types of high-quality insulation. Perforated metal doesn't come with insulation, and it's difficult to apply highly effective insulation to it. Of course, most designs don't need insulation, but if yours does, keep this in mind.

Tensioned-wire stators have all the advantages of wire stators, are flattest of all, and you can use insulated wire. Their most interesting feature is that you can build them with long, unsupported stretches of small wire while retaining flatness. If all else is equal (rarely the case), they should have the highest output.

Their disadvantage is that they are extremely labor intensive—you must attach hundreds or thousands of wires to a supporting structure and place them under tension. Their supporting framework must be very strong to withstand the tension of a multitude of stressed wires.

Rigid-wire stators have disadvantages, too. They are *much* more labor intensive and expensive than perforated-metal stators. You won't build a set of wire ESLs in a day. More likely, the project will take weeks.

Rigid-wire stators require separate, strong frames to which you glue rigid wires or string flexible wires. This complicates construction. Simple perforated-metal stators are self-supporting.

Wire stators are much heavier than perforated aluminum ones. An 18″ × 36″ aluminum panel weighs around 4 lbs. A steel wire one weighs nearly 30 lbs.

What is the performance difference between small-hole perforated metal, with simple construction, and the best wire ones? The output difference will probably not be audible. At worst, it will likely measure less than a couple of decibels. The high-frequency response difference based on stray capacitance will probably be no more than a few hundred hertz.

The performance difference between wire and perforated metal is small while the work, time, and expense is great. Therefore, most builders opt for perforated-metal construction.

Before getting into building cells, I'll review the perforated-metal stator process because they are the simplest to build. I'll follow this with very detailed instructions on how to build them. This base information can then be expanded into the other construction methods.

BUILDING PERFORATED-METAL STATORS: AN OVERVIEW

Figure 10-1 shows the basic construction of this simplest-of-all stator designs. You just take a sheet of perforated metal, cut it to the size needed, glue spacers to it, and it's done!

To complete a cell, glue a conductively-coated diaphragm to one stator, heat shrink it, and glue another stator on the other side of the diaphragm. Your cell is finished.

Here are the details: cut strips of plastic that you will use as spacers from a sheet of acrylic (Plexiglas) or polycarbonate (Lexan). The plastic's thickness determines the basic D/S spacing. Glue the spacer strips together to form the frame to which you can attach the perforated metal.

You must make the spacer frames identical. When they are glued to each other with the diaphragm sandwiched between them, the spacers must face each other, which is easy to do if you draw your spacer pattern on the underside of a sheet of glass. By building each spacer frame over this pattern, they will be virtually identical.

With the spacer frames complete, glue on the perforated metal. Do all assembly work on the glass sheet so everything is flat and accurate.

Lay some Mylar film over the glass. Sprinkle powdered graphite and grind it into the Mylar by rubbing it hard with a paper towel.

Put epoxy glue on the stator spacers and set it on the Mylar. When the epoxy has cured (ten minutes to an hour depending on the type), lift it from the glass and heat shrink it with a heat gun to get it tight as a drum and wrinkle free.

Lay the finished stator/diaphragm upside down on the glass so the diaphragm is facing up. Put epoxy on the spacers on the other stator assembly, and lay it on the diaphragm. When the epoxy has cured, trim the excess diaphragm from the edges and attach electrical contact bolts to finish the cell.

Now that you have the general idea, I'll go into the construction sequence in greater detail and describe a generic cell—one without specific dimensions. In the chapter on Systems, I present detailed drawings for specific designs and cell sizes.

PMS CONSTRUCTION. Start by making a drawing of the cell. Dimension everything and carefully consider the way you will mount it in its frame. You can change things on paper more easily than you can after the parts are cut and the cell is partly assembled.

Chapter 4 on output has the information you need to select the specifications for the perforated metal—thickness, hole/slot patterns, and cell dimensions. The only question is whether to make one big cell or several small ones.

For example, if your speaker is to be 6-feet tall and 1.5-feet wide, you can make one cell that size, two cells $3' \times 1.5'$, four cells $1.5' \times 1.5'$, or any combination that fits the dimensions.

Large cells are slightly more difficult to make, but you are better off with one or two large cells rather than many small cells. The easier construction of the small ones is far out-weighed by the extra parts you must assemble.

Mounting many small cells to make a single speaker is much more difficult than mounting a single large one. Making very narrow borders around any ESL is difficult. When you have two cells together, their borders cause a significant void in the speaker output at that point. This adversely affects the sound. Also, the borders of an ESL usually have stray capacitance.

In short, large cells are easier to make and have better performance than small ones. You can use internal spacers to support the diaphragm in a large cell if necessary (it usually is).

On the other hand, at some point the cell becomes too large to manage. I think a single cell $2' \times 6'$ falls into that category.

A reasonable compromise is to make cells no more than $3'$ in any dimension, which also happens to be a common size for materials like perforated metal or welding rod that you probably will use to build the cells. If you want cells larger than $3'$, you will incur additional expense buying oversized material.

For this discussion, I will assume that you wish to use a D/S spacing of around 70 mil. Therefore, the holes or slots in the metal should ideally be around 1/16-inch diameter, and you will use 1/16" Plexiglas spacers.

If you were going to use 90-mil spacing, you could use Lexan "unbreakable windows," which is 80-mil thick. If you wanted to use 130-mil spacing, you could use 1/8" Plexiglas spacers.

If you used 130-mil D/S spacing, the holes in the perforated metal could be larger—up to 1/8". Smaller holes produce higher field density and more output even with large D/S spacing. You might as well use perforated metal with 1/16" holes for any D/S spacing 70 mil or larger.

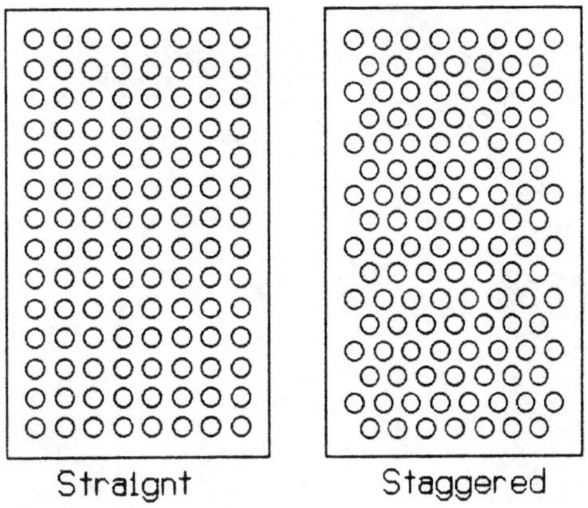

PERFORATED METAL. From your reading of the chapter on output, you know that the

FIGURE 10-3: Perforating patterns.

percentage of open area in a stator should be around 50%. Perforated metal with round holes comes in two patterns: straight and staggered (*Fig. 10-3*).

To get 50% open area requires that the holes be very close together. You must use the staggered pattern and a very close hole spacing dimension. A suitable pattern would be 1/16" perforations on 3/32" staggered centers.

What metal should you use? Aluminum is lighter than steel, doesn't rust, and can be anodized in various colors. It comes in very thin sheets that you can cut with scissors, yet is adequately strong. The ideal perforated metal for most ESLs is 20-mil-thick aluminum with 1/16" holes on 3/32-inch-staggered centers.

Where do you get it? Metal processing firms will perforate almost any metal and thickness you desire, but this can be costly if it's not a commonly stocked item. Look in the "Yellow Pages" of any large city telephone directory under the heading "Metal Perforators."

Fortunately, a good substitute is available. It comes as decorative aluminum sheets and is available from any ACE® hardware store.

Although it doesn't come in the exact pattern described above, it comes in a pattern called Lincaine, which is nearly as good. Lincaine has 1/8" holes surrounded by 1/16" holes.

With Lincaine, you have a choice of finishes—plain aluminum or gold anodized. I urge you to buy the gold anodized version. The plain aluminum type is coated with an oily film that is difficult to remove. The gold type is spotlessly clean, aesthetically more pleasing, and costs only a few dollars more.

This material is 20-mil thick and comes in $2' \times 3'$ or $3' \times 3'$ sheets. By contrast, metal perfo-

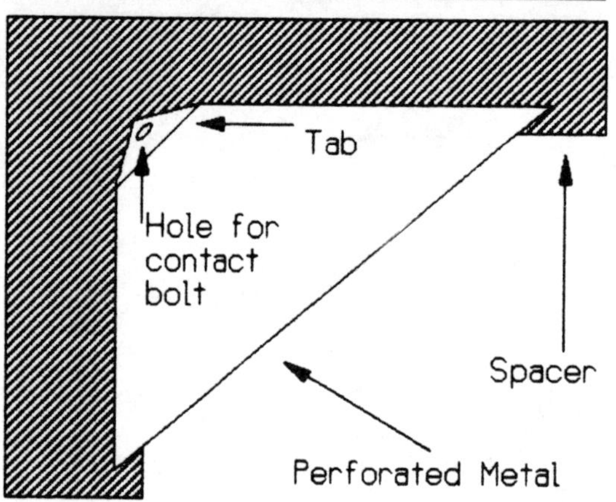

FIGURE 10-4: Detail of stator contact.

rators usually supply metal in $2' \times 10'$ sections and charge to cut it to smaller sizes. Shipping also is likely to be a problem for perforators, while Alcoa supplies Lincaine in cardboard shipping containers.

The hardware store is unlikely to have this in stock. You can find it in their catalog and special order it.

Such soft, thin material can easily be damaged in transit. Commonly, a corner is bent, which you can easily straighten. Beware of a crease across the sheet; you can never get it flat.

The sheets will have a modest bow or warp to them from the perforating process, but this is normal and will not usually be a problem. The laminating process will flatten it.

You can cut a sheet in half with a pair of scissors. Depending on the cell size, you may get two stators out of one sheet. If you want a perfect cut, take it to your local sheet metal shop and have them cut it on a sheet metal shear. It will take them only a moment.

You will note that the holes have rounded edges on one side and sharp ones on the other. During construction, be certain to face the rounded holes toward the diaphragm to minimize the possibility of developing a corona.

Drill a 5/32″ hole in one corner of each sheet, which will later accept a 6-32 nut and bolt used for the electrical connection to the stator. Remove the burr left by the drill by twisting a much larger drill bit or a countersink in the hole. Twist it by hand for good control.

You will bend this corner at right angles to the diaphragm *after* you complete the cell, and put a bolt through it for the electrical connection. If you bend it now, you can't lay the metal flat for construction purposes.

The tab need only be about ½″ (*Fig. 10-4*). If you prefer to have all the connections in the same speaker corner, remember that the stators will be mirror images of each other. Be careful to pick the appropriate corners.

SPACERS. These must have very good insulating qualities. Clear plastics work well. You can buy sheets of acrylic: one brand is Plexiglas, another is Lexan, also known as "unbreakable windows," in glass shops or home improvement centers.

Normally, you cut the sheet into narrow strips and glue them together to make a spacer frame. Some builders cut openings in a solid sheet to avoid gluing strips together. They cut openings with a hand-held router, milling machine, punch press, or computer-driven laser.

Few of us have access to such machinery. Routing wastes most of the expensive material, so most of us glue the strips together.

Some glass shops will cut the plastic into strips for you, but it's not difficult to do and you can save money. Although there are many ways to cut plastic, no way is perfect, but they all work. You can even score and break off strips as you would for glass, but it is very difficult to do for long, thin strips, even with a jig. Sawing the plastic into strips is most practical.

The best tool for this job is a band saw with a very fine blade. Clamp a rip fence to the table to guide the plastic, and you can cleanly cut narrow strips. A 24-tooth/inch hacksaw blade works beautifully.

Unfortunately, band saws are not common. You more likely have access to a table saw. Table saws work, but there are problems. First, nobody makes fine-toothed blades for them. You can use a plywood blade, but this will cause chipping along the edge of the cut. Fortunately, the chips, although unsightly, don't degrade performance.

You will also discover that the plastic sheet wants to climb the saw blade instead of being cut. If you try to stop this by raising the blade, it will tend to catch and shatter the plastic.

A solution is to clamp a piece of wood on the side of the rip fence about ¼″ above the table. With the saw running, move the blade up into this wood about ¼″. The block will prevent the plastic from climbing up the blade (*Fig. 10-5*).

DIAPHRAGM CONTACT. You must somehow connect the high-voltage polarizing sup-

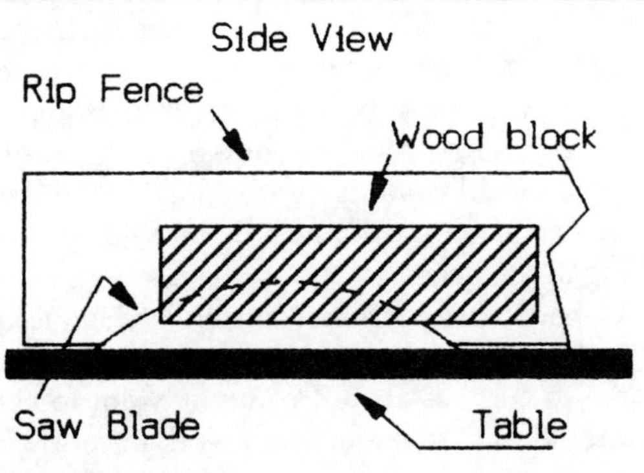

FIGURE 10-5: Table saw modifications.

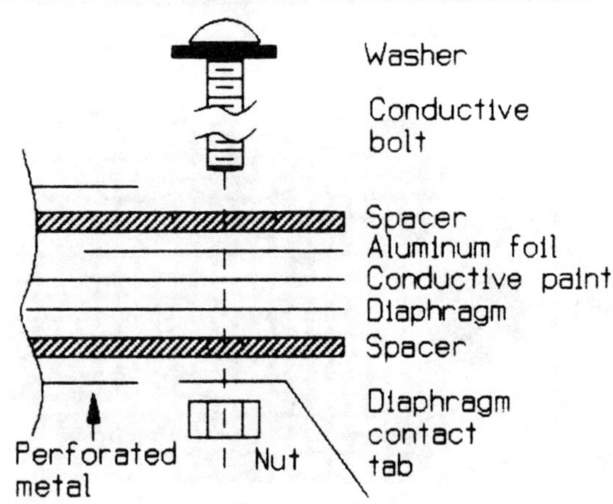

FIGURE 10-6: Detail of diaphragm bolt contact.

ply to the diaphragm. This contact is potentially very fragile, and you must take care to make it durable.

The method I prefer is to put a metal bolt through the spacers so it makes contact with a strip of aluminum foil, which is in contact with the diaphragm. The polarizing voltage connects to this bolt (*Fig. 10-6*). I call this a **bolt contact**.

Another good method is to glue a tab to the stator that contacts the coated side of the diaphragm. This tab will stick out from the edge of the cell after assembly. Clamp it between the stators with a nylon bolt (a good insulator) and it makes a very secure contact. You can also use the nylon bolt to hold two tabs that make up the stator contacts. The result is a rugged and compact set of contacts (*Fig. 10-7*). I call this a **tab contact**.

The tab contact's disadvantage is its interference with mounting cells by their edges. The bolt contact sticks out at right angles to the spacers and does not interfere with edge mounting the cells, but it has the disadvantage of being more fragile and complicated to build.

To make either contact, start by clamping two spacer strips together so you can accurately drill a hole through them for the bolt. Select the strips based on the contact location on the completed cell. The contact's position doesn't affect performance, so you can put it anywhere you like.

Drill a 5/32 hole (the shank size of a 6/32 bolt that you will use for the diaphragm contact bolt) through both strips. Be careful—conventionally shaped drill bits tend to shatter the plastic just as they exit the hole!

One way to prevent this is to stop drilling just as the point starts to exit the hole. Turn the stack over and finish the hole by drilling from the other side. The other solution is to

regrind the tip of the drill bit so it has a much steeper angle—60° works well.

If you use the bolt-type contact, separate the two pieces and increase one hole to ½". This allows the head of the screw and the washer that make up the contact point to pass through one strip and make contact with the diaphragm. From now on, keep the strips together in matched pairs. The holes must match closely when you glue the stators to each other.

If you use narrow-perimeter spacers, you may need to cut a notch in the perforated metal so it clears the diaphragm contact (*Fig. 10-8*). Do this after you have drilled the strips so you can easily see the notch's correct location.

GLUING SPACERS. You will need a piece of glass slightly larger than your completed cell

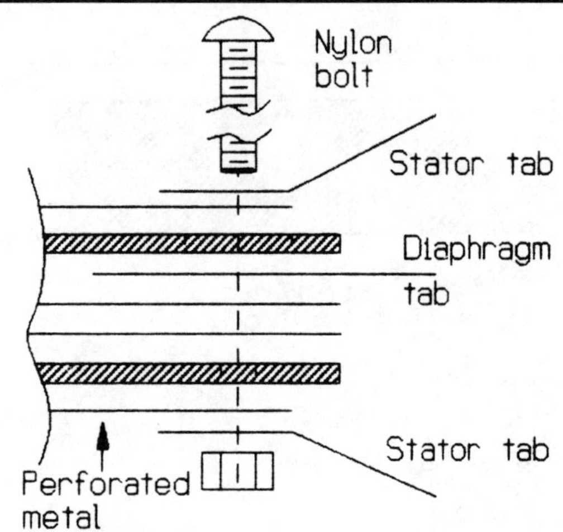

FIGURE 10-7: Detail of diaphragm tab contact.

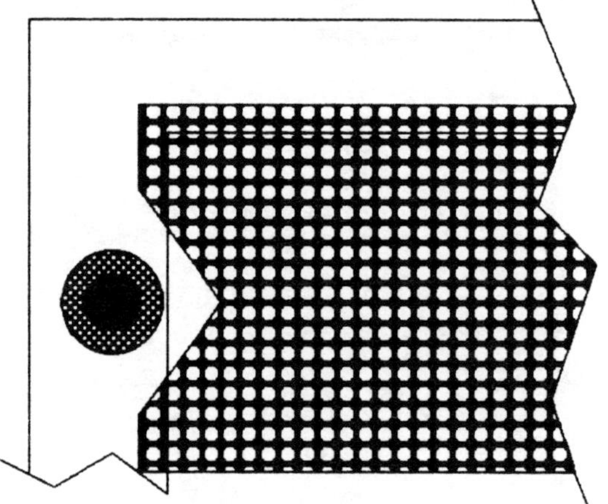

FIGURE 10-8: Notch perforated metal to clear diaphragm contact.

for building ESLs. Draw your spacer pattern on the underside of the glass with a felt marking pen. Alternatively, you can use masking tape to lay out the pattern.

All the stators must be identical, so that when you glue them together on opposite sides of the diaphragm, the spacers match. This is particularly important for the center strip spacers since these will be very narrow.

ADHESIVES. You will need two types of glue for this project: epoxy and cyanoacrylate. Get these from a hobby shop or hardware store.

You will need a small amount (¼ oz. or 15 ml.) of cyanoacrylate for gluing the plastic strips into a spacer frame. You may recognize this as "instant" glue.

Brand names include Krazy Glue and Hot

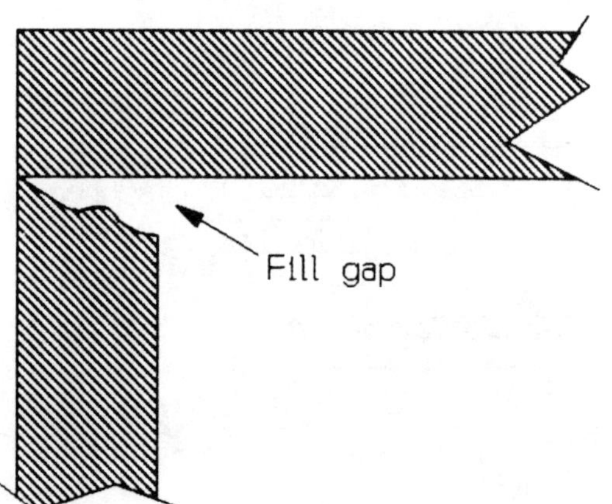

FIGURE 10-9: Gap in spacer frame.

Stuff. The gap-filling type is better than the water-thin type because the watery type requires very close fitting joints. If your joints have very large gaps, as shown in *Fig. 10-9*, you can fill them with baking soda. Cyanoacrylates work much better with something in the gap.

Cyanoacrylates are not "instant" when used on plastic. They may take five minutes or more to cure (catalyze). You may want to get some "kicker" to make them work instantly if you are the impatient type.

To use the "kicker," you first put glue on the joint, then spray a little "kicker" on it. It speeds the cure rate so much that the glue actually smokes from the heat of the reaction.

You *must* use epoxy for gluing the diaphragm to the spacers. Other types of adhesives require contact with air to cure. Air cannot enter a joint made with air-tight materials like plastic and Mylar, so air-cured adhesives won't work.

Incidentally, some builders use double-sided tape instead of epoxy. I discourage this for two reasons: first, the tape tends to "creep" over time and the diaphragm loses tension; second, trying to remove gummy old tape years later to change a diaphragm is a nightmare. Epoxy comes off easily.

To cure epoxy, you mix catalyst with it—it doesn't need air. Epoxy also works well to join the perforated metal to the spacers, but you may wish to use silicone rubber here since air is available.

You will need several ounces of epoxy. To save money, get it in 9 oz. squeeze bottles from hobby shops rather than little tubes from hardware stores. I like Devcon brand epoxy best because it has relatively low viscosity and is very high quality. But other brands also work.

Epoxy is manufactured with different cure rates. Some frequent ESL builders use 5-minute epoxy exclusively, but if you choose this method you must apply it with a syringe and work quickly. However, 30-minute epoxy is a more sensible choice. It gives you more time to work and you need not apply it with a syringe—although a syringe is the ideal applicator.

Epoxy cure time is strongly affected by temperature and film thickness: low temperatures and thin films greatly slow curing. For example, if the temperature is below 50°, a thin film of 5-minute epoxy may take hours or even days to cure.

When epoxy cures, it gives off heat. This

heat increases the temperature of the epoxy and speeds the reaction. A couple of ounces of 5-minute epoxy in a cup on a hot day will cure so rapidly that it will actually boil and be too hot to touch—in about two minutes. Avoid this exothermic reaction by mixing small amounts and avoiding excessive heat.

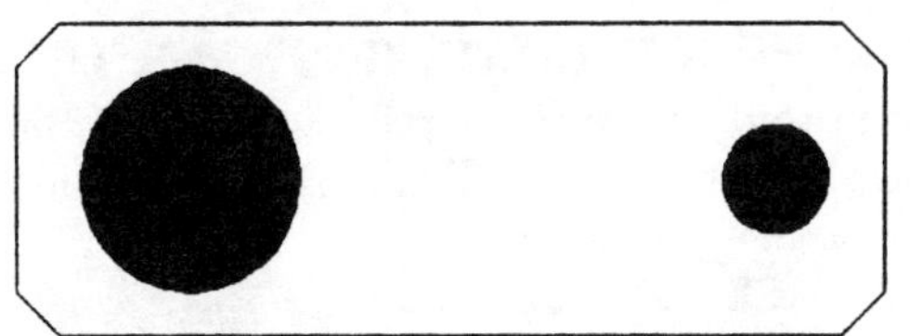

FIGURE 10-10: Typical brass tab contact.

Clean your glass, aluminum, and spacers with a vacuum cleaner, or wipe them with a damp rag. Protect the glass by coating it with two layers of car wax, or by putting a layer of plastic film such as Saran Wrap® over it. Wax paper is porous to epoxy—don't use it.

Lay out the spacers and tack them together with cyanoacrylate. Lightly sand the glued areas to remove any high spots of glue. If you have used baking soda to fill large gaps, sand extra well to make sure there are no lumps of soda that will prevent the cell from lying flat.

Wrap the sandpaper around a small sanding block. A large (1″ × 1″ × 2″) art gum eraser works well as a sanding block.

GLUING PERFORATED METAL TO THE SPACER FRAME. Place a spacer frame on the glass and spread a thick film of epoxy or silicone rubber where you will place the perforated metal. Avoid putting epoxy in the area of the contact tab so you can bend it up later.

Lay the metal on the spacer frame, remembering to keep the rounded holes toward the spacers. Be certain the stator contacts are in the desired position. Cover the stator with Saran Wrap.

Place flat weights on this assembly until the epoxy cures. You must have a flat stator. If the weights are not flat, neither will be the resulting assembly. Books are the usual weights, but you must select them for flatness.

A second sheet of glass to place on top of the assembly is a little more expensive, but well worth the cost. You may put books on that. Even better is a sheet of ¼-inch-thick steel, which won't require extra weight. Putting a flat sheet on top will assure a truly flat stator.

When you lift the assembly from the glass after the epoxy has cured, it will still be warped. Don't worry: it will be flat after you sandwich it to the other stator at final assembly.

If you are using a tab contact, now is the time to install it. Make a tab from thin brass shim stock about 0.01-inch thick. You can get this from hobby shops and auto parts stores.

The size and shape of the tab isn't critical. Something as in *Fig. 10-10* works fine.

Attach a tab to one stator from each matched pair with some epoxy. Be sure to align the hole in the spacer with the hole in the tab. A good way to be sure it's aligned and firmly held in place while the epoxy cures is to put a bolt through it.

CONDUCTIVE PAINT. You may use conductive paint to help get a secure diaphragm contact, but it isn't essential. A piece of aluminum foil clamped between the stators usually works adequately. Still, the paint is good insurance. I use LocTite Quick Grid (for repairing rear window defrosters in cars) because it's common—but any type will work.

You should arrange your stators in pairs, remembering that the diaphragm-contact holes are matched pairs. You will paint around the diaphragm-contact hole on one stator from each pair. If you're using a diaphragm-contact bolt, take the stator from the pair with the large hole. If you're using a tab, pick the stator with the tab.

Paint from the area of the hole to the inside edge of the spacer: a line about ¼-inch wide for a distance of about 6″. Go in whatever direction is necessary to *avoid* the joint in the corner spacers (*Fig. 10-11*).

I discovered the hard way that the joint usu-

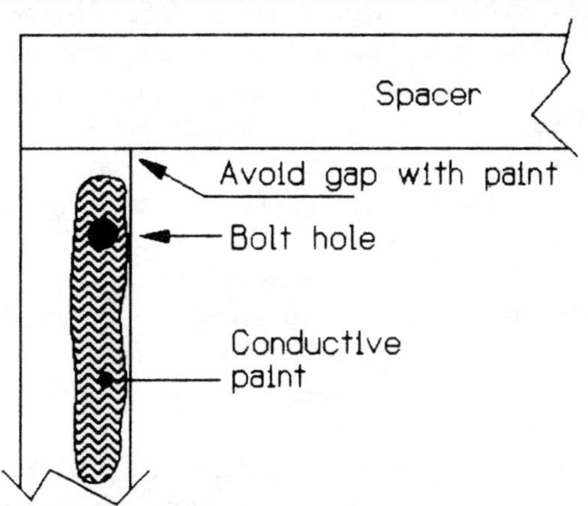

FIGURE 10-11: Typical paint pattern.

ally has a void in it which allows the conductive paint to run down and touch the stator, causing an electrical short in the cell. Stay close to the inside edge of the spacer, and don't allow the paint to run over the spacer edge where it can touch the metal stator.

This reminds me of an old saying: "Experience is the toughest teacher. She gives the test first—then the lesson!"

Be particularly careful that the paint does not run inside the hole. Don't paint near the outside edge of the spacer because you will need that area for gluing the cell together.

If the paint migrates to where it's not supposed to be, you can clean it off with a cotton-tipped applicator soaked in acetone or paint thinner. Or, you can wait until it's dry and sand it off with sandpaper.

When you are finished, take an ohmmeter reading between the stator and the paint, using the meter's "megaohm" position. *Any movement of the meter represents a short that is easier to fix now than later.*

The paint can almost, but not quite, touch the perforated metal. This could happen if it ran part way down the side of a spacer. The speaker will arc at this point when under high voltage, but an ohmmeter test won't find this problem because an actual short isn't present.

To test for a "near short," connect one wire from your polarizing supply to the paint and the other wire to the perforated metal. Turn on the power supply. If the stator arcs, you know you have a problem, and you know exactly where it is. If it doesn't, you know your contact paint is satisfactory.

STATOR INSULATION.

Most builders believe they should insulate their ESLs, but as I've said earlier, this is unnecessary in most designs. If you wish to use insulation, I must say that I have not found a spray-on type which will not arc when you apply high voltages. When foreign objects (like insects) get into the speakers, this hazard is particularly likely.

Uninsulated cells work fine and have higher output than insulated cells if you use high impedance diaphragms. Neither will be damaged by an occasional arc or a collapsed diaphragm. Persistent arcing by a foreign object will cause a hole in both. Fortunately, a few holes do not affect the sound.

The insulation most builders use is GLPT varnish, also known as Glyptol or High Voltage Insulating Varnish, available from electronics supply firms.

Another technique is to have your perforated aluminum anodized black. Unlike gold, black anodizing insulates the cells.

For cosmetic reasons, you may wish to paint the stators flat black so they are nearly invisible behind a moderately sheer grille cloth. On the other hand, you may prefer some gold visible through a black grille cloth.

DIAPHRAGM COATING.

Use only Mylar. Imported imitations will not hold high diaphragm tensions. Experiment with other material if you wish, but for a proven product, stick with DuPont's Mylar.

Mylar comes in many formulations, but the ¼ or ½ mil are the only two choices: type S and type C.

I have always used type S with splendid results, but type C has the same characteristics. The only difference between the two is that the more expensive type C is made to tight dimensional tolerances for use in capacitors. This is unimportant for ESLs.

Mylar is available from large plastics houses. Although it is inexpensive, in recent years it has become difficult to purchase in small quantities. You will probably find it easier to get some through the ESL Clearinghouse, where someone always seems to have a large roll on hand that he or she will share for a modest price.

Mylar comes in a variety of widths, but 48″ is usually best for our purposes. You can place a 3′ cell across the Mylar and have 6″ on either side with which to work.

Mylar film is an insulator which must be made slightly conductive so it will accept an electrostatic charge. Coating it with graphite, as described below, produces diaphragms with impedances between 10–100kΩ/inch. These work beautifully despite opinions of some that the resistance must be higher. Recall that aluminum-coated diaphragms with only a few ohms/inch work fine.

You may have read that you must achieve a particular resistance. This serves no useful purpose and may be ignored. This also means that even a haphazard coating job won't degrade the ESL's performance. In short, this job is easy and noncritical.

ESL manufacturers usually use relatively hazardous solvents and uncommon materials to coat their diaphragms. Fortunately, this is unnecessary.

You can use many substances to coat diaphragms. Almost anything works—for a limited time. For example, spray-on antistatic coatings work very well initially, but as the adhesives evaporate, vibration loosens the coating and the cell dies. One reader says he mixed a cloth fabric softener (the type used in clothes dryers) with water. By stirring it, he got a slightly soapy solution which he simply painted on the diaphragm. He claims that in three years it is still working fine. I haven't tried it. Experiment if you wish, but if you prefer a cheap, safe and reliable coating, use graphite. Fine-powdered graphite is available from any hardware store. Manufacturers sell it as a lock lubricant. Pencil lead rubbed against fine sandpaper is also an excellent source of fine-powdered graphite.

Rather than relying on some type of adhesive, few of which will adequately bond to Mylar, you're going to grind graphite into the Mylar's surface. Then you know it won't come off.

Many builders are intimidated by this simple job of rubbing graphite into Mylar. They constantly look for another method, like wiping on a liquid or using a spray-on coating.

I'm puzzled by this, but have had so much feedback on the subject that I have made the following directions very detailed to cover all possible questions. Let me assure you that the process is simple, quick and *easy*, despite the length of the directions.

Rubbing graphite is not the problem. The problem is tearing the film on grit caught between the glass and the Mylar when you rub hard. The key to success is cleanliness. If you wet-clean the glass with acetone on a paper towel, you should have no problems.

Acetone is a superb cleaning agent. It cuts grease, will not dissolve Mylar, evaporates without a trace, is an effective epoxy solvent, and is safe to breathe in small quantities.

Acetone is a normal body waste product so, unlike most other solvents, it's not toxic in small quantities. This is particularly important, since you will be working indoors.

Acetone is available at minimal cost from any hardware or paint store. I cannot stress too strongly that acetone is EXTREMELY FLAMMABLE. Treat it accordingly.

The only other solvent I can recommend is distilled water. Although not as good as acetone, because it evaporates slowly and will not cut grease or epoxy, it is nontoxic, odorless, and will not burn.

Keep windows closed and fans off during cleaning to avoid stirring up dust. Damp mop-

ping the floor and work table to remove dust and grit is a good idea, but not essential. Avoid vacuum cleaners at this stage of construction: they blow grit everywhere.

Wipe your glass with an acetone-wetted paper towel. If you feel lumps of epoxy or anything else on the glass, shave them off with a single-edge razor blade, and wipe again with acetone.

Cut a piece of Mylar a little larger than your glass. Fingerprints do not accept graphite very well, so try to keep your hands off the surface.

Some builders use cotton or thin latex gloves to prevent fingerprints. This isn't necessary, since fingerprinted cells work fine, but I thoroughly wash my hands before handling Mylar and try to minimize contact with it. You can always remove fingerprints with acetone.

Unless your glass is much larger than your cell, allow a couple of inches of border to overhang the edge.

With a dry paper towel, smooth the Mylar lightly (about 1 lb. of pressure) from the center outward while gently pulling on the film edge with your other hand. The object is to rid the Mylar of major wrinkles, adhere it to the glass, and find the grit.

CAUTION

ACETONE IS
EXTREMELY FLAMMABLE

When the wrinkles are out, wipe harder (about 5 lbs. of pressure). A strong light will reveal any little "tents" in the Mylar caused by grit. Lift the edge and wipe away any grit which forms a tent larger than about 1/8" across. Smaller ones are OK, unless they are very sharp or pointed.

Incidentally, you will probably notice creases which appear as fine lines. Ignore these, as heat shrinking will remove them (mechanical stretching will not). Anyhow, they don't cause problems.

The process of finding and removing grit, although simple, is the hardest part of coating diaphragms. If the glass and Mylar are clean initially, there is nothing to it.

If you somehow have so many grit tents that you are overwhelmed, you can get a head start by wiping the *top* of the Mylar completely clean with acetone. Fold the Mylar in half on itself so the surface you just cleaned faces another clean surface. Carefully remove the Mylar from the glass and set it aside.

Clean the glass with acetone, then put the *clean* side of the Mylar against it. Now you should have almost no tents.

Tape down the Mylar every 4–6″ around its perimeter. Tape down the corners first, then tape halfway between the corners, then tape halfway between that, then halfway between that, and so on until you have a piece about every 6″. Pull it moderately tight. Use 3-inch long pieces of masking tape. There should be no wrinkles in the Mylar.

Sprinkle a small amount of graphite onto the Mylar. The exact amount is difficult to describe, but for a 2′ × 3′ cell the needed amount would fill only about half a pea. More graphite reduces diaphragm resistance. Therefore, if you want to make very high resistance diaphragms, use less.

Gently spread it evenly around the Mylar with a clean, dry paper towel. A light-colored tablecloth under the glass enables you to see where the graphite is on the diaphragm—it is a light gray color. If you are using very small amounts of graphite to achieve a very high resistance coating, make sure you disperse it uniformly.

Now rub hard to grind the graphite into the Mylar. I place the heels of both hands—one on top of the other—on a paper towel. Make sure you rub everywhere, particularly around the area of the diaphragm contact.

Rubbing hard is important. I don't hold back, but lean on my hands with most of my weight. I don't care what resistance the diaphragm has, and don't even bother to measure it anymore. If the diaphragm looks gray and I've rubbed hard, I know it will work perfectly and last indefinitely. It takes less than a minute to coat a diaphragm.

Next, rub off all the excess graphite with a towel. Again, rub hard. If you can rub off any of the coating, you didn't rub it in hard enough. Repeat a couple of times with clean paper towels until the last one remains clean.

If you don't clean off the excess, the cell will hiss for several hours until it burns it off. Eventually, it will be silent.

You can measure the resistance of the diaphragm coating with an ohmmeter. Just place the probes about 1-inch apart on several areas of the diaphragm.

Some builders believe they must coat and make contact on *both* sides of the diaphragm, but this is not so. You need a conductive coating on only one side.

The Mylar will now be lightly stuck to the glass and may have a few wrinkles. Peel the tape from one side, and lift one edge enough to get some air under it and release it from the glass.

The next step is to get the diaphragm uniformly tight before gluing it to the stator. Lay the Mylar on the glass, trapping a little air under it. Don't rub or wipe it anymore, but start taping it to the table again. Work your way around the table while lifting, pulling, and sticking the tape onto it.

Add more pieces of tape and keep pulling until you have the Mylar tight and wrinkle free. You will need tape about every 3″ around the perimeter. Getting the diaphragm tight, uniform, and free of wrinkles is surprisingly easy. Put the diaphragm under fairly high tension, but don't get carried away—heat shrinking will finish the job.

Cut a piece of aluminum foil about ¾″ × 2″. You will place this under the diaphragm-contact-bolt hole.

GLUING DIAPHRAGM TO STATOR. With a vacuum, clean the inner surface of the stator (the side with the spacers), which you will glue to the diaphragm. If you are using the bolt contact, this stator will be the one with the *large* diaphragm-contact hole in it. If you are using a tab contact, pick the stator with the tab.

Next, put epoxy on the spacers and position the stator on the Mylar. Epoxy is easiest to mix in 1 oz. plastic cups from doctors' offices, hospitals, or hobby shops. Small plastic cups or the tops from spray-paint cans also work well. Do not mix epoxy in waxed paper cups because it penetrates the wax and becomes contaminated.

You will need about 15 ml. of epoxy (½ oz. for a 2″ × 3′ cell). Mix equal amounts of resin and catalyst (7 ml. of each) for at least 30 seconds. Popsicle sticks are perfect for this task. You can use a heat gun to heat the epoxy and reduce the viscosity, making it easier to handle, but this increases the reaction rate and reduces working time.

The epoxy is best applied with a 10 or 12 ml. disposable syringe without a needle. Suck the epoxy into the syringe. You can't get it all, but you can get about 11 ml. Wipe any epoxy from the outside of the syringe with a paper towel.

Lay a *small* bead (about 1/8-inch wide) down the center of each spacer. When you reach the

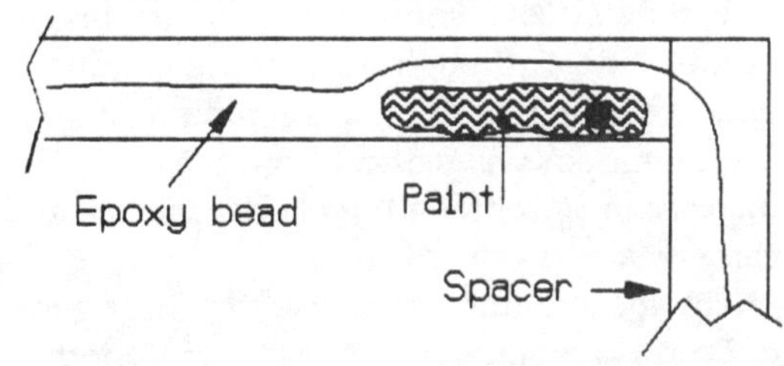

FIGURE 10-12: Avoid getting epoxy on diaphragm contact.

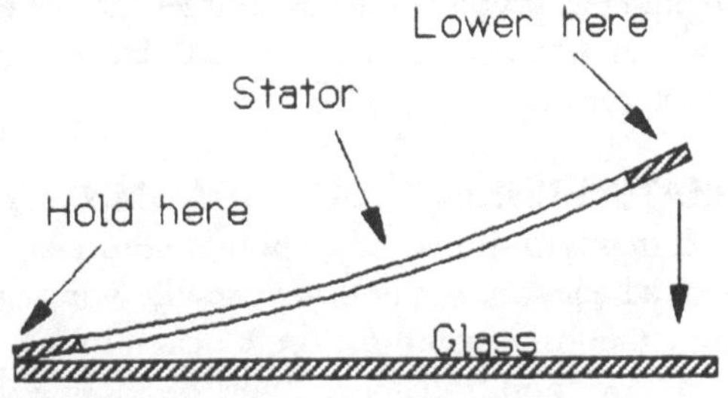

FIGURE 10-13: How to place stator with wet epoxy on diaphragm.

diaphragm contact area, move the bead to the outside edge so the paint can contact the diaphragm, rather than being coated with an insulating film of epoxy (*Fig. 10-12*).

Avoid a thick bead of epoxy, or it will form an uneven film which causes nonuniform D/S spacing. It will also tend to run down the edges of the spacers. Since epoxy is an insulator, this won't damage the speaker, but it is tough to remove if you must change a diaphragm. You need a thin film only on the spacer surface facing the diaphragm. Run your finger along the spacer edge as a guide for the syringe.

If you are using a bolt contact, put the aluminum foil over the diaphragm-contact hole. It should cover the width of the spacer and will lie on the epoxy bead. Do not let the foil extend beyond the inside edge of the spacer, where it might touch the stator and short the cell.

Turn over the stator and carefully place it on the diaphragm. The idea is to position the stator on the Mylar without smearing epoxy everywhere. Put one end of the stator in place first, and while using your fingers to hold it in place, lower the other end (*Fig. 10-13*).

Press hard over all the spacers to squeeze the epoxy into a thin film. Avoid sliding the stator and smearing the epoxy.

I place my thumbs over the spacers and press as hard as I can every 2″. At first, I merely weighted the stator, but this didn't squeeze the epoxy into a thin film. Even several hundred pounds of weight won't do the job if it's spread over the entire cell.

You can easily put 50–100 lbs. of force onto your thumb. That much pressure concentrated over a square inch can really squeeze the glue. Now put your sheet of glass or steel on top to keep the stator flat and let the epoxy catalyze.

Epoxy cures much faster in a cup or syringe than in a thin film, so don't remove the cell from the glass just because the epoxy in the cup has hardened—wait about twice that

long. To free the Mylar from the glass, cut it with a razor blade rather than trying to remove the tape. After cutting it, carefully lift one end of the assembly to get air under it, as it will tend to stick to the glass.

The stator/diaphragm assembly will still be floppy and warped. Although your Mylar was free of wrinkles when you glued the stator to it, you will probably find that it now has a few as the stator bends. Don't sweat it—you're in good shape.

With your thumb, gently press the Mylar at the diaphragm-contact hole. Clearly outline the hole's shape and location.

This next step is critical: take a pencil, awl, or other sharp object, and gently puncture the aluminum foil at the diaphragm contact. Puncture from the diaphragm toward the spacer, so you don't tear them apart (*Fig. 10-14*). Recall that you did not put epoxy between the Mylar and the aluminum foil, so this area is vulnerable to disruption if you are not careful.

The hole should be only as large as the shank of your diaphragm-contact screw. If you are using the bolt contact, locate the puncture in the exact center of the stator hole.

Keep it small. If you make it the full size of the stator hole, you will have no place to contact the diaphragm after you assemble the cell. If this happens, you must install a new

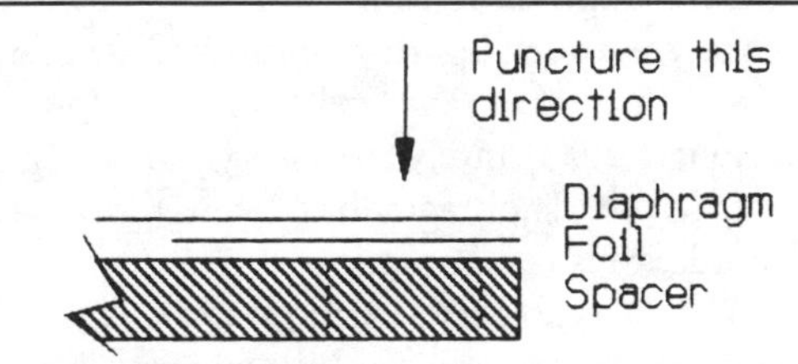

FIGURE 10-14: Puncture diaphragm toward spacer.

diaphragm. If you use a tab contact, however, the puncture should be the full size of the tab/stator hole.

HEATSHRINKING DIAPHRAGMS.

The next step is to shrink the diaphragm drum-tight with a heat gun. A hair dryer usually won't get hot enough, so you must use a heat gun.

Hobby shops sell cheap ones for about $20. If you ask around, you may find a modeler who will let you borrow one. Hardware stores sell them for stripping paint.

Heat guns vary in power and heat. The common hobby type is rated at 900W. Many have vents on the back that you can adjust to control the heat.

Test your gun by having someone lightly stretch a scrap of Mylar between their hands while you heat it. The object is to determine how much heat you can apply without melting the Mylar.

Effective heat shrinking requires that you almost, but not quite, melt the Mylar. I generally find that leaving the vents wide open while holding the gun about 1–2″ from the film is about right. When you hold the gun stationary, the Mylar should not melt. By adjusting the distance from the film and the vent openings, you can readily determine the "safe zone," which is not as difficult as it sounds.

Hold your stator/diaphragm assembly vertical while passing air from the heat gun over the diaphragm in a slow, uniform motion. To help give you the "big picture," I move the gun about 4″/second and cover a 2–3″ path.

This takes less than a minute—one pass will do the job unless you have huge wrinkles. Feel free to heat and shrink repeatedly, or in troublesome areas until you get the diaphragm smooth and tight.

Mylar will shrink a maximum of about 10%, which is plenty for all but the sloppiest of construction. The diaphragm may relax slightly after the first heat application, and it is a good idea to reheat it about a day later to obtain maximum tension.

You can touch up the diaphragm by heating it anytime after the cells are completed, although this is rarely necessary. Once a genuine Mylar diaphragm has been thoroughly shrunk, it remains tight.

Problems can occur during shipment or accident where the cell is stressed, causing the diaphragm to slacken. Being able to shrink the diaphragm again is wonderful.

To reshrink an assembled cell if you use steel-wire stators, simply heat the diaphragm on one side as though no stator is present. If you are using perforated-aluminum stators, the aluminum will expand from the heat. The increase in stator length will bow the cell and can break glue bonds.

The ideal method is to use two heat guns: one on each side and opposite one another. But you can do it with one gun by briefly heating one side, then quickly switching to the other side, and then back again. Keep shifting sides about every five seconds to keep the expansion even. Both techniques will expand the aluminum evenly enough to prevent broken glue bonds.

When you're satisfied with the diaphragm, lay the assembly on the glass with the diaphragm up. Put epoxy on the spacers of the other stator in the matched pair. Set it very carefully on top of the just-finished assembly. Again, be careful not to smear the epoxy.

Examine the diaphragm-contact hole. You must center the hole you made in the Mylar directly under the one in the second stator. If they aren't concentric, you won't be able to insert the diaphragm-contact screw later.

As before, squeeze the epoxy into a thin layer over every inch of spacer. Double check the alignment of the diaphragm contact to be *absolutely sure* you have centered the two holes, and that everything is flat and aligned.

This lamination locks the cell into its final, rigid shape. If you mess up now, you must tear down the cell and put in a new diaphragm to correct any problems. Few things are as frustrating as having a seemingly perfect panel—then finding that the diaphragm-contact holes are misaligned, or you left out the aluminum foil, or overlooked some other detail which ruins the cell. Pay attention to the details. When you're sure everything is perfect, put the glass or steel sheet over the assembly and weight it.

When the epoxy has cured, gently lift the assembly. Nothing sticks to graphite-coated Mylar very well, not even epoxy. The glue joints are adequate, but they may break if abused. The assembly will be amazingly rigid and flat, with no give. The only way to bend the cell now is by breaking a glue bond.

Take a sharp, single-edge razor blade and trim the Mylar. Mylar is tough and will quickly dull a razor, which can tear your diaphragm. Razor blades are cheap—replace them often.

You may remove the last fragments of diaphragm by sanding the edges with medium-grit sandpaper, but this is not necessary. Bend up the stator-contact tabs and insert the brass or steel 6-32 bolts for your electrical contact.

If you're using a bolt contact, place a small washer under the head of the diaphragm-contact bolt, and insert the bolt and washer through the hole. Don't omit the washer: it protects the delicate aluminum foil from tearing if you twist the bolt. This goes through the large hole first, so the head and washer lie against the aluminum foil. Add the nut and gently tighten.

Be gentle when connecting wires to these contact bolts. The contacts can be damaged by excessive force.

If you're using a tab contact, place one stator tab on a nylon bolt and push it through the hole. Position the other stator tab and the nut, and gently tighten.

You can now test them with music. When first energized with the polarizing voltage, they may make a faint frying or hissing sound which will eventually stop.

If you are making a hybrid system, and you test a cell with music—alone and without equalization or woofers—you will be disappointed in the SPLs and frequency response. They will sound like a cheap transistor radio (albeit a very clear one).

Hang in there; nothing is wrong. Yours is a *system*, and when all the parts are working together, the sound will be splendid.

TROUBLESHOOTING. The above instructions have 17 years of mistakes behind them. I've tried to cover all the pitfalls involved in constructing your cells, and they will work fine if you followed the directions.

Writing a troubleshooting section suggests that the speakers are going to be difficult to build, unreliable to operate, and that you are going to have problems. Actually, the opposite is true, but, realistically, things sometimes go awry. I know how frustrating it is to have a problem and have no idea of how to fix it, so I'm including this section.

If the cell doesn't work, only three things can be wrong. Like checking for gas in your lawn mower, do the obvious first—make sure you have polarizing and drive voltages at the connections.

If so, then two possible problems exist with the cell itself. The most common is a short between the diaphragm and a stator.

SHORTED DIAPHRAGM. Usually, this is caused by a foreign object within the ESL. You can remove it by vacuuming the cell with a soft brush or blowing compressed air into it.

You may have let conductive paint run down the spacer edge or into the diaphragm-contact hole. If so, the only cure is to tear down the cell and start over. Hence, the importance of using care when you applied the conductive paint, and why I advised you to test it before assembly.

If you painted the stators with insulation, you may have some conductive object embedded in the paint. You may also have something conductive on the outside. Keep in mind that although you trimmed the diaphragm, it's possible to make electrical contact somewhere along its edge. Although this is usually not a problem, because it's about 1″ from the edge of the cell to the conductive stator, don't overlook the possibility.

Look closely for something touching the edge of the cell which is also touching a stator. It can be quite insignificant and still cause problems at high voltage.

Figuring out which stator has the short is easy. Connect the stators to your step-up transformer. Do *not* connect the polarizing supply to the diaphragm. You do not need to play music for this test, but it doesn't hurt if you do. You don't even need to connect the amplifier, although you may.

With the polarizing voltage on, bring the diaphragm-contact wire close to the contact bolt or tab. Try to almost, but not quite, touch the diaphragm contact with the wire.

In a properly operating cell, a small and momentary arc will occur as the current charges the diaphragm. In a shorted cell, the arc will be much larger and will persist. Also, you will usually hear a pronounced "pop" when you connect the polarizing supply.

You can tell which stator has the short by removing one connection and again bringing the polarizing-voltage wire near the diaphragm contact. Alternately connect and disconnect each stator. The shorted stator will cause persistent arcing between the polarizing-supply wire and the contact.

You can usually find the short by connecting an ohmmeter between the diaphragm contact and a stator contact. Keep in mind that the short may be several megohms and still disable the cell.

Ohmmeters don't always find the problem, as the foreign material may not actually touch

TYPES OF CELL PROBLEMS

1. Shorted diaphragm
2. Nonconductive diaphragm

the diaphragm or stator. The same potential problem exists as when you were applying conductive paint. It need only be very close to arc to disable the cell. The meter measures with low voltage and will see an open circuit. When you apply high voltage, the speaker will arc at that point, preventing high-voltage operation.

You may have a badly warped stator where the aluminum is not flat, or a wire stator where one or more wires have bent or broken away from their support structure. The problem of warped metal is why a glass or steel plate is placed over the top of the cell.

These problems are evident in the form of persistent arcing at that location whenever you apply the polarizing voltage. Turning down the room lights will help you see the arc more easily.

You can solve the problem by bending the perforated aluminum away from the diaphragm. Insert a small punch or nail in the 1/8″ holes along one edge and pry gently. If you used wire, remove the offending wire or fix its attachment.

Usually, the short is fairly easy to find and correct. If not, the only cure is to tear down the cell and put in a new diaphragm.

OPEN DIAPHRAGM. The second problem is a failed diaphragm contact or lack of diaphragm conductivity. It can be easily identified using the test above, where you bring the polarizing-voltage wire near the diaphragm contact. This time, there will be no arc, pop or noise, because no current is flowing to the diaphragm.

A variation is a cell which works but has surface dead spots, caused by areas where you

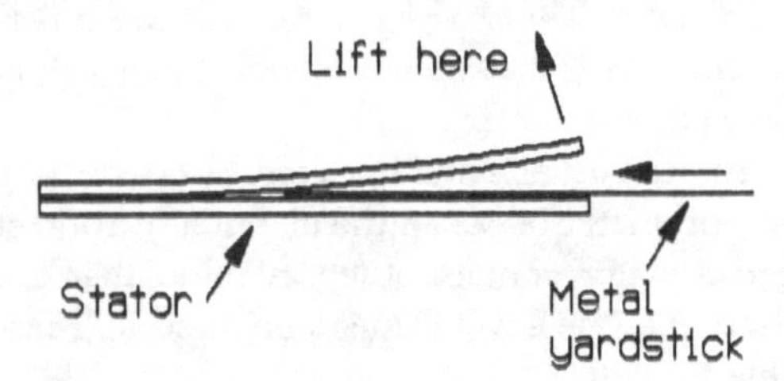

FIGURE 10-15: Disassemble cells.

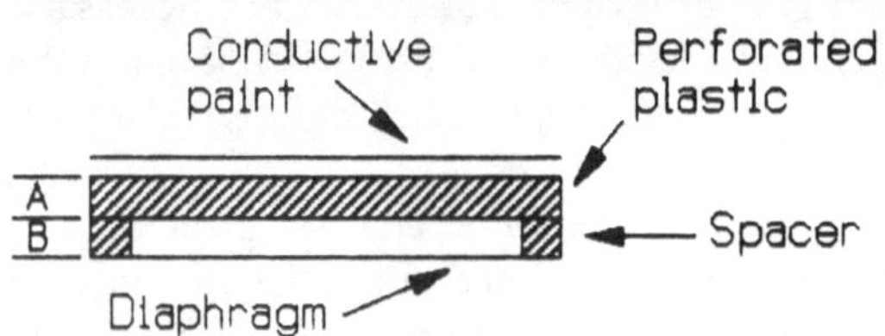

FIGURE 10-16: Perforated plastic stator design.

either missed applying graphite or where a coating has started to come off. This won't happen with properly applied graphite if you rubbed it hard, but is common with many other coatings.

DIAPHRAGM REPLACEMENT. Replacing a diaphragm is not difficult. You need two tools:

- A sharp putty knife
- A metal yardstick (or other long, thin, metal object) with one end sharpened

After removing any bolts, run the putty knife around the cell's perimeter to separate the spacers. Slide the yardstick down any center spacers, while lifting one end of the top stator (*Fig. 10-15*).

With the stators separated, run the putty knife under the epoxy on the spacers. It generally separates easily.

You will have difficulty if you applied too much epoxy, which ran over the edges of the spacers. If so, just peel and chip away as best you can.

Correct the problem which made disassembly necessary. Vacuum the cell to remove all debris. Clean everything, and you are ready to install a new diaphragm.

PERFORATED-PLASTIC STATORS. Plastic is not conductive: it must be coated with conductive paint. This offers two advantages.

You can apply the conductive coating to the *outside* of the stator (the side away from the diaphragm). This technique uses the stator itself to form a very high quality, effective, and uniform insulation.

You can make the stator conductive only in certain areas, depending on where you apply the conductive paint. This makes it easy to eliminate stray capacitance.

Against these advantages, you must weigh

the serious disadvantages compared with every other stator design. Plastic stators are relatively thick, which seriously degrades output for two reasons.

- The holes/slots must be larger than with a thin stator. This reduces field density.
- Applying the conductive coating to the outside increases the D/S spacing by the thickness of the plastic. Commonly, it doubles it, and this is a tremendous penalty (*Fig. 10-16*).

Applying the coating to the inside has no deleterious effect on output, but why would you do so? Not only is it a major chore, it offers no advantages over thin-metal stators, while the disadvantages of plastic stators remain.

Plastic has other problems as well: it is weak compared to metal; it must be thicker than metal to have adequate strength; since it is weak, it is difficult to obtain high open-area ratios.

Perforated plastic does not come ready-made to specifications which are suitable for ESL construction. It must be custom perforated, which is expensive, time-consuming, and presents shipping problems.

Because of these problems, I think plastic stators are best reserved for ESL designs with large D/S spacing. Such designs may require voltages high enough to require stator insulation. Also, wide spacing means the relatively large perforations are not such a liability.

Suitable plastic must be at least 80-mil thick, and 120 mil (1/8") may be necessary unless you use elaborate external support structures. Refer to *Chapter 4* on output for perforation patterns.

Construction techniques for perforated-plastic stators are similar to perforated metal, but with a few modifications and additional tasks. You can build the spacer frames as usual, but you need not glue the frames together before gluing on the perforated sheet. If you are accurate, you can assemble the frames by gluing the strips directly onto the perforated plastic.

Unlike a perforated-metal stator, you can run the plastic sheet all the way to the spacer frame's edge. To avoid the penalty of stray capacitance, don't apply conductive paint all the way to the edge. Remember that stray capacitance is caused by a *conductive* part overlying an immovable section of diaphragm.

When gluing the spacer frame or strips to the perforated plastic, don't use epoxy. Use water-thin cyanoacrylate instead.

Apply the cyanoacrylate in a bead beside the joint. As it wicks its way inside the joint by capillary action, it welds the two plastics together to form an extremely strong joint.

You can do clever things with conductive paint. For example, I've already mentioned that you can reduce stray capacitance by omitting paint where a spacer is present. Another idea is to use the paint to turn a single cell into a segmented speaker (*Fig. 10-17*).

To make contact with the stator's conductive paint, apply it on the desired area near the cell's perimeter. Drill a hole through both the perforated plastic and its spacer. Cut a brass shim stock tab and secure it with a nylon bolt. The tab contact used in perforated-metal cells is similar to this.

If you use flat-head nylon bolts, you can recess them into a countersunk hole anywhere inside the stator to make a contact point. This could come in handy if you have conductive strips you wish to contact (*Fig. 10-18*).

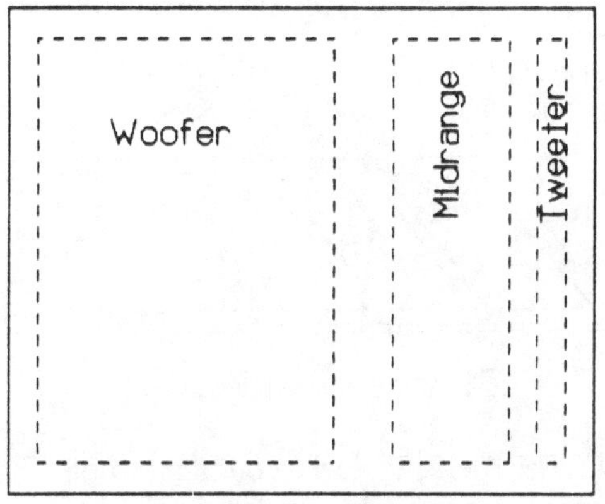

FIGURE 10-17: Segmenting a single cell with conductive paint.

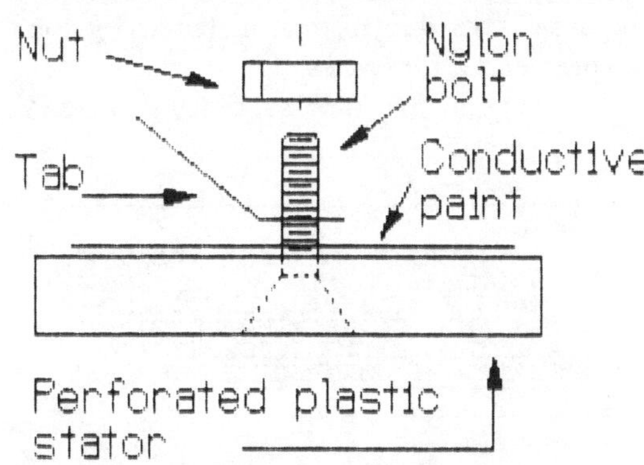

FIGURE 10-18: Detail of stator contact.

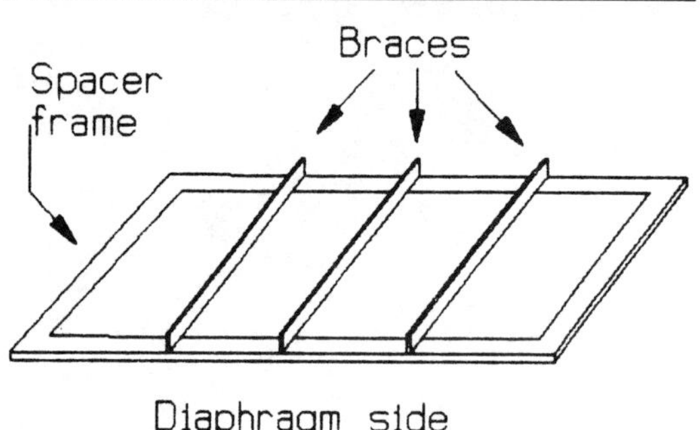

FIGURE 10-19: Typical external bracing.

Do not use a spray gun to apply conductive paint. If you do, some will get into the holes and compromise the otherwise excellent insulating quality. Using a fine-nap paint roller works much better. Rollers as narrow as 2″ are available in hardware and paint stores.

When using a roller, lightly coat it with paint. Apply only a thin coat of paint to the plastic. If you overdo the coat, it will run inside the holes and compromise your insulation. Also, press lightly on the roller. If you roll too heavily, you will squish paint into the holes.

Unless your cells are very small, you must brace them to support the flimsy plastic. You have a choice of ways to do this. Most builders use some type of plastic or wood strips, which are glued across the cell in several locations (*Fig. 10-19*).

Be sure whatever material you use is straight. Any curvature in the brace will affect the stator and cause nonuniform construction, with adverse effects on output.

Use a dimensionally-stable material. Wood is unsuitable because it absorbs moisture, and tends to warp and change length depending on humidity.

At the voltages we are using, wood is a fairly good conductor. Because of this, and because of its dimensional instability, I don't recommend it for braces.

Use a good insulator as bracing material,

since it will cross conductive areas and could short one section to another. A noninsulator would be a shock hazard, since it would carry the full charge applied to the stators.

Metal braces are a poor choice because they are conductive. Plastic best meets the requirements. The problem with plastic braces is they aren't very strong, and are hard to cut perfectly straight. Still, 1/8-inch-thick plastic, 1-inch wide and laid on edge, works well if you can cut straight pieces.

A good solution to this problem is to glue the perforated plastic to ½- or ¾-inch-grid fluorescent-light diffusers, as shown in *Fig. 10-20*. They are inexpensive, flat, and available at home improvement centers, they have a limited size selection and the resulting cell is very thick. With a little creative thinking, however, you can get them to work.

When gluing plastic braces to your stators, use epoxy, since you will be gluing them to the conductive paint and not to the plastic stator. If you use cyanoacrylate, it cannot contact the plastic in the stator in order to weld the two together.

RIGID-WIRE STATORS. An easy way to visualize the construction of a rigid-wire stator is to imagine that you removed the metal from a perforated-metal stator and laid rigid wires side-by-side in its place (*Fig. 10-21*).

While this concept appears valid, in reality problems with structural strength exist. Unlike a metal plate which is rigid in two directions, a series of wires is rigid in only one direction. Also, an array of wires is *much* heavier than a thin aluminum sheet.

A stiff-wire stator needs a stronger framework. With perforated-metal stators, the

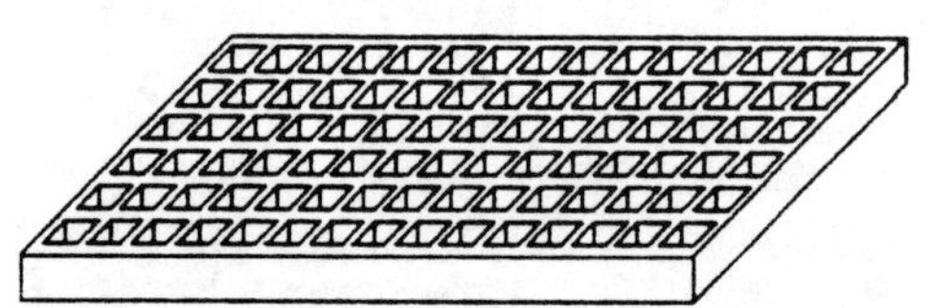

FIGURE 10-20: Typical fluorescent light grid/diffuser.

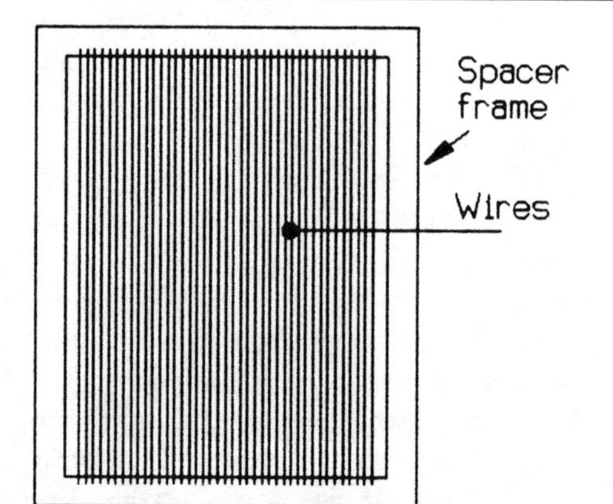

FIGURE 10-21: Rigid wire construction.

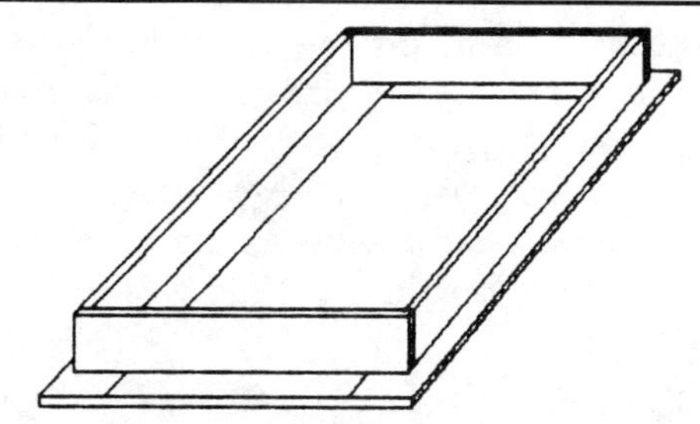

FIGURE 10-22: Bracing for rigid wire stators.

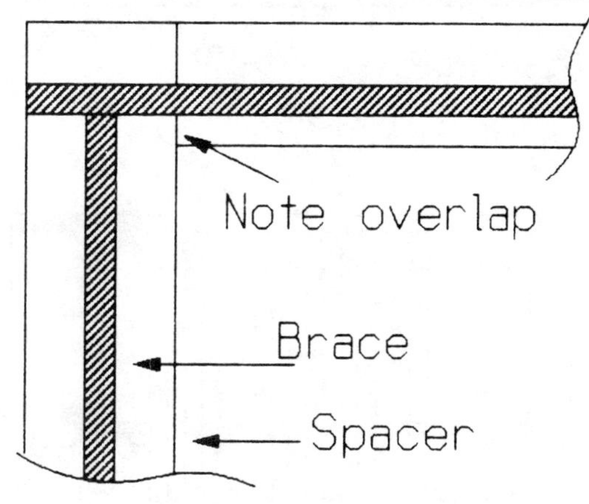

FIGURE 10-23: Overlap corner braces.

metal holds everything together and the spacers separate the diaphragm from the stators. In rigid-wire designs, the wires do *not* hold things together. The spacer framework must both support the wire load and withstand the considerable stress of the diaphragm tension.

The technique commonly used to achieve the necessary strength is to make the perimeter cell spacers stronger. Do this by laminating a brace of heavy plastic onto the spacers that form the cell perimeter. *Figure 10-22* shows one such design.

The glue bonds for this type of frame must be very strong. A butt joint between plastic strips isn't stout enough. *Figure 10-23* shows the proper way to overlap parts so they are adequately robust. Use cyanoacrylate glue. Make sure the parts are straight and in tight contact so you get solidly welded joints.

When designing wire stators, several questions commonly arise:

- What is optimum wire size?
- How far apart should I place them?
- What material should I use?
- How important is stray capacitance?
- Should I use exterior supports to reduce stray capacitance?

Conceptually, the wire size is easy to decide: it should be as small as possible. You need small slots and large open space for best performance, and this means the wires should be extremely small.

WIRE SIZE. Ideally, the wires should be so small that they are deflected by electrostatic drive under high-output conditions. In practical terms, however, the limiting factor is that small wire is never straight and stiff enough to span the distance between spacers.

The smallest practical size, therefore, is 40 mil. Since 1/16" (63 mil) wire is most readily available, it is most often used.

You will probably make your wire choice based on practicality: what size wire is commonly available? Like perforated metal, anything is available if you are willing to sacrifice time, effort, and money. But it's much easier to find and use "off-the-shelf" wire which is locally available, and which you can buy in small quantities.

Quantity is relative, however. I had quite a surprise when I bought wire for my first ESLs (the large ones shown in this book). At first glance, it doesn't appear that many wires are involved. A little thought and math proved me wrong.

I designed the ESLs with 12 wires/inch. Each cell was 18-inches wide, and I needed eight stators for four cells. I expected to reject 20% of the wires as excessively crooked. It turned out that I needed *2,000* wires!

I had to clean each one, test it for straightness, position it, space it, glue it, and solder an electrical contact wire to it. Even a little task consumes a lot of time, effort, and money when you repeat it 2,000 times. When working with quantities of this size, specialized orders from wire manufacturers are more practical.

WIRE TYPE. The wire material determines whether it is commonly available. Since it must be strong to be rigid in small diameters, steel is the preferred material.

The many different steel alloys vary greatly in strength and stiffness. The most readily available, and least expensive, is mild-steel wire.

Any welding supply shop sells gas welding rod. Several diameters are suitable for ESLs, including 1/16", 3/32", and 1/8".

These rods are usually 3-feet long, which is ideal for most ESLs. The material is electro-

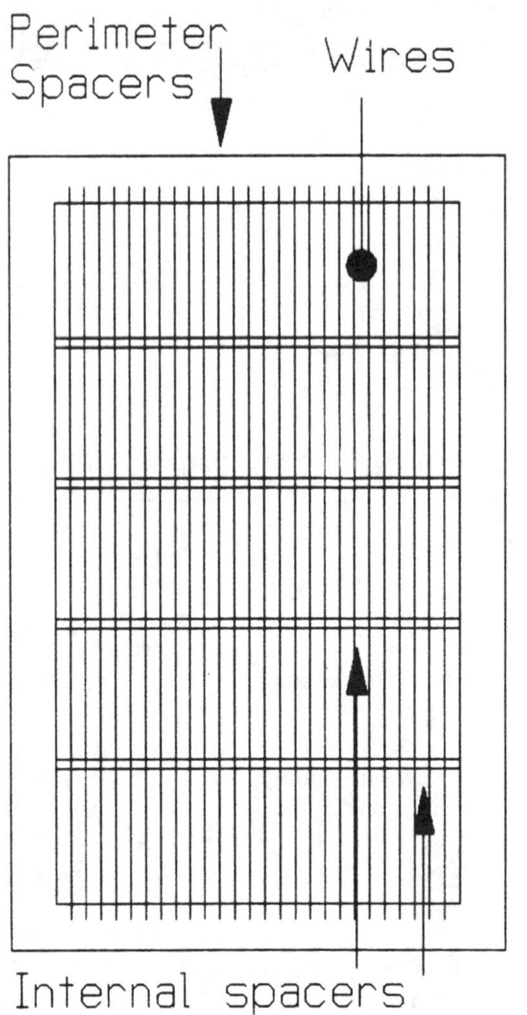

FIGURE 10-24: Simple rigid wire stator design.

plated with either nickel or copper to prevent rusting. As in anodizing aluminum, these coatings require that the wire be spotlessly clean. If it has been stored in a closed container, it will be ready to use when you buy it. Welding rod is not the cheapest material, but having ready-to-use wire is worth the extra cost.

For the quantities you will be purchasing, you should negotiate a low price. Look up a wholesaler or distributor of welding supplies who is willing to work with you directly. Also, check the "Yellow Pages" of large cities under "Wire."

Welding rod is not perfect. It doesn't come smaller than 1/16″, is soft and, for steel, not very strong.

"Music wire" is much stronger and comes in smaller sizes. One of a family of high-carbon tempered steels, it is so hard and strong it's used to make springs. You can use it in smaller

diameters than mild-steel wire because of its strength. Unfortunately, it's not electroplated.

You can purchase music wire from hobby shops in 3′ lengths, but it's expensive. You can get more reasonable prices from a wire supply house. Again, look in the "Yellow Pages."

Manufacturers supply wire to wholesalers in large rolls. They cut it to custom lengths after running it through a wire-straightening machine. Ask for "cut-and-straight" wire.

If asked, the supplier will tell you the wire is perfectly straight when it comes out of his machine. To him, it probably is. For your use, however, it isn't. Expect to discard at least 20% of the wire as excessively crooked.

Plan to paint the wires to prevent rust. They have an oily coating to assist with fabrication and resist oxidation in storage, and you must thoroughly clean them. Electroplating is an expensive alternative.

WIRE CLEANING. A slight oily film may not sound like a big deal, but how do you clean oil from a thin wire 3-feet long? More to the point: how do you remove oil from several thousand such wires? It simply isn't practical to wipe each wire with a solvent-soaked rag.

How about washing them in a pan? A piece of rain gutter makes a cleaning pan when lined with an 8-mil plastic sheet. Raise the plastic at the ends so the solvent can't escape. Or dig a trench in the ground and line it with the plastic.

Several thousand wires are very heavy. My bundle weighed over 100 lbs. You must use a stout pan. The solvent is toxic and highly flammable, whether it be gasoline, acetone, or paint thinner. Safer, less volatile solvents, such as kerosene or stove oil, leave a slightly oily residue which won't accept paint. Some solvents (such as acetone) will dissolve your plastic "pan."

SELECTING STRAIGHT WIRE. To determine whether a wire is straight, roll it on a piece of glass. You will immediately see any corkscrew shape left by the wire-straightening machine. If the wire rolls freely, keep it.

WIRE ORIENTATION AND POSITION. How far apart should the wires be? If you use 1/16″ welding rod, 10 wires/inch gives a good open area percentage. If you use 50-mil music wire, place 12 wires/inch.

Stray capacitance and supporting structure complexity for rigid-wire stators are interrelated.

Most builders accept about 10% stray capacitance as a fair trade-off for ease of construction.

By crossing the wires over the D/S support spacers, you can get the necessary support without building any structure outside the cells. The same technique is used for building perforated-metal stators. Refer to *Chapter 4* on output for further discussion of stray capacitance.

If you wish to reduce the stray capacitance, arrange the wires so they don't cross any spacers. You must then add external bracing to support the wires. You can use the same techniques as in the section on perforated-plastic stators.

Figure 10-24 shows the most common and easiest-to-build construction layout. The disadvantage is that the stray capacitance is higher than in *Fig. 10-25*.

Don't get too excited about stray capacitance. While it matters, it isn't a serious problem in most designs. The difficulties involved in building near-zero-stray-capacitance ESLs aren't worth the performance gain.

An exception is when the panels are very large. Then 10% stray capacitance becomes significant, particularly when the audio drive system is already straining to meet the high capacitance demands of a large cell.

Another extremely important consideration in the orientation of the wires to internal spacers is diaphragm tension. You will recall that the diaphragm puts a lot of tension on the perimeter spacers.

Look again at *Fig. 10-25*. Note that the perimeter spacers which are parallel to the wires have no way to resist the diaphragm's tug. They will bow inward, and the diaphragm will not develop full tension.

Now look again at *Fig. 10-24*. See how the internal spacers brace the sides of the perimeter spacers, and keep them from bowing inward from diaphragm tension.

The low-stray-capacitance design in *Fig. 10-25* can be strengthened by using external bracing perpendicular to the wires. You must do this, anyway, to support the wires. A lot of extra work will be necessary to avoid a small amount of stray capacitance.

MOUNTING WIRES. Place the wires parallel to each other, and a small but precise distance apart. A good way is to lay a fistful of wires on the frame, and roll them around until they are all lying flat beside each other.

Obtain a piece of threaded rod as wide as your cell. Hardware stores carry threaded rod in 3′ sections. It comes in different diameters

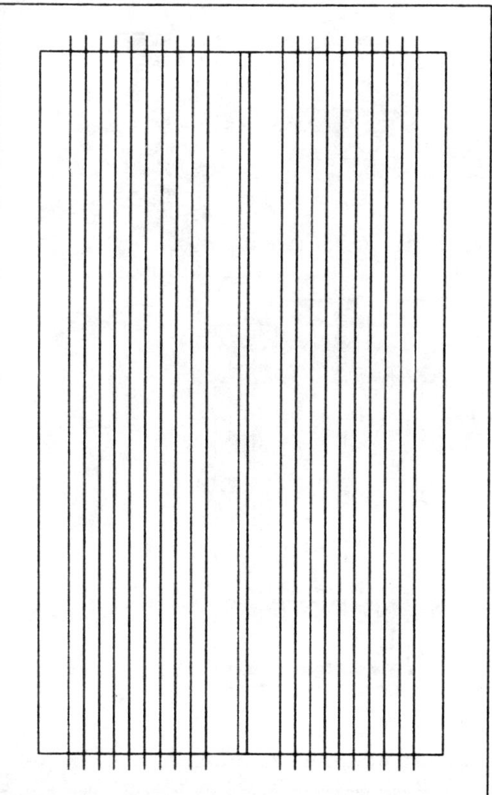

FIGURE 10-25: Low stray capacitance stator design.

and threads-per-inch, so you can usually find a suitable size.

For example, if you are using 1/16″ welding rod placed 10/inch, buy threaded rod with 10 threads/inch. Get the largest diameter available with the threads-per-inch you need.

If you can't find a rod with the exact number of threads required, have a machine shop make one for you. A metal lathe can easily cut threads on a rod to your specifications, and this is such a quick and easy procedure that the cost should be minimal.

To parallel the wires, use two threaded rods. Place the first rod at one end of the wires. Work the wires so one lies in each groove (*Fig. 10-26*).

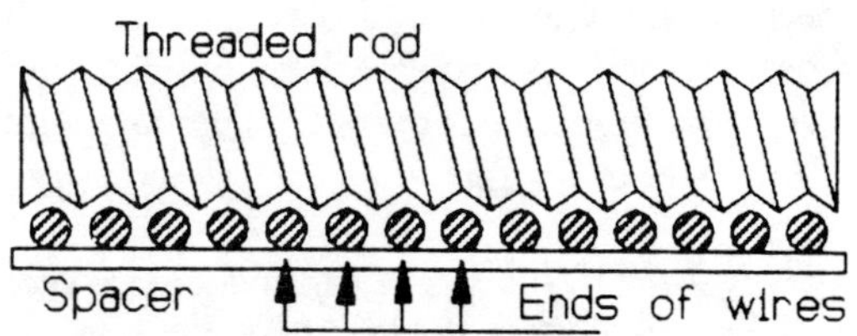

FIGURE 10-26: Position rods.

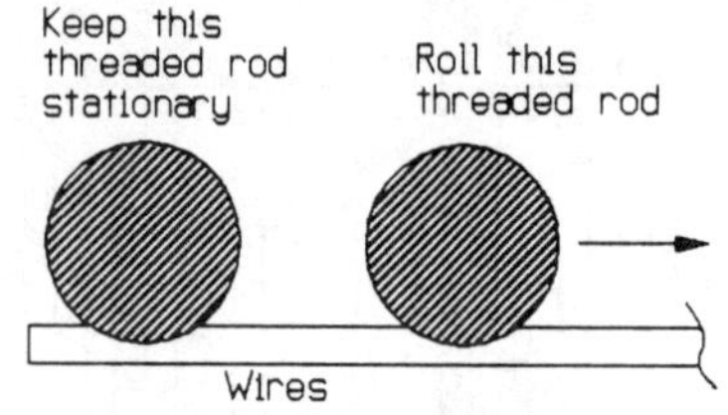

FIGURE 10-27: Spacing rods.

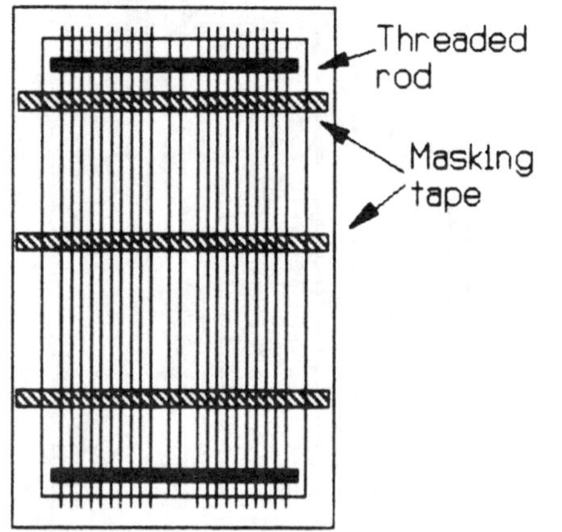

FIGURE 10-28: Anchor rods.

Place the second rod close to the first. While you hold the first in place, roll or comb the second rod toward the opposite end of the wires (*Fig. 10-27*).

When you have positioned the wires, secure them with a few strips of masking tape (*Fig. 10-28*). Glue the wires to the frame by pouring epoxy over them. If you prefer, you can use silicone rubber.

WIRE GRID. Connect the wires to form a uniform electrical grid. I do this by soldering a copper or steel connecting wire across them prior to painting the stator (*Fig. 10-29*).

An alternative is to pour a conductive paint, or conductive epoxy, along one end of the wires. Bend up one end for the electrical contact (*Fig. 10-30*).

If you use conductive paint, embed a solder tab or a brass strip in the paint for a contact. If you rely on the paint to hold the tab, it will be mechanically unsound and may later get torn loose. A small nylon bolt will hold it firmly in place. An alternative is to solder a contact wire to one wire in the grid.

TENSIONED WIRE. You can make wire very straight and stiff by holding it under tension. Tensioned wires allow you to make an extremely flat stator with very small diameter wire.

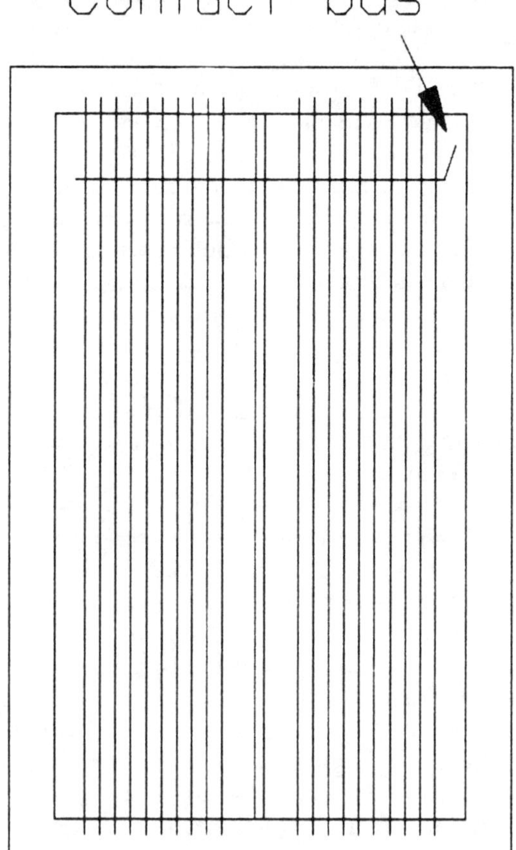

FIGURE 10-29: Make electrical connection to rods.

Of the different types of construction, tensioned-wire stators can potentially produce the highest output. If you use high quality insulated wire, their insulation properties can be as good as perforated-plastic stators. Unfortunately, they are the most difficult to build.

Use the same design guidelines for tensioned-

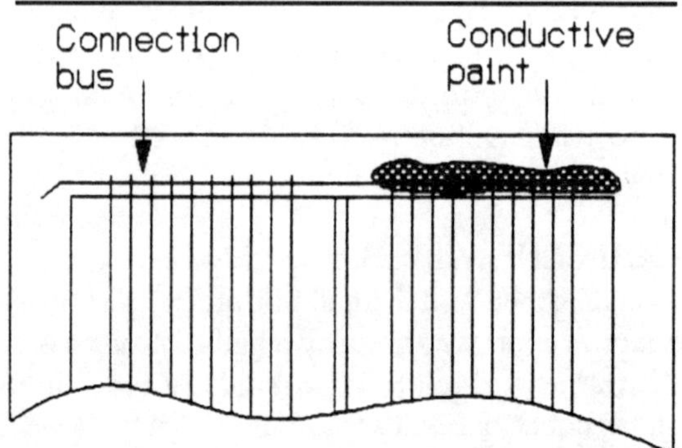

FIGURE 10-30: Alternative method of connecting rods.

wire formats that you used for rigid-wire stators. In short, to generate a high field density, many small wires placed close together are better than a few large wires spaced far apart.

WIRE TYPE. Many types of wire are suitable for tensioned-wire stators. Copper is the most common, and you may either use it bare or with various coatings.

These include tin, and silver, as well as many types of insulation. Most builders use insulated wire. The literature that discusses insulation and its use with ESLs is outdated, and there is little data on modern wire insulation.

For example, Teflon-coated wire probably offers the highest insulation quality for the thinnest coating. I expect it is the highest quality stator possible, yet it's not mentioned in the literature.

Magnet wire (used to wind motors, electromagnets, and transformers) works well. Not only is it inexpensive, it has excellent insulation and comes in very small diameters.

The older references give magnet wire a poor rating, however. I believe this is because the enamel insulation of that day was not nearly as good as modern magnet-wire insulation. Today's insulation is far superior to the older types.

You can find magnet wire in hair-thin sizes. A stator made from it could suffer from deflection caused by the electrostatic drive force. Don't use extremely small wire unless you support it at close intervals.

So what constitutes a "close interval"? This is highly dependent upon wire diameter, since it determines both the maximum tension and the mass of the wire.

Recall that the combination of wire mass and stiffness prevents deflection of the stator. The stiffness of a tensioned wire is determined by how much tension you apply to it and the distance between support structures.

The atomic weight of the elements used to make the wire times the wire's diameter defines its mass.

Copper is relatively dense and heavy, with much higher mass and inertia than aluminum wire. If mass were the only variable, copper would be ideal for tensioned stators.

Unfortunately, it is soft and has poor tensile strength. You can only bring it to moderate tension before it stretches and breaks.

Small-diameter steel wire would be much better. A high-carbon alloy such as music wire would be better still.

If you are looking for the best combination of strength and inertia in a hair-thin wire, it would have to be tungsten. Extremely strong and available in very small diameters, tungsten is relatively massive. It isn't insulated, but again, most ESLs don't need it.

TENSIONED-WIRE FRAMES. True tensioned-wire ESLs are very rare. I have yet to see a design I like, and consider them experimental. If you are the creative type and have access to a machine shop, I encourage you to explore the possibilities.

The major problem with tensioned-wire stators is that you need a very strong, rigid framework upon which to string the wires. Most of these designs use some form of plastic "egg-crate," fluorescent-light diffusers to serve this purpose.

Like all stator designs, you must support the conductors at reasonable intervals. Egg-crate construction supports the wires at very close intervals, usually every ½".

Mounting the wires is a major problem. Drilling holes along the ends of the crate for the wires is one way (*Fig. 10-31*). Running a long, continuous wire through a series of holes is a major task, however. You need a better way.

Another way is to put pins along the ends, over which you can hook the wire (*Fig. 10-32*). An alternative technique was used in the 1950s

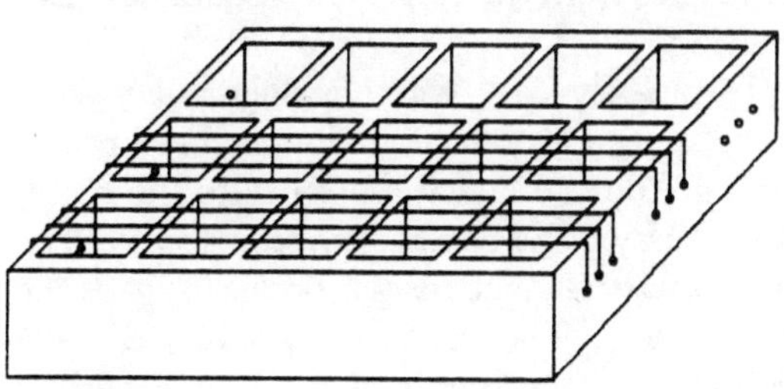

FIGURE 10-31: Tensioned wire strung through holes.

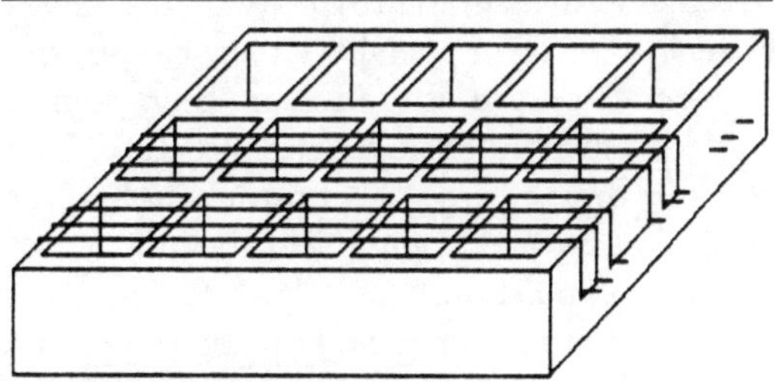

FIGURE 10-32: Tensioned wire strung on pins.

by the late Arthur Janszen in his commercial ESL tweeters. He wrapped the wires around both sides of a thick plastic grid (*Fig. 10-33*).

Egg-crate construction has its problems. Since plastic isn't strong, you can't apply high tension to the wires. A multitude of small wires under even moderate tension produce an amazing amount of pressure on the support structure.

Think about a stator 18-inches wide, having 20 wires/inch, each wire with a tension of *only* one pound. One pound isn't much tension. Even so, the force trying to collapse this egg crate is 360 lbs.

If each wire had 10 lbs. of tension, there would be over 1½ *tons* of pressure on this flimsy plastic grid. Look inside a piano to see the support structure necessary to handle many wires in tension.

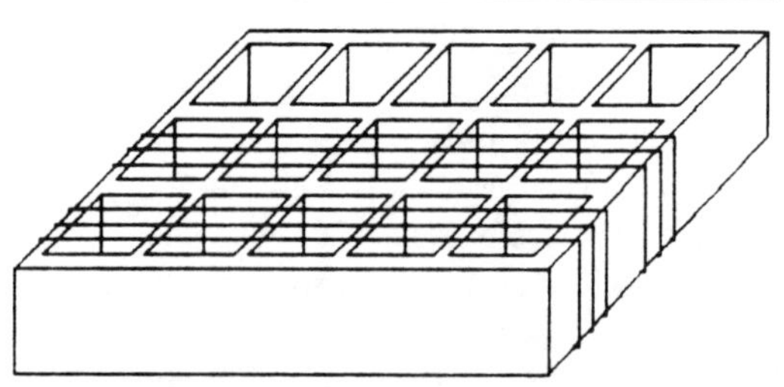

FIGURE 10-33: Wrapped wire stator.

Of course, the wires in an ESL don't have to be that tight. My point is that plastic light diffusers aren't strong enough to do the job.

Builders use them successfully because they glue the wires to each crossrib in the crate. With the wires supported every ½", the light tension just keeps them in position until the glue cures. Such a stator design is actually "quasi-tensioned."

One builder made an interesting variation on the "egg-crate" wire stator by gluing aluminum window screen to the crate. He did this by placing the screen on glass, which he covered with aluminum foil with the edges turned up to form a shallow pan. He put the "egg crate" on top of the screen, then poured liquid plastic cement over the structure. The cement drained to the bottom, where it pooled on the aluminum foil at the junction of the screen and crate.

The cement softened and melted the plastic crate slightly, then evaporated. By weighting the crate, the screen welded to the plastic. The aluminum window screen has very small

wires with 1/16" openings, and an open area higher than 50%. Performance is reportedly similar to Lincaine perforated metal.

A close-to-perfect design would be a truly-tensioned-wire stator. It would need a steel, aluminum, or carbon-fiber, composite framework to manage high forces without significant distortion. A series of pins along opposing sides could accept the wires, each of which would be strung with several pounds of tension.

If you wanted insulation, you could use 20-mil magnet wire or small-gauge, Teflon-coated wire. Putting 20 magnet wires/inch would give a very high field density with an open area of 60%.

I would forego insulation and use 10-mil music wire or tungsten. For the ultimate stator, I would use 40 wires/inch for an open area of 60% and extremely high field density.

The ultimate stator would need to have near-zero stray capacitance, so I would use external supports. Because the tension would be high, I would use one only every 4". It would be fascinating to compare this high-performance design with a more conventional stator.

EXTERNAL BRACING. You must take special care when using external bracing. If positioned improperly, it will deflect the wire from perfect straightness.

Presumably, if you take the trouble to build a tensioned-wire stator, you want it as perfectly flat as possible. A brace which touches a wire will deflect it, and it will no longer be straight.

The braces must be positioned so they almost, but not quite, touch the wires. The problem then becomes how to glue the wires to them.

A thick adhesive, such as epoxy, is unacceptable unless the wire is under very high tension, when it can overcome the epoxy's viscosity to find its naturally straight position. Also, the epoxy tends to form a glob, which protrudes from the stator and reduces the D/S spacing.

Fortunately, each wire has only a very weak force applied to it, so the adhesive need not be strong. You can attach the wires to the braces using spray paint. Several light coats sprayed on the brace/wire junctions will bridge the gaps and bond them. The light liquid spray will not deflect the wires from their straight position.

Another consideration is the expansion and contraction which temperature changes will cause in the wires and frame. If the wires are

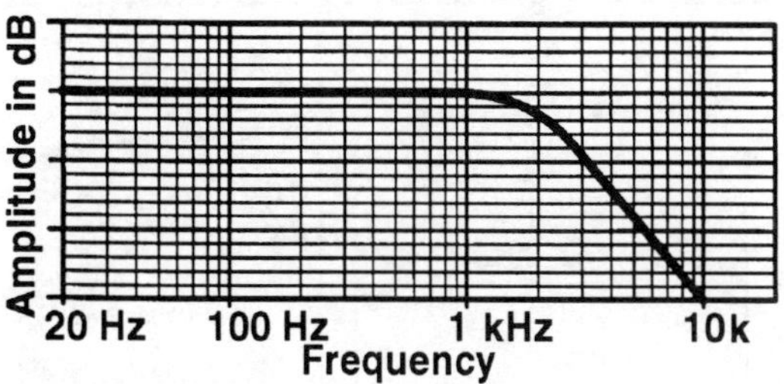

FIGURE 10-34: Incorrect frequency response.

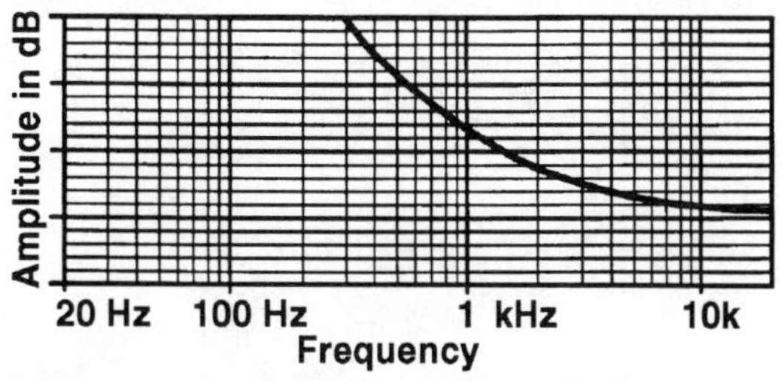

FIGURE 10-35: Ideal frequency response.

pulled tightly around pins or through holes, they will loosen and tighten with temperature, adversely affecting performance.

A problem will occur when winding the wires onto the frame because no framework is perfectly rigid and dimensionally stable. The first wires you attach will be tight. As more are wound on, however, they will slightly deform the frame and the original wires will loosen.

The framework needs to incorporate some type of tightening device, which can be adjusted after the stator is completed, so the wires are held at high tension. Alternatively, you can use an arrangement of strong springs to hold the wires in tension.

As with rigid-wire designs, tensioned-wire stators require you to connect all the wires electrically. Tensioned-wire stators present a different dilemma: insulated wire. The insulation prevents you from conveniently soldering to every wire in the grid.

Often, only one wire is used to string the entire stator. It may be more than 1,000-feet long. The resistance of such a length of wire can adversely affect high-frequency response.

If there is significant resistance in the stator, areas which are far from the amplifier contact point won't have time to become fully charged at high frequencies. The result is reduced high-frequency response at locations far from the contact.

While this is unacceptable for most designs, some builders deliberately compensate for phase cancellation this way. However, the frequency response curve is wrong, as in *Fig. 10-34*. Recall that you need frequency compensation as appears in *Fig. 10-35*.

You can calculate the high-frequency response at any stator location with the formula:

$$f = \frac{159{,}155}{RC}$$

Where:

  F = Frequency in hertz

  R = Resistance in ohms

  C = Capacitance in microfarads

Obviously, quite a bit of resistance is needed for this to be a problem. Still, it isn't reasonable to connect only the ends of a very long wire to the amplifier. Ideally, you should connect every wire to a bus which connects to the amplifier, as you did for a rigid-wire stator.

Connecting the ends of every wire to the bus is unnecessary and a lot of work. A reasonable compromise would be to make a connection about every 100' along the wire. Give thought to this when you are winding the stator. You will have to tie off the wire at these points, so the insulation can be stripped and connected.

Wires should not be allowed to make electrical contact with a metal or conductive frame. If they do, the frame will be at the drive voltage, producing stray capacitance and a shock hazard.

When using insulated wire, prevent bare parts from contacting the framework. If you use bare wire, insulate the pins by slipping Teflon "spaghetti tubing" over them. A notched-plastic strip insulator can be made to accept the wire.

Precisely cutting the notches is difficult to do freehand. The same applies to drilling holes and installing pins. A milling machine does this type of work quickly and accurately. While this is not the type of machine you are likely to have in your garage, it would be worth having the holes drilled or slots cut by a machine shop because of the precision it requires.

Consider a set of ESLs consisting of two cells/channel. These four cells require two stators each for a total of eight stators. Each stator has a set of pins or slots at each end to which you wrap wire, for a total of 16 strips. If each cell is 18-inches wide and has 20 wires/inch, the total number of pins or slots required is 5,760!

Another interesting figure is the amount of wire used to wind these ESLs. A speaker that

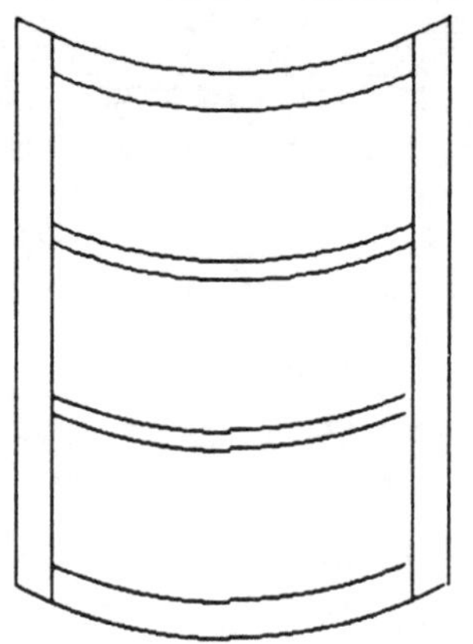

FIGURE 10-36: Spacer orientation in curved cell.

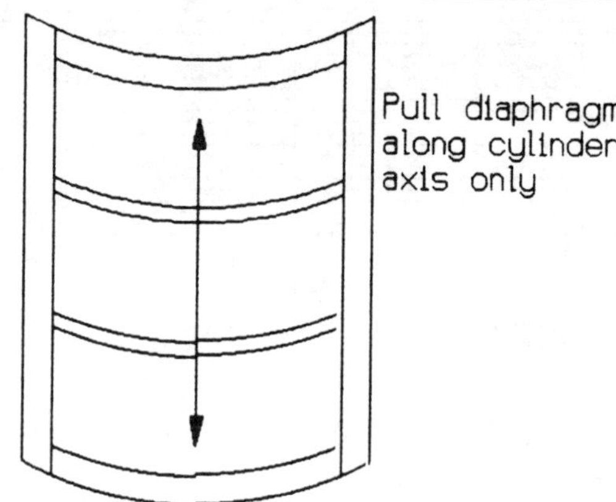

FIGURE 10-37: How to prevent diaphragm collapse in curved cell.

wires/inch, would require a total of 8,640′ of wire. Suddenly, the compromises of perforated metal don't seem so bad.

RESONANCE IN CELL STRUCTURES.

Some builders worry that their ESL structure will resonate, buzz, or rattle. They try all kinds of things to prevent this anticipated problem.

They pack foam or felt along the edges of cells, mount them on springs or rubber bands, and put silicone rubber along the edges of diaphragms. Yet I've never heard any resonance or other coloration caused by an ESL structure to justify this.

My suggestion: *don't worry about it.*

CURVED ELECTROSTATIC CELLS.

Curved ESLs follow all the basic guidelines, design parameters, and general rules as planar cells with one important exception: any internal spacers must be horizontal. These help support the diaphragm in a curve (*Fig. 10-36*).

Two major building differences between curved and flat cells are:

- Curved cells are built on at least one, and preferably two, curved tables.
- The diaphragm must be tensioned in only one direction, along the axis of an imaginary cylinder. If the diaphragm is stretched along the circumference, it will be pulled into the inner stator and short the cell (*Fig. 10-37*).

Wire-stator construction becomes troublesome when building curved cells. Since the wires are straight and rigid, they must run in the direction of the cylinder axis. The framework and external bracing (if any) must run in the direction of the cylinder circumference.

Thus, you must cut curved braces or heat plastic braces until they almost melt, while you mold them into the correct curve around some sort of jig. While possible, it is difficult.

Positioning the wires is an added problem. You can't roll a threaded rod down them because it would have to be curved. You might use a combing action, but keeping the entire rod in contact with all the wires would be difficult.

All things considered, it makes more sense to use perforated metal, which you can easily bend into the required curve. Also, curving perforated metal produces a perfectly flat surface in the direction of the cylinder axis, and prevents any problems with warped stators.

You should *not* glue the spacers to the perforated metal on a flat surface, and later try to curve it into the cell. The perforated metal and spacers form a lamination, and laminations are very stiff. Therefore, the stator resists being forced into a curve.

If the glue bonds don't break at the stator-to-spacer joint, they likely will at the diaphragm-to-spacer joint. Cells built on a flat surface must include nylon screws to help protect the glue bonds.

Building the stators on a curved table, so they will naturally follow the correct curve, is much better. Then the glue bonds will not be under stress in the finished cell.

Ideally, you should laminate the spacers to the perforated metal on a table which has a little *more* curve than you wish. When you release the stator from the table, it will spring back slightly toward the flat position. If you then finish the cell on a table with about 15% less curve, the cell will place nearly zero stress on the glue bonds.

Unfortunately, this requires two tabletops: one for gluing the spacers to the metal, and a second for gluing the diaphragm-to-spacer laminations. Making one curved tabletop is trouble enough. Why make two, when you can get by with just one? The glue bonds will be stressed, but not badly. If you are a perfectionist, make two.

The diaphragm can move more easily toward the inner than toward the outer stator. As the diaphragm moves outward, it must stretch tighter to move to a larger circumference. Moving inward reduces tension (*Fig. 10-38*).

The diaphragm's inward movement also produces a certain amount of nonlinear distortion not present in planar cells. Fortunately, this distortion is subtle. Acoustic coupling is also a problem.

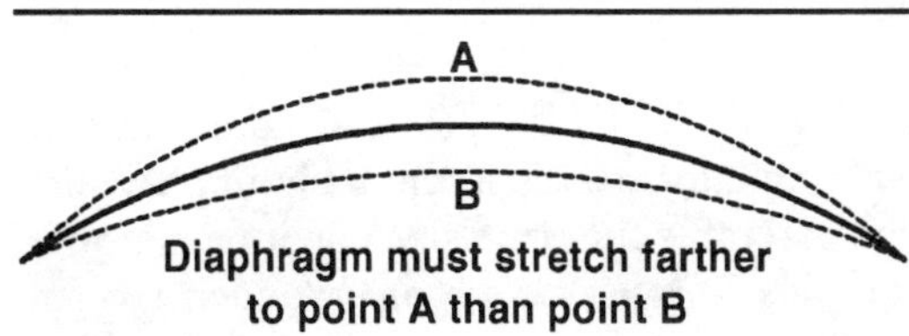

FIGURE 10-38: Non-linear diaphragm motion due to curvature.

A way to minimize these problems is to use a larger D/S spacing on the inner stator. For example, you could use 70-mil spacing (1/16 acrylic spacers) on the outer stator and 90-mil spacing (80-mil polycarbonate) on the inner stator. This will not cause distortion. Remember that the force field is the same anywhere between the two stators.

The inner stator must be slightly smaller because its circumference is less. The more curvature in the cell, and the wider it is, the greater the size difference between the inner and outer stator.

This difference is usually great enough that you must make the stators slightly different sizes. If you choose a very flat curve, and the cell is narrow, you can safely make them the same size. They will not match perfectly, but they may be close enough. For all others, different-sized stators are essential.

Determining the size difference between stators is surprisingly easy. You need only know the degrees of dispersion required, and the D/S spacing for both stators. For this

example, assume the D/S spacings are 0.07″ and 0.09″, and the dispersion desired is 20°.

The perforated metal will not give, so the circumference difference should be determined at the center of each piece of sheet metal. The thickness of the metal is small enough to be insignificant. For calculations, you may assume it is infinitely thin.

The difference between the circumferences of the stators is the difference of the circumferences of two circles whose radii are the difference times the sum of the D/S spacing. By taking the angle of dispersion as a percentage of a circle, you can find the size difference. A simple formula for this is:

$$W = D\,2\pi\,\frac{\text{Angle}_D}{360}$$

Where:
   W   = Difference in width between inner and outer stators
   D   = Sum of D/S spacers
   Angle$_D$ = Dispersion angle

Solving this equation for the above example, the difference between stator widths is 0.055″, which is almost 1/16″. That may not seem like much, but remember this cell does not have much curve.

Consider the same cell with 60° of dispersion. Now the error becomes 0.168″, or nearly 3/16″. About ¼″ difference would exist if the cell had 90° of dispersion. If you wanted 90° of dispersion in a cell using 140-mil D/S spacing, the difference would be 440 mil, or nearly ½″.

Another decision involves your desired dispersion angle. Against the desired amount of dispersion, you must balance reduced detail and output, nonlinear distortion, and construction difficulties. I recommend you use the least amount of dispersion consistent with your needs.

The main problem with highly curved cells is correctly fitting the diaphragm. When stretched in one direction, a diaphragm tends to form a slight hourglass shape, rather than remaining rectangular (*Fig. 10-39*). The diaphragm also tends to develop wrinkles along its edges, and the problem becomes more pronounced as you increase the curvature.

You can remove the wrinkles by gently pulling the edges in the direction of the cylinder's circumference. This cure is severely limited, however, because tension in that direction collapses the diaphragm into the inner stator. A better solution is for the diaphragm

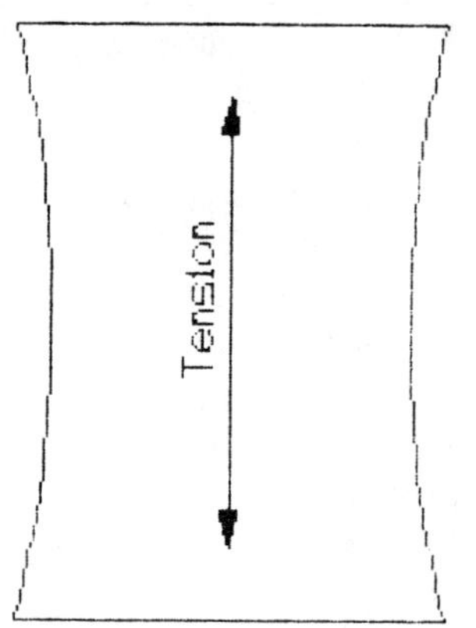

FIGURE 10-39: Stretched curved diaphragm shape.

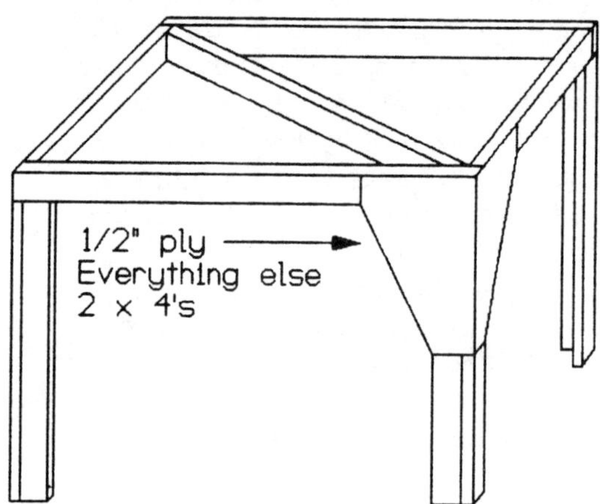

FIGURE 10-40: Simple table for building curved cells.

to be significantly longer than the cell itself, so you only use the middle 50–70% of it.

If you use a modest amount of cell curvature with an oversized diaphragm, you can get excellent results. Dispersion angles up to 20° are relatively easy to make. Beyond that, things get difficult.

For greater dispersion, several cells side by side, each having minimal curvature, work better than one large, wide cell with severe curvature. In planar ESL designs, many small cells are generally more trouble and introduce more stray capacitance than one large cell. In the specialized case of curved ESLs with large

dispersion angles, however, it is the best solution to a difficult problem.

CURVED TABLE. An important part of building curved cells is the table upon which you assemble the panels. While it need not be complex, it must be strong. Wooden two by four construction works fine for the basic frame and legs. For simplicity, I've shown only the basic framework in *Fig. 10-40*. Across this, you will attach a curved top.

Figure 10-41 shows an easy method of making rigid legs. Diagonal bracing will be needed for adequate stability, but it won't be much if you use the leg design shown.

Make the tabletop frame out of 2 × 4s and 1 × 8s (*Fig. 10-42*). Over these ribs you will attach a top to form the smooth curved surface upon which you build your cells. The number of ribs you need depends on the stiffness of the material you put over them. For most materials, a rib every 6″ is adequate, but using more won't hurt.

You need an accurate building surface to produce accurate ESLs. Since the ribs define the surface, you must cut them carefully so they closely match. Make a rib template out of

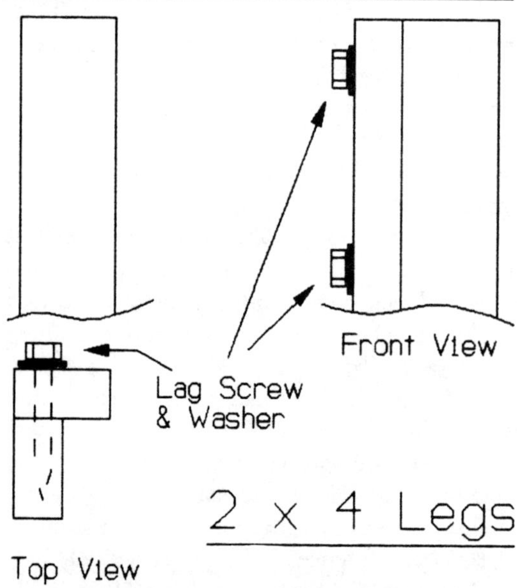

FIGURE 10-41: Strong table leg design.

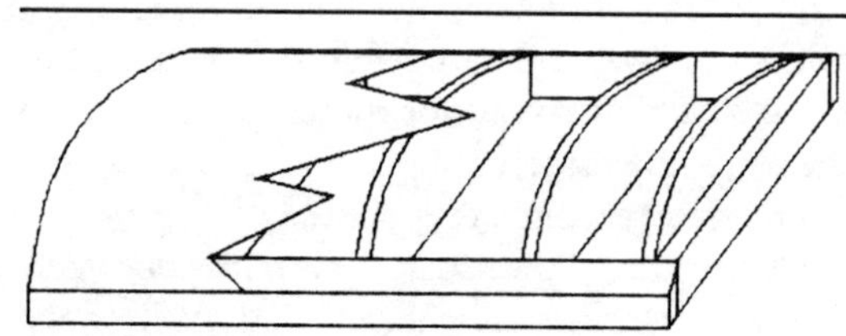

FIGURE 10-42: Curved table top.

thin cardboard, such as a file folder. Trace around this template for each rib.

A band saw is the best tool for cutting the ribs, but you can also use a hand-held jig saw. Use as wide a blade as possible for the smoothest cut. When using a band saw, keep light pressure against one side of the blade for control.

The curved surface may be made from sheet metal, sheet plastic (acrylic, polycarbonate, or Formica™), or smooth wood products such as Masonite™. The thicker the material, the better (only not so thick that you can't bend it): 1/8–¼″ plastics, 1/16–1/8″ metals, and ¼″ Masonite™ work satisfactorily.

You must attach this material to the frame without distorting it. I find that steel bars laid over the edges of the curved surface help hold it flat. The table will need to be several inches wider than the finished cell so these parts don't get in the way (*Fig. 10-43*).

I suggest you make two tabletops with different curves. Use the deeply curved one for laminating the perforated metal to the spacers, and the shallower one when installing diaphragms.

You will need some type of hold-down to secure the stator/spacer assemblies to the curved surface. Strong nylon straps are ideal. Attach one end to the side of the table, and the other to a sliding buckle or some other mechanism that you can tighten. You can also fasten a heavy weight to their loose end.

Alternatively, you can build a second curved surface which is a concave image of the building surface. Think of it as a lid, and build it as you would the tabletop (*Fig. 10-44*).

Glue a piece of ½″ foam to the lid's underside. You can hinge it to the table or leave it free. Whatever clamping device, belts or

FIGURE 10-44: Table top "lid".

weights you use, be certain the lid can produce enough pressure to hold the stator firmly to the shape of the tabletop. If each rib is cut from a rectangular piece of sufficient width, the scrap can become a rib for the lid.

DIAPHRAGM STRETCHER. Heat shrinking pulls the diaphragm in all directions. To make a curved cell, you must stretch the diaphragm in only one direction—along the cylinder's axis.

Most builders use a one-way mechanical diaphragm stretcher instead of heat shrinking. In the next chapter, Barry McClune shows a clever way to heat shrink a curved diaphragm.

If you use a stretcher, it must be very strong. A surprising amount of force is needed to stretch a sheet of Mylar.

To tension the diaphragm, attach one end of the Mylar directly to the table or its extension. Connect the other end to a movable section made from 2″ × ¼″ steel for adequate strength (*Fig. 10-45*).

I use Scotch® double-sided tape to hold the Mylar to the stretcher. Others use more complicated clamping devices. If you use tape, the steel's surface must be very clean and smooth, so the tape will stick well. Grind or sand it to bare shiny metal. If there is any oxide or roughness to the surface, the tape will pull loose and your stretcher won't work.

You may find bending the steel to the tabletop curvature difficult. One easy way is to bend it around a large cylinder (like a garbage can) while it is still a long bar, which gives you leverage. After it's bent, you can cut it to the desired length.

A very good reason why the pivots are so far away from the tabletop is in the drawing. The steel surface should move so all its points are equidistant from the Mylar's stationary attachment.

Pivoting attachments don't achieve this because the curve's middle moves farther than its ends. This stretches the center of the

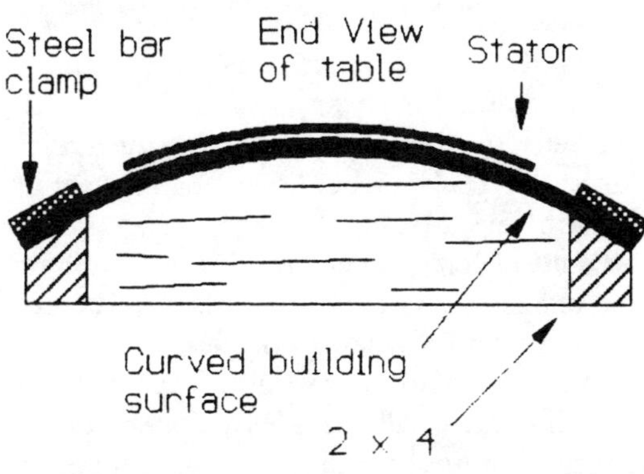

FIGURE 10-43: Detail of curved table top.

diaphragm tighter than the sides, and aggravates its tendency to assume an hourglass shape.

Making the steel stretcher surface move the same everywhere is conceptually easy. All you must do is make a parallelogram mechanism, although this requires much metal fabrication. It's easier to use a pivot, but extend its length so the difference in motion between the stretcher's middle and its ends is insignificant.

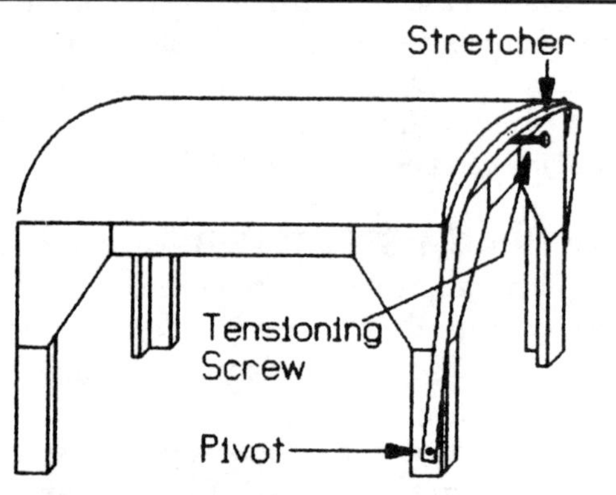

FIGURE 10-45: Stretcher table.

CELL CONSTRUCTION. You still need a piece of glass for making spacer frames and coating diaphragms. Construction follows the general principles for as perforated-metal, flat ESLs, with a few modifications.

Cut plastic strips for the spacer frames, taking care to make the sizes for the inner and outer stators slightly different. Glue the spacer frames together on the glass. Transfer them to the curved table, where you will glue them to the perforated metal.

Be certain to make both an inside and an outside stator! The outside one uses thinner spacers, larger sheet metal, and a wider spacer frame, which will lie against the curved table with the perforated metal on top.

The inner stator's qualities will be just the opposite: thicker, smaller and narrower, respectively. The perforated metal will be placed against the curved table with the spacer frame above it.

One point bears repeating: the outer stator is built *face down* (spacer frame against the table), and the inner stator is built *face up* (perforated metal against the table).

Also, pay special attention to the diaphragm contact details. The *larger* diaphragm-contact hole will need to be in the outer stator if you use a bolt contact.

Be particularly cautious with the parts: do not let them slide around and become misaligned when you are compressing them onto the curved tabletop. Use masking tape or cyanoacrylate glue to tack a couple of corners into position, so they cannot move just prior to compression.

Prepare a diaphragm by cutting it to fit the stretcher. I coat it on the glass prior to attaching it to the stretcher, because it's easier to get under it there to remove grit than it is on the curved tabletop.

Coat the diaphragm as you would for a flat cell, but with one change: minimize the pieces of tape. Tape complicates diaphragm installation. You can probably get by with tape only in the corners.

STRETCHING THE DIAPHRAGM. Prepare the stretcher for the diaphragm by cleaning its contact surface with acetone. Apply double-sided, thin Scotch tape. Do not use the thick foam type because the diaphragm tension will rip it apart.

The tape won't stick well if the metal is cold. Warm it to room temperature with a heat gun if necessary, so it will adhere well. Don't get carried away and make it so hot it's gummy.

Having an assistant when you transfer the Mylar to the stretcher is very helpful. With the two of you holding the Mylar in a tensioned rectangle (keeping the graphite side up), press the diaphragm to the tape on the stationary end.

Check that it's aligned, and that there are no big wrinkles. When you are satisfied, press the other end to the moveable steel stretcher. If it's misaligned, peel the Mylar free and try again.

Double-sided tape adheres amazingly well to Mylar. It tends to pull away from the steel first, so it is important that the steel is perfectly clean and smooth. When the tape is properly applied, it will adhere tenaciously to both the steel and the Mylar.

Stretch the diaphragm by turning the tensioning screw. The stretcher has enough strength to deform the Mylar, although an impressive amount of force is necessary to do so. Stretch the diaphragm as tightly as possible. The optimum tension is where it just starts to deform.

With the diaphragm tight, look for wrinkles. If a few small wrinkles are present at the edges, gently smooth them by wiping the Mylar with a paper towel. Wipe from center to edge; there should be enough suction between the Mylar and the tabletop to hold it in place.

You may use tape, but try to avoid it. Use as little tension at right angles to the stretched direction as possible, since this will tend to collapse the diaphragm toward the inner stator. Use enough tension to remove the wrinkles and no more.

Glue the outer stator to the diaphragm. Take care to get the diaphragm contact right. If the curve of your stator matches the tabletop so it lies flat on the diaphragm, fine. If not, you must hold it in place with straps or a lid. If you use straps, you can press on the spacers to squeeze the epoxy, taking care not to disrupt the alignment.

After the epoxy has catalyzed, slacken the stretcher and cut the diaphragm free. Get some air under it, and lift the assembly from the table.

Examine the diaphragm carefully. You should not see wrinkles, although you can live with a few small ones. If large wrinkles compromise the D/S spacing, install a new diaphragm.

Place the inner stator on the table face up, and glue it to the outer diaphragm/stator assembly. Carefully check the diaphragm contact alignment. Place the assembly under compression until the epoxy has cured.

Handle the finished cell more carefully than you would a planar cell if the glue bonds are under tension. Store it vertically—not horizontally, where you could accidentally push on it. Very little pressure is needed to straighten the curve and break the glue bonds.

Mount your cells using a method which helps hold them in an arc. Some builders insert nylon bolts through the spacers to take some stress off the bonds. If you made curved stators at the start and handled the finished cells with care, this is not necessary. The bonds are not that fragile, but why take risks?

MOUNTING ESLs. You'll need to mount your ESLs in some type of frame to make a finished speaker. Aesthetics play a very important role here. Many choices are available, and it's such a personal decision that I won't present specific designs. Instead, I'll give you a few ideas and you can take it from there.

Most cell designs have insulating spacers at their edges, which you can use to mount the cells. The two spacers are typically only 1/8". By cutting a slot in a wooden or metal frame, you can just slip the cells into the frame.

Remember, at ESL voltages, wood is *not* an insulator. It may not be a good conductor, but if you short the stators and diaphragm to the wooden frame, the speaker won't work. Don't let the audio-drive voltage to the stators touch or come close to any part of the frame.

The diaphragm-polarizing voltage can be in contact with the wood, and in fact is hard to avoid. The diaphragm comes to the edge of the spacers around the perimeter in most designs. Both the spacers and the diaphragm will touch the wood. If the stators don't contact the frame, this is not a problem.

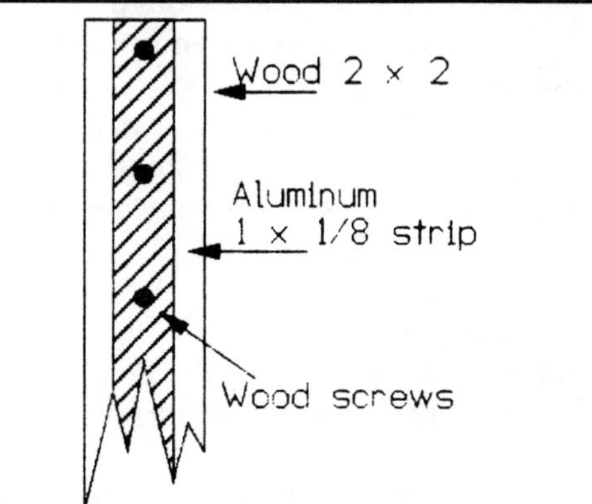

FIGURE 10-46: How to stabilize a thin wood frame.

If you don't want the diaphragm voltage to contact the frame, sand the edges of the spacers to remove any diaphragm scraps which may be hanging out. Then seal and insulate the edges with a coat of epoxy.

Nylon bolts can be useful, as they are strong enough for mounting the cells and are excellent insulators. You can drill right through the stators and their spacers, and put bolts through them without worrying about shorts if you don't introduce foreign matter into the hole.

Another useful idea is to use plastic for the frame. Since it is an outstanding insulator, you needn't be concerned whether speaker parts touch it. Acrylic or polycarbonate comes in 1/4" and greater thicknesses, in round or rectangular tubes, and solid rods. You can make very *avant garde* frame designs with it.

If you wish to use thin frames, be aware that long, thin, wooden beams are not dimensionally stable over time. The frames in my original speakers were oak 2 × 2s, but they warped. Laminating them would have stiffened and stabilized them.

Alternatively, you can screw a metal strip to

a solid-wood beam to stabilize it. Attractive, clear, anodized-aluminum strips and bars are available from hardware stores. A strip one or two inches wide and 1/8- or ¼-inch thick gives great stability to a long 2 × 2 (*Fig. 10-46*).

GRILLE CLOTH TESTING. Grille cloth is a problem. Putting *any* obstruction in front of a speaker we wish to sound very clear is illogical. Grille cloth degrades sound quality, yet we use it for cosmetic and safety reasons.

Grille cloth can sound as though you put a blanket over the speaker—except some blankets sound thinner than others. The worst types are close-weave cloths and thick foams. The best have open weave and smooth plastic fibers.

Mellotone™, usually available in electronics parts houses, is called "flame proof" and is not cheap, but it's the best I've found. Not surprisingly, it is very sheer and doesn't visually hide the speaker.

Many builders refuse to compromise—they don't use grille cloth. If you decide to use bare cells, give careful thought to cosmetics so they are aesthetically acceptable.

If you use grille cloth, consider testing different types to find the least obtrusive. Start by mounting one on a simple frame. Have an assistant alternately place it in front of the speaker and then remove it, while you listen to wideband complex source material. White noise (hiss), such as FM interstation noise, is an exceptionally good test.

You must perform this test "blind." If you can see the grille cloth in front of the speaker, your visual bias will trick you into thinking the sound is bad. No matter how objective you think you are, you cannot override your unconscious biases with logic.

You *must* perform the test blindfolded. If you close your eyes, you will surely peek. If you can see the grille cloth, the test will be void.

So put on a blindfold, plug your ears, and have an assistant place the grille cloth for the start of the test. It doesn't matter whether they choose on or off. All that matters is that you don't know the starting position.

Now unplug your ears and listen. At your signal, the assistant switches the grille cloth position. You should readily and reliably hear when the grille cloth is on. If you can't, you have incredibly good grille cloth.

ELECTRICAL SAFETY. I presented this section at the end of the electronics chapter.

Thinking that some of you might not have made it that far in the chapter, I'm repeating it here.

Several high-voltage sources are present on ESLs. Despite this, they are not very hazardous when you analyze the situation. Still, caution is warranted.

Current kills, not voltage. High voltages only make it possible to drive current through a person. Voltage itself is not dangerous.

The way to kill someone with the least amount of electricity is to drive electric current through his heart. About 50mA will disrupt the heart's electrical system. Any other electrocution method requires large amounts of current to literally burn and coagulate human tissues.

The maximum current a large ESL system can deliver is only a fraction of an amp. Therefore, high-power electrocution doesn't apply to ESLs.

The polarizing voltage is the source of most shocks. Most builders occasionally get zapped by the diaphragm voltage. While unpleasant, it isn't hazardous—more like touching a door knob after having walked across a new carpet on a dry day.

The charging resistor isolates you from the polarizing power supply. This prevents significant current transfer.

Assume, for example, that the polarizing voltage is 3.5kV and there is a 22MW resistor in series with the diaphragm. Using Ohm's Law, the maximum current which can flow through this resistor would only be slightly more than 1/10mA. Even ten times the voltage would deliver less than 2mA, which is far below the 50mA required to stop a heart.

The diaphragm coating stores electricity even when the polarizing supply is disconnected or switched off. In fact, it is when working with a disconnected cell that most shocks occur. Again, the diaphragm has a high resistance which prevents significant current flow.

While an ESL diaphragm is not an electrocution danger, this is not true of the polarizing power supply. Although these usually have a rating of only a few milliamperes, they may have enough capacitance to momentarily deliver enough fatal current. Treat the power supply with caution, and mount it so neither people nor pets can touch it.

A practice used by technicians who work with high voltages is to keep one hand in a pocket. For current to travel through the heart, it must travel from one side of the chest to the other. The obvious path is from arm to arm.

By using only one hand, the technician makes it nearly impossible for current to pass through the heart.

If you don't do it this way, you will forget where your other hand is while probing the innards of a high-voltage source. Most likely, it will end up resting on the chassis, where it can complete the electrical circuit if the other hand touches something "hot."

The main risk is the audio-drive voltage. It can pass several hundred milliamperes because there is no resistance in series with the voltage source.

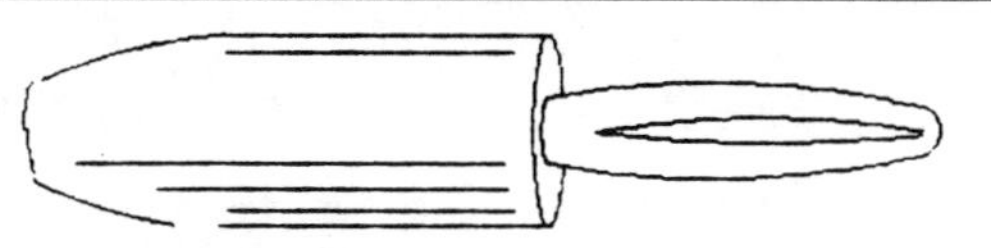

FIGURE 10-47: Banana plug.

You can't easily get shocked by the stators. Since they are "floating" and aren't referenced to ground, you can't readily get shocked by just touching one. You have to touch one with one hand, and the other stator with the other hand, while playing loud music.

Even this does not guarantee a fatal shock, but I wouldn't want to try it. How likely is it that somebody will touch both stators while loud music is being played? Not likely, but then people do weird things. At low output levels or when the music is off, there is no hazard.

The greatest risks are the connections between the cell and amplifier, which are close together, easily reached, and often worked with while testing.

I recommend the following precautions. First, use plastic grille cloth. It offers significant protection from touching the stators.

Second, be thoughtful when designing your connections between the ESLs and the step-up transformers. Do *not* use screw terminals, since they leave exposed contacts and wiring.

Safer and more convenient connectors are banana plugs. Besides being well-insulated, they give a solid plug-in contact, and you can color code them. If you mount them on a piece of insulating plastic, and the connections between the plugs and the stators are protected from contact, they are quite safe (*Fig. 10-47*).

One final caution: direct-drive, high-voltage amplifiers are extremely dangerous. Not only are high voltages present with high current capacity, but have large storage capacitors in the high-voltage power supplies. This setup can be truly lethal. A conventional amplifier and step-up transformer setup is much safer.

CHAPTER 11:
CURVED ESL CONSTRUCTION

By Barry McClune

The advantages of curved electrostatic loud-speakers are the elimination of the "head-in-a-vise" syndrome and expansion of the listening "sweet spot." The syndrome causes the stereo image and frequency balance to shift with slight side-to-side movement of your head.

While your head is in the "vise," the stereo image is amazing, but I like to shift in my chair without destroying the image. Curvature of 10–15° will cure the syndrome and may be sufficient to enlarge the "sweet spot" for two or three consenting adults. Using my techniques, you can build an ESL of any size and with any degree of curvature.

CURVATURE. How much curvature should you use? You already know the advantages, so let's review the disadvantages. The greater the curvature: the harder it is to make the curved-metal stators perfectly parallel; the harder it is to tension the diaphragm in only one direction; the greater is the tendency for the diaphragm to be pulled onto the inner stator.

To deal with these problems, I suggest:
- Large D/S spacing
- Low spacer ratios (50:1) for modest curvature, even lower for severe curves
- Minimize the amount of curvature

If you just want to eliminate the "head-in-a-vise" syndrome, 10–15° should be more than adequate. If you want to increase the "sweet spot" to encompass two or three adults, 10–15° may still do the job depending on your distance from the speakers.

Peter Baxandall, in his *Headphone and Speaker Design Handbook*, mathematically modeled the dispersion characteristics of a flat panel and a curved conventional speaker. Even if you lack faith in the limited experimental evidence supporting mathematical models, the simple equation below will give you a conservative approximation. Note that the accuracy of the predicted horizontal dispersion pattern *decreases* the further you are from the speaker.

$$W = (Tan\ C)(D)$$

Where
D = the distance from the speaker

C = the degrees of curvature
W = the width of the listening window

So, if your speaker has 30° of curvature and your listening position is 10′ away:

D = 10′
C = 30°

$$W = Tan\ 30 \times 10$$

$$W = 0.577 \times 10$$

$$W = 5.7'$$

CONSTRUCTION. Five steps are included in building a curved ESL of my design:
- Building the supporting frames for the stators
- Attaching the perforated metal to the frames
- Attaching the spacers
- Heat shrinking the diaphragm
- Final assembly

BUILDING SUPPORTING FRAMES. You must choose the style of frame you wish to build. I built the frame pictured in *Fig. 11-1* for a speaker with 30° of arc.

The frame in *Fig. 11-2* should work just as well, and if you have access to a table saw, you can cut the angle right into the frame. The angle cut into each support equals one-half of the degree of curvature desired (for 30° of curvature, cut a 15° angle along the length of each main support of the frame).

I used a jig saw to cut the curve of the speaker into the base and top of the frame in *Fig. 11-1*. I don't think it's necessary, and I've omitted it from *Fig. 11-2*.

I cut the outside edges of Frame A at a 45° angle in an attempt to eliminate the sharp transition at the edge of the speaker. The main supports in Frame B in *Fig. 11-1* were made from 1¼″ particleboard 7-inches wide.

I chose this width in order to extend the path from the front of the speaker to the back, not for structural support. This slightly lowers the phase cancellation frequency. If you are building a speaker to match Roger's designs, a 3-inch width and 1-inch thickness for these pieces would be adequate for structural support.

Make the width of the frame from outside edge to outside edge (B1 to B2 in both frame designs) equal to the width of the stator, less a fraction of an inch to compensate for the stator's curve. To help you calculate the necessary dimensions, see the table in *Fig. 11-3* or the formula in *Fig. 11-4*.

If you know your way around a carpentry shop (or have a friend who does), please feel free to ignore the following directions on how I assembled the supporting frame. I have no doubt there are easier ways to do it. For example, give *Fig. 11-1* or *11-2* to your local carpentry shop with the appropriate dimensions and have them do it. (OK, this is a book on amateur construction, but sometimes it pays to have someone with the right tools and knowledge do the job.)

Once you've cut the frame supports to the dimensions required, you must assemble them as accurately as possible. Glue alone, or with a few screws, isn't strong enough.

For the design in *Fig. 11-1*, I used plenty of glue and two wooden dowel rods for the joints in Frame B. In Frame A, I used glue, one dowel rod, and a screw in each joint.

The assembly of Frame B is the same for either *Fig. 11-1* or *11-2*. Carefully measure and outline the position of each support on both the bottom and top end pieces of the frame. Line up each support with one end piece, and drill holes for the screws and dowel rod(s). Secure it in position.

When you have both supports firmly attached to *one* end piece, take the other, and with the help of a friend, hold it in position. Double and triple check your measurements between B1 and B2 for both the top and bot-

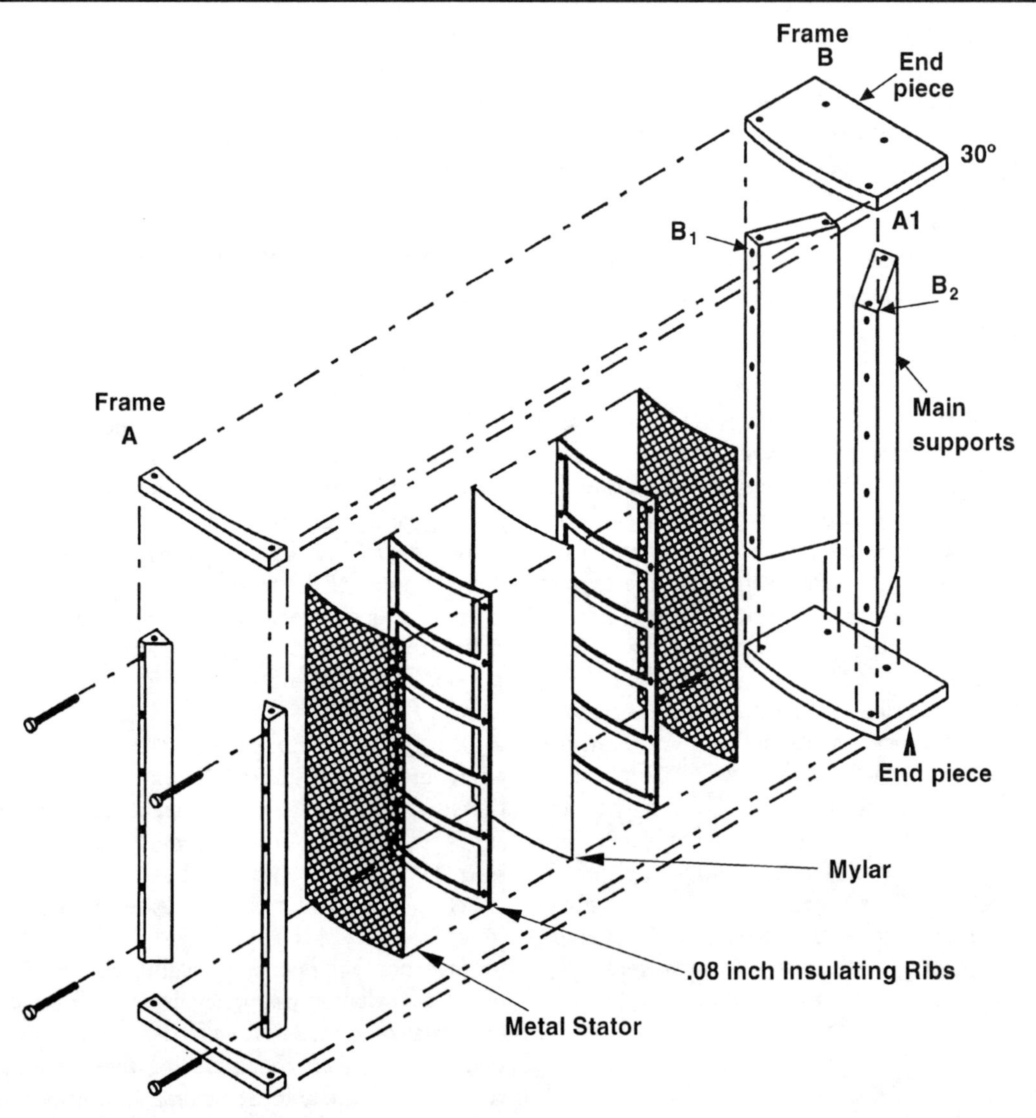

FIGURE 11-1: Support frame.

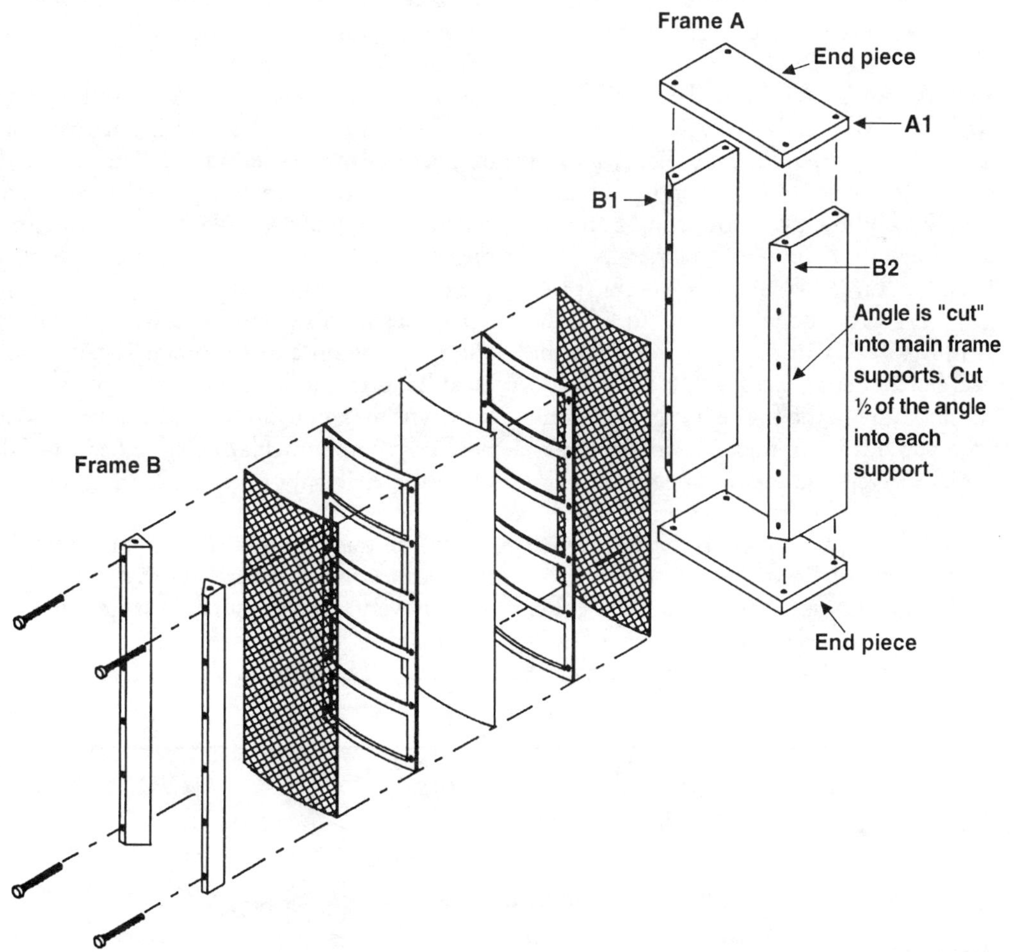

FIGURE 11-2: Alternate support frame.

tom. Also, check the angle made by the two supports, if you are building the frame in *Fig. 11-1.*

The measured distance between point B1 and B2 at the top of the frame should be exactly equal to the measurement at the bottom of the completed frame. Once you're satisfied all is well, attach the last piece. Assemble Frame A in *Fig. 11-2* with the same degree of accuracy, by resting the end pieces and supports on Frame B and drilling holes for the screws and dowel rod. In the *Fig. 11-2* version, place Frame A over Frame B and drill ¼" holes about 6-inches apart on both sides. Drill from the front of Frame A through Frame B.

After you drill each hole, pass a 3/16" × (needed length) bolt through the hole. This will ensure that the next hole you drill will be aligned with the previous hole. The procedure is the same in the *Fig. 11-2* version, except you may need to use greater care in drilling the holes to ensure that they pass through the *center* of both Frame A and Frame B.

The bolts will contact both the perforated metal and diaphragm. If they are conductive, they will short out the cell. Nylon bolts are best, but you can use metal bolts if you insu-

---

IMPORTANT

Be sure to make provision for connecting the polarizing power supply to the diaphragm. In this design, the coated side of the diaphragm is on the outside of the curve. You may make the connection in the *Fig. 11-2* frame on the bottom curved spacer, if the spacer is wide enough to extend at least ½" beyond the width of the particleboard. Please refer to Roger's methods of making a diaphragm contact for more ideas.

late them with a couple of layers of electrical tape or two sleeves of heat-shrink tubing.

CUTTING AND ATTACHING PERFORATED METAL.

Let's review a couple of points so there's no confusion. The width of your metal stators will be equal to the distance between B1 and B2, plus a fraction of an inch to allow for the curvature of the metal.

The angle formed by the two supports (A1 in *Fig. 11-1*) sets the degree of curvature for the frame in *Fig. 11-2*. Cut the curve into both the top and bottom end pieces.

The curvature in the frame in *Fig. 11-2* is cut into the supporting frames (A1 in *Fig. 11-3*). I've already given you the math needed to figure out the width of the stators and angle A1. Before cutting metal, it might be a good idea to review your calculations.

When you cut your metal, you must either leave a tab of metal outside the frame for electrical connections, or decide how and where to make the electrical connection to the stators. In building the speaker in *Fig. 11-1*, I left tabs on one side near the base.

Don't expect to solder a connection to an aluminum stator without trying it first—aluminum doesn't solder easily. You can easily cut aluminum stators with metal shears, large scissors, or a table saw with a fine wood-cutting blade. Be sure to allow for the width of the blade in your measurements.

Unquestionably, the best way to cut the aluminum is on a sheet metal shear at your local heating and air conditioning shop. This tool will make absolutely flawless cuts with perfectly perpendicular sides. Except for the travel time, it only takes a moment, and the cost is insignificant.

Cut the metal accurately, and take care that the outer edges are parallel. You can remove any unevenness along the cut by gently hammering it on a flat surface.

Attach the perforated metal to Frame B

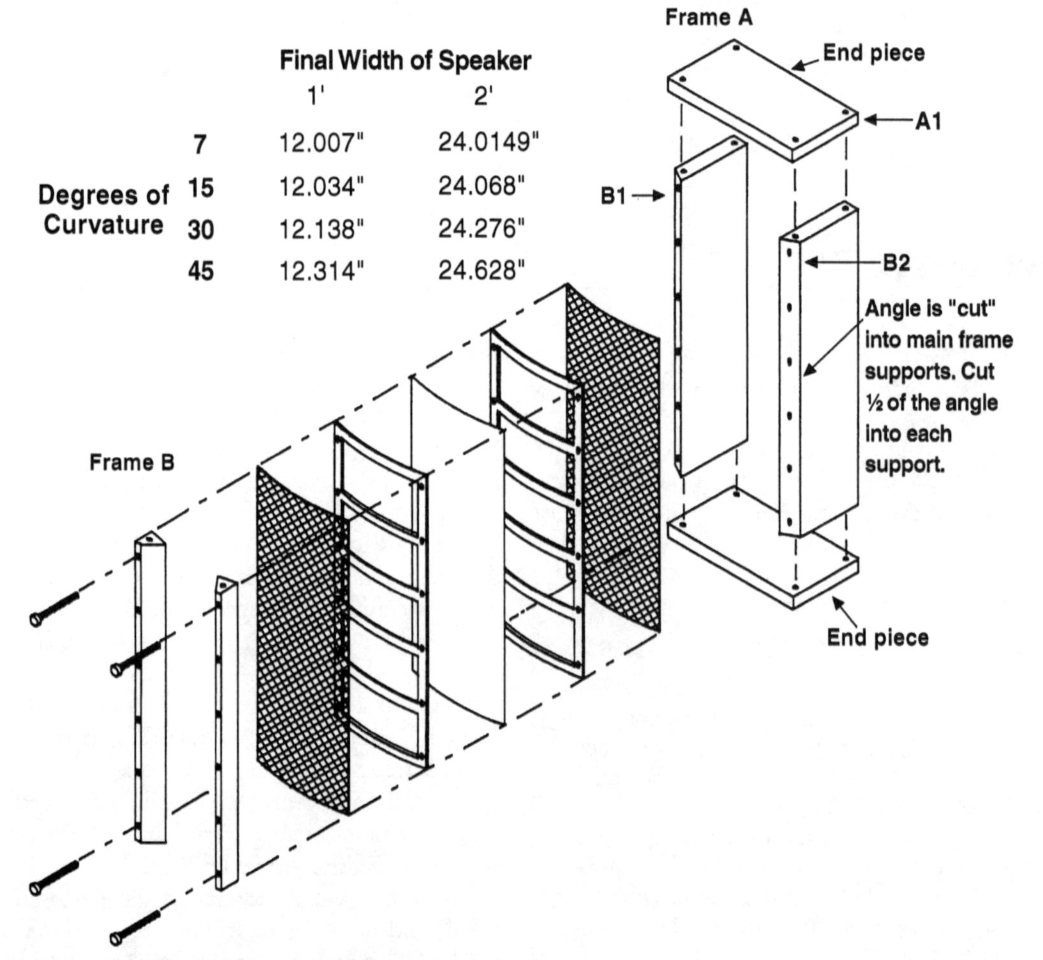

FIGURE 11-3: Final speaker design.

using epoxy. Be certain the holes with the rounded edges are facing toward the diaphragm side.

I relied on several extra pairs of hands to align the metal with the outside edges of Frame B. With 20/20 hindsight, I would place a generous amount of adhesive along one edge of Frame B, carefully position the metal, then countersink small, flat-headed nails along the outside edge to secure the metal to the frame. Repeat the procedure on the other edge, aligning the parallel surfaces of the metal with the outside edges of Frame B.

Next, make a sandwich consisting of:
- Frame B and the attached metal
- A single layer of wax paper or Saran Wrap (to prevent bonding the metal to both Frames B and A) *(Epoxy can ooze through wax paper; I recommend Saran Wrap—Roger)*
- Frame A

Take a drill with a ¼" bit and pass it through the holes in the frames to create a hole in the metal. Pass your bolts (with washers) through the sandwich, tighten snugly, and let the epoxy cure.

After the epoxy is solid, carefully remove all the bolts, Frame A, and the wax paper. Please resist the temptation to pull the metal to see how well it's attached to Frame B.

Now, carefully lay the second sheet of aluminum over the first and align the edges. Ideally, you should also have the small holes in the aluminum aligned. Position Frame A over Frame B, and again pass a ¼" drill bit through the sandwich, placing the bolts as you go.

Place a new layer of Saran Wrap over the Frame B metal, and position the Frame A metal, being sure that the ¼" holes line up. Don't use nails; just apply a generous amount of epoxy to Frame A and position the whole frame (*Fig. 11-1*) or the separate pieces (*Fig. 11-2*) on the metal.

Line up the ¼" holes, and make sure there are no gaps in the metal-to-Saran-Wrap-to-metal sandwich. Pass the bolts (with washers) through the sandwich and tighten.

You can use considerable pressure to ensure a good bond between Frame A and the metal. After the epoxy cures, remove the bolts, *slowly* remove Frame A, and finally the Saran Wrap. Your stators are complete.

ATTACHING THE SPACERS.

Cut the 0.080-inch-thick, plastic spacers to the dimensions as the edges of Frames A and B. If you used 1" particleboard and built the *Fig. 11-2* frame, the spacers should be 1-inch wide. The long spacers on the outside edges should extend the full length of the frame. The shorter "curved" spacers should just fit inside the long outside-edge spacers.

Make the curved bottom and top spacers the same width as the outside-edge spacers. Cut the other curved spacers as narrow as possible (¼–½"). All the spacers should have both surfaces lightly sanded to improve the bonding qualities.

You must drill the outside spacers to permit passage of bolts. Position all the Frame A and B outside spacers along Frame B, and set Frame A into position, lining up the holes. Make sure that the outside spacers are lined up. It helps if Frame B is sitting horizontally.

Drill through the sandwich with a ¼" bit, placing bolts as you go. When finished, remove the bolts and Frame A. Epoxy the spacers you just drilled and align them on their respective edges. Put wax paper or Saran Wrap between them. Again, position Frame A to make a sandwich, pass bolts through, tighten, and let set.

Attaching the top and bottom curved spacers can be tricky if you have used a high degree of curvature. Check that you cut the pieces to the proper length with no more than a 1/16" gap on either side of the spacer. Apply epoxy

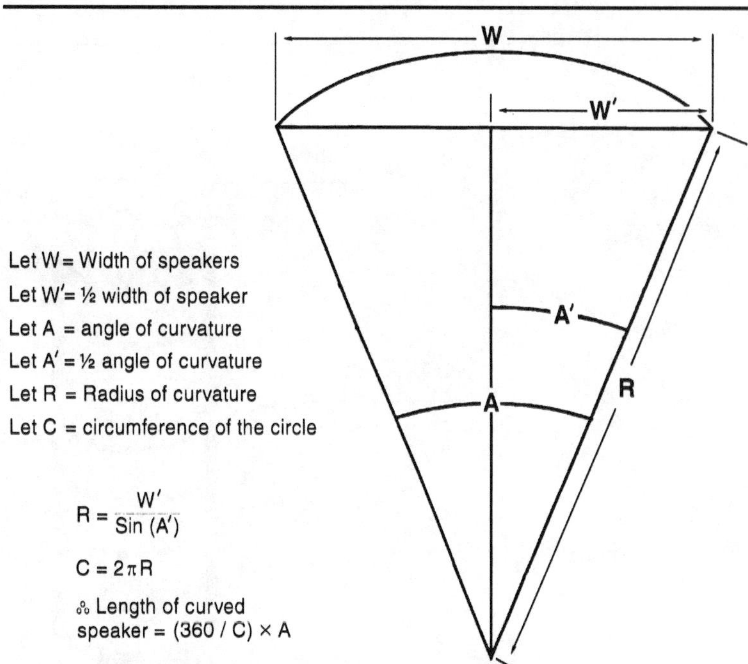

$$R = \frac{W'}{\text{Sin}(A')}$$

$$C = 2\pi R$$

$$\text{∴ Length of curved speaker} = (360 / C) \times A$$

FIGURE 11-4: Formula to determine the dimensions of your speaker frame. (See also *Fig. 11-3*).

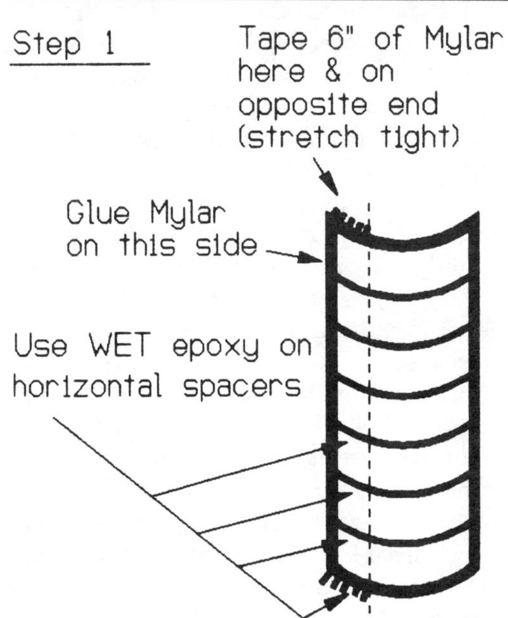

FIGURE 11-5: Tape first 6 inches of Mylar.

and use masking tape to hold the pieces temporarily in position, while again bolting together a sandwich of Frames A and B.

Position the remaining spacers horizontally along the speaker. You may need to hold the center spacers in place just like the curved end pieces—by tape—after applying the epoxy. Bolt Frames A and B together as before, and allow the epoxy to cure.

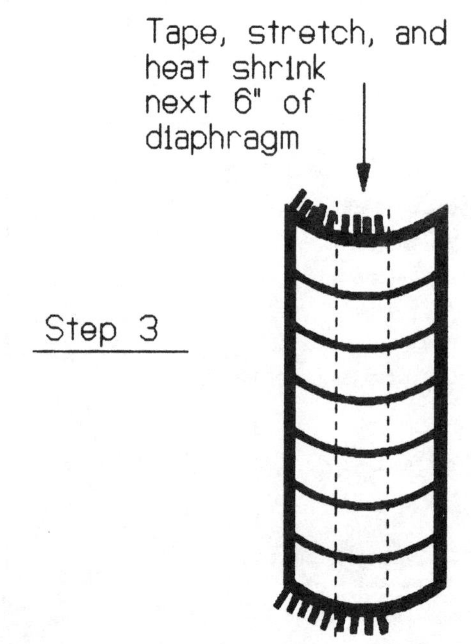

FIGURE 11-7: Tape next 6 inches of Mylar.

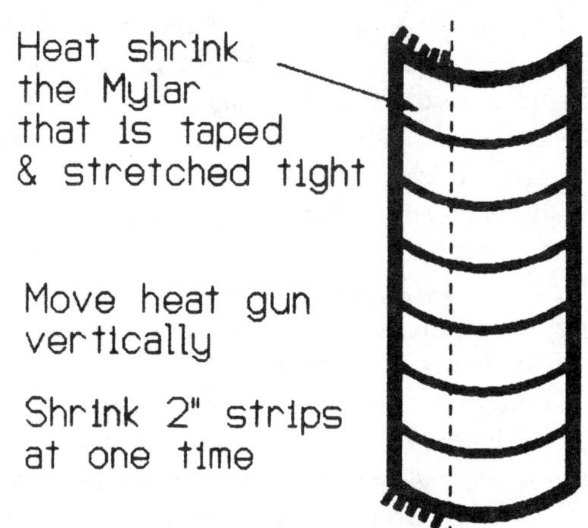

FIGURE 11-6: Heat shrink diaphragm.

FABRICATING THE DIAPHRAGM.

Following Roger's directions in *Chapter 10* on ESL construction, cut a Mylar diaphragm to size and coat it with graphite. I will now explain how to mount and heat shrink a diaphragm, so you don't have to build a stretcher.

Use 5-minute epoxy to glue the uncoated side of the diaphragm to the flat outside-edge spacer of Frame B's stator. I've found that if you first apply a bead of epoxy along the spacer and then run the edge of a razor blade along its length, you can get a very thin, uniform coat of epoxy. Attach the diaphragm and let the epoxy cure.

Apply a thin layer of 45-minute (or longer) epoxy to the top and bottom 1″ spacers on Frame B. *Do not put epoxy on the flat perimeter spacer on the opposite side yet!*

Attach the Mylar as uniformly as possible over the rest of the frame. Stretch it by hand to get the big wrinkles out, and try to avoid smearing the epoxy on the Mylar.

Place 2-inch wide (or wider) strips of masking tape on the diaphragm at the top or bottom ends of the frame. Anchor half the width (or about 6″) of the Mylar on the side of the frame to which you have already glued the diaphragm.

If you are building the *Fig. 11-2* frame, pull the masking tape tight and attach it to the bottom or top end pieces. The *Fig. 11-3* frame may require that you temporarily position a piece of wood between the end piece and the unsupported curved metal. This will permit you to apply sufficient force to the Mylar

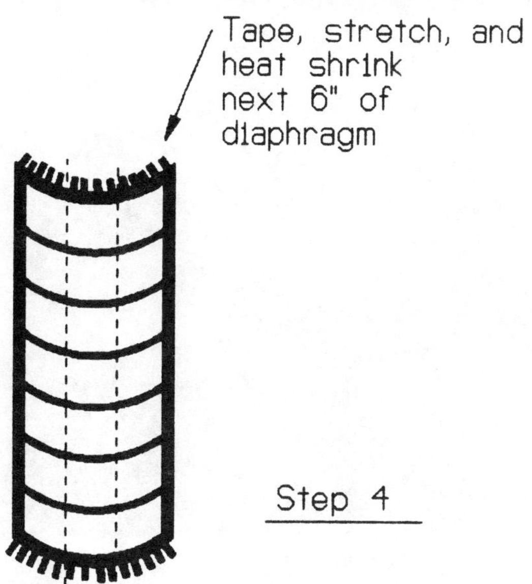

FIGURE 11-8: Heat shrink diaphragm.

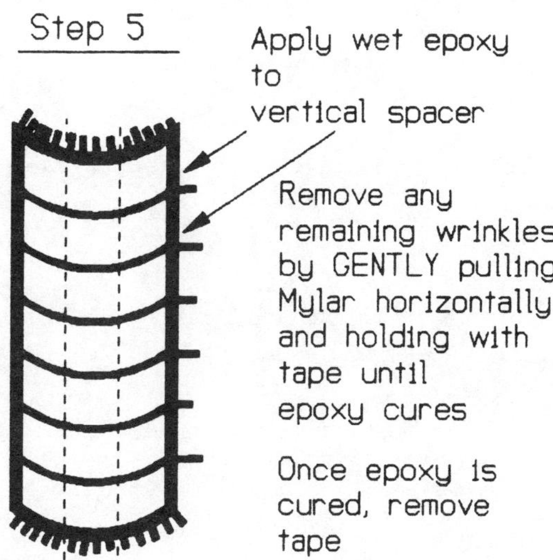

FIGURE 11-9: Glue edge of diaphragm.

without distorting the curved metal. Go to the other end and do the same, stretching the Mylar vertically along its length as much as you can without tearing it (*Fig. 11-5*).

Take a heat gun, and starting at the attached side, partially heat shrink the Mylar. Follow a vertical line from the top to the bottom of the section of Mylar that you have anchored with masking tape (*Fig. 11-6*).

You will see the Mylar stretching tight vertically (because you've anchored it vertically with masking tape), but it will *not* shrink tight horizontally. Instead, it will slide over the smooth plastic supports. This relieves the horizontal tension and prevents the Mylar from being pulled onto the inner stator.

Attach the Mylar to the second horizontal half of the stator (or the next 6″) the same way (*Fig. 11-7*). *Don't heat shrink yet!*

Now, fully heat shrink the diaphragm beginning at the very edge of the side that you already partially heat shrunk. Heat shrink 2″ vertical strips, but stop short of heat shrinking the last several inches that are tensioned by the masking tape. You will fully heat shrink the last several inches after attaching the next 6″ of diaphragm. Repeat the taping and heat shrinking for each 6″ or so (*Fig. 11-8*). When

you've heat shrunk all but the last inch or so, you are ready to finish the job.

Apply 5-minute epoxy to the remaining outside edge and heat shrink it. Now is the time to smooth out any remaining small wrinkles along the edge. If necessary, you can use a little Scotch Magic® tape to pull the diaphragm horizontally. Use as little tension as possible—just enough to remove any remaining wrinkles (*Fig. 11-9*).

FINAL ASSEMBLY. After the epoxy has cured, carefully remove the tape. You may glue Frame A into place, but this is not necessary. Place Frame A over Frame B, and line up the holes through the entire assembly.

Using a sharp instrument like a nail or awl, puncture the Mylar to allow passage of the bolts. Install and tighten the bolts, and make your electrical connections. You're ready to test the speaker.

COMMENTS. Building a curved ESL is *not* an easy project. Fabricating the wood frames accurately is the most time-consuming and difficult part. The only problem I've encountered is tearing the Mylar by initially applying too much vertical tension with the masking tape. Feel free to call if you have questions.

CHAPTER 12:
TRANSMISSION LINE DESIGN

ESLs can't produce linear bass and high output. The solution to this problem is to avoid using them in the bass. By now, you should be saying that in your sleep. But if you aren't going to use ESLs in the bass, what *are* you going to use?

Like ESLs that are fundamentally better than magnetic drivers at high frequencies, magnetic woofers are naturally better than ESLs at low frequencies. Despite an ESL's inherent potential, you have to design them well to reach it. The same is true for magnetic woofers, since it's hard to make one provide great bass. Yet superb bass is the key to the ESL equation. An ESL without bass is not much more than a very clear transistor radio. Making woofers produce bass like ESLs produce treble is a true challenge.

This book is about ESLs, not magnetic speakers. But a book on ESLs is not complete if it doesn't tell you how to build magnetic woofer systems which will really make your ESLs perform. This chapter fills that void, but is necessarily brief—a comprehensive book on magnetic systems would be voluminous. It concentrates on transmission-line (TL) woofer systems, because I've found they work best with ESLs.

Before diving into the theory and construction of TLs, an issue must be settled: is it possible for a hybrid ESL/TL system to sound as clear and detailed as a full-range ESL?

Audiophiles argue this point *ad nauseam*. I wasn't sure, so I set up a test to find out. I matched a full-range ESL against a hybrid system, and asked a panel of "golden ear" listeners if they could hear any *difference* between the two. Note carefully that I didn't say we tried to hear which one was better. All I wanted to find out was whether a hybrid system could sound the same as a full-range ESL.

We tested several different types of woofer systems, including bass reflex, horns, transmission lines, infinite baffles, and acoustic suspension types. This type of testing demands rigorous control of all variables, as I explain in the chapter on testing. Suffice to say here that the output levels for the two systems were identical, as were the frequency response, amplifiers, source material, connecting wires, and other variables. The only variable which I allowed to affect the sound was woofer presence.

And the results? The panel detected differences between the systems—except when transmission lines and horns were used. The horn and TL woofers had as much clarity and detail as the full-range ESL.

Several listeners commented that they found the results hard to believe. One even concluded that I rigged the test. His bias was so strong he could not accept that a hybrid system could be as good as a full-range ESL, although he heard it with his own ears.

The results proved beyond a doubt that a hybrid system can equal an ESL in the critical area of clarity and detail, while simultaneously providing the deep bass and high output of which ESLs are incapable. It's not easy to get a magnetic woofer system to perform this well, but it is possible with transmission line and horn technology. Why this is so is the subject of the next section.

WOOFER THEORY. Magnetic woofer systems suffer from many of the same problems as ESLs. Phase cancellation in particular is an even greater problem.

Magnetic woofers are smaller than electrostatic woofers. Recall that when the length of the sound wave reaches about a quarter of the minimum dimension of a driver, the output falls. Even a moderately large magnetic woofer 10" in diameter will start to roll off the bass at around 5kHz.

Phase cancellation beginning this early causes severe loss of bass output. The problem is so serious that the only practical way of dealing with it is to put the woofer in an enclosure, which isolates the front wave of the woofer from the back so that phase cancellation cannot occur.

Enclosures introduce many problems, the most significant being resonances. You must eliminate, or at least minimize, resonances for high quality sound, which is difficult to do and requires careful attention when designing enclosures.

Magnetic woofers are small compared to ESLs and operate at very low frequencies. Because of their small size, they experience another problem: loss of bass due to the falling

radiation resistance of air. Radiation resistance is the resistance air has to being pushed.

Air behaves differently at different frequencies. At high audio frequencies, it behaves like a soft solid—somewhat like Jello. As the frequency falls, it behaves progressively more like a liquid. At very low frequencies, it behaves like a gas.

The speaker must have something to push against to make sound. At high frequencies, the air presents a high resistance to being pushed because of its inertia and general stiffness. At lower frequencies, there isn't as much resistance because the air has less inertia. It acts more like a liquid that gets out of the way of the woofer. It has less resistance, so the output falls.

A useful analogy is to substitute a hammer for the speaker. Hitting a wall (something stiff with much resistance) with the hammer produces loud sound. The same hammer stroke striking a puddle of water makes very little sound. If the hammer just hits air, there is no sound. Falling radiation resistance is a serious problem below 100Hz and gets progressively worse with decreasing frequency.

Like ESLs, the output from magnetic woofers falls dramatically below fundamental resonance. Unlike ESLs, the better magnetic woofers have very soft suspension systems and are much more massive. Therefore, their fundamental resonances may be very low. Often, the fundamental resonance is so low that the output loss below it is not a problem, as in ESLs.

Because magnetic woofers are small and massive compared to ESLs, they couple poorly with air. Their fundamental resonance is not affected by the mass of the air seen by the speaker, as in ESLs. In summary, the three factors which limit low-frequency performance in magnetic woofers are:

- Phase cancellation
- Radiation resistance
- Fundamental resonance

The usual solutions to these problems are:

- Enclosures
- Increased bass output at very low frequencies
- Low fundamental resonance

You can solve the radiation resistance problem by increasing the driver's output to compensate for the air's falling radiation resistance, which is no different than dealing with phase cancellation in a dipole ESL. The speaker's output must be boosted at the frequencies where the falling radiation resistance reduces output.

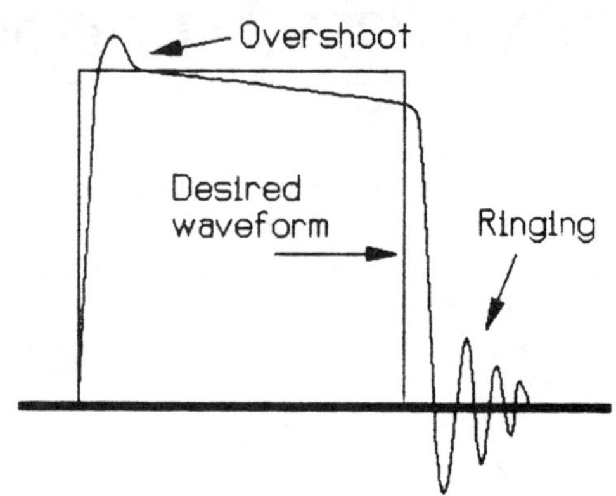

FIGURE 12-1: Overshoot and ringing.

You can get this output boost with equalization before the power amplifier, as in an ESL, but this is a poor technique in magnetic woofers for several reasons. Equalization forces a woofer's magnetic motor to push the cone harder. This accelerates the cone faster and increases output, but the electrical drive signal does not *stop* the cone.

An equalized magnetic driver *overshoots* the desired waveform and *rings* after voltage stops changing. **Ringing** refers to the production of sound from small vibrations of the cone that continue after the music stops (*Fig. 12-1*).

By comparison, the air mass damps (stops) an ESL's diaphragm, so equalization works very well with them. But equalizing a magnetic woofer exaggerates overshoot and ringing. A better method of dealing with the problem of falling radiation resistance is with clever enclosure designs which add output at lower frequencies.

ENCLOSURES. Enclosures for magnetic drivers usually address two problems: first, to stop phase cancellation; second, to increase deep bass output to compensate for radiation resistance losses.

Stopping phase cancellation is easy, since all enclosures isolate the front and rear waves. Designing an enclosure which increases the output in the deep bass without introducing other problems is a challenge. Engineers have designed many such enclosures. Not surprisingly, they all have faults that require compromise.

No perfect or ideal enclosure exists for magnetic drivers. Consequently, there are many types in use, and there are strong opinions regarding which one is best.

What follows is an overview of some common magnetic-woofer enclosure designs. This overview is only a general outline of the concepts involved with each design, and is not a comprehensive discussion. I've omitted details and present only generalizations.

Infinite-baffle or **closed-box** enclosures are the simplest enclosure designs. They consist of a magnetic woofer mounted in a closed box (*Fig. 12-2*).

Except for fundamental resonance, these designs do not make any special effort to support the deep bass, but only stop phase cancellation. They have a single fundamental resonance whose amplitude and frequency can be influenced by the box dimensions.

The fundamental resonance in an infinite-baffle system has a narrow bandwidth. Unfortunately, the decreasing radiation resistance of air is a gradual phenomenon and behaves much like the roll-off caused by phase cancellation. The narrow bandwidth of a single resonance fails to produce the necessary mirror-image frequency response to compensate accurately for radiation resistance losses.

Figure 12-3 shows the generic frequency response of a woofer experiencing radiation resistance losses. *Figure 12-4* shows this same generic woofer's fundamental resonance. You can see that they are a very poor mirror-image match. Mixing them produces a frequency response like that shown in *Fig. 12-5*.

Infinite-baffle enclosures usually do not produce truly deep bass unless they are enormous. The deep bass gets some help from the system's fundamental resonance, but the response is neither deep nor linear. Infinite-baffle woofer systems using large drivers and having deep bass response are very large. This

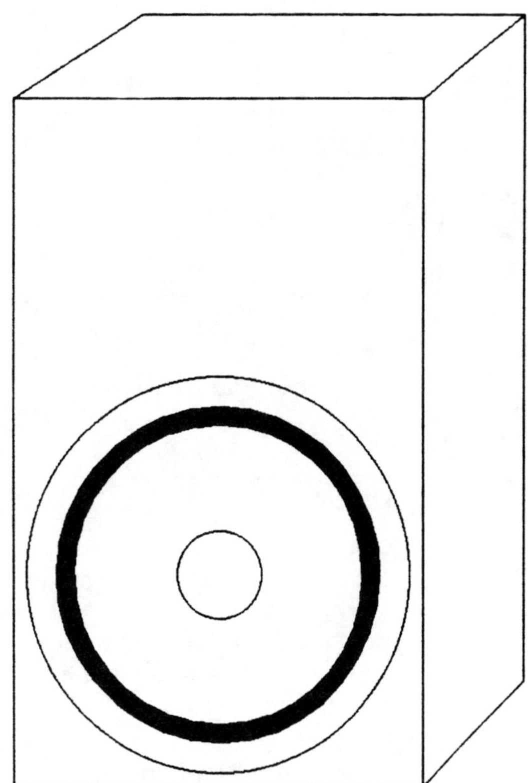

FIGURE 12-2: Closed box enclosure.

has limited their popularity among current commercial designs.

Rectangular box-type enclosures produce severe resonances. They have three pairs of parallel walls (one for each dimension). When the wavelength of the sound wave is equal to the distance between the walls, it produces strong resonances. Less intense harmonics of these resonances are also present.

These resonances vibrate the air in the

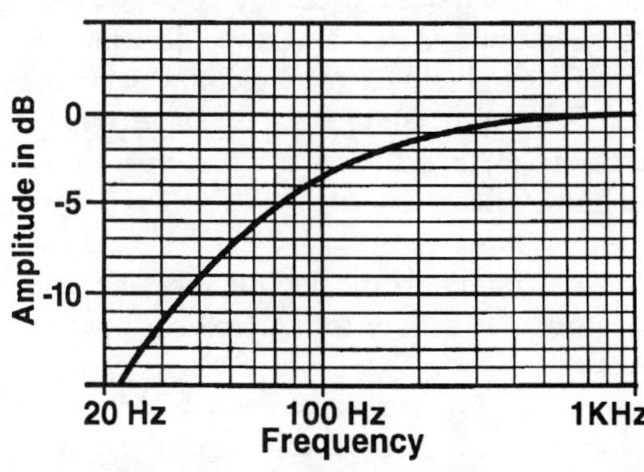

FIGURE 12-3: Poor low frequency response caused by falling radiation resistance.

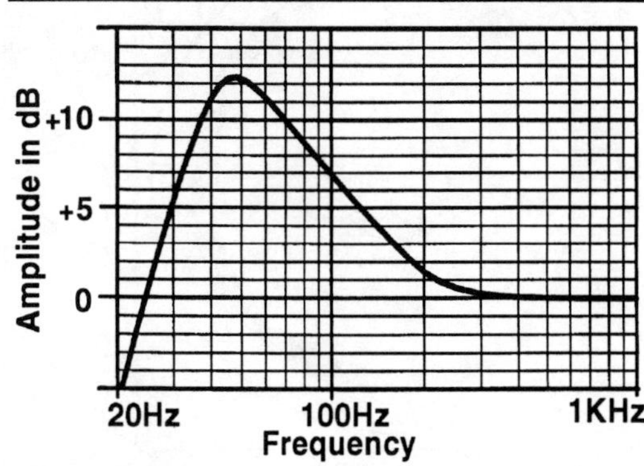

FIGURE 12-4: Poor low frequency response caused by fundamental resonance.

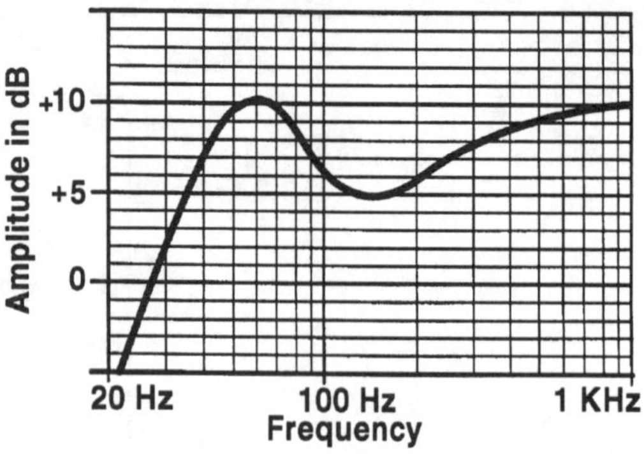

FIGURE 12-5: Poor low frequency response caused by combination of radiation resistance and resonance.

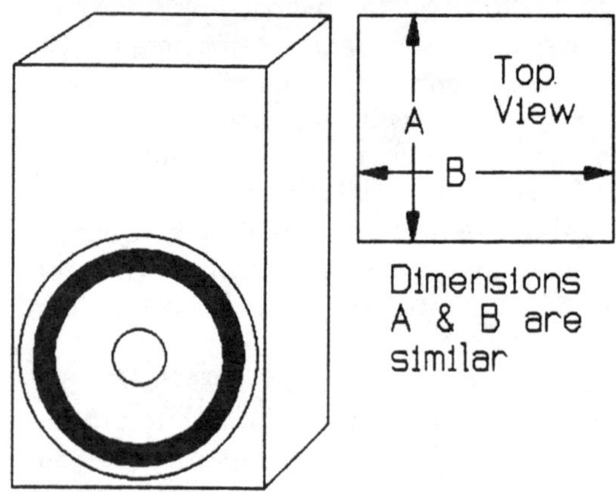

FIGURE 12-6: Similar dimensions cause large resonances.

room by flexing the walls of the enclosure, or by escaping through the woofer cone. This sound from the enclosure is not supposed to be there—you should only hear the output from the driver—and adversely affects the frequency response, detail, and clarity of the woofer system.

Resonances are worse when all three dimensions of the enclosure are the same. For

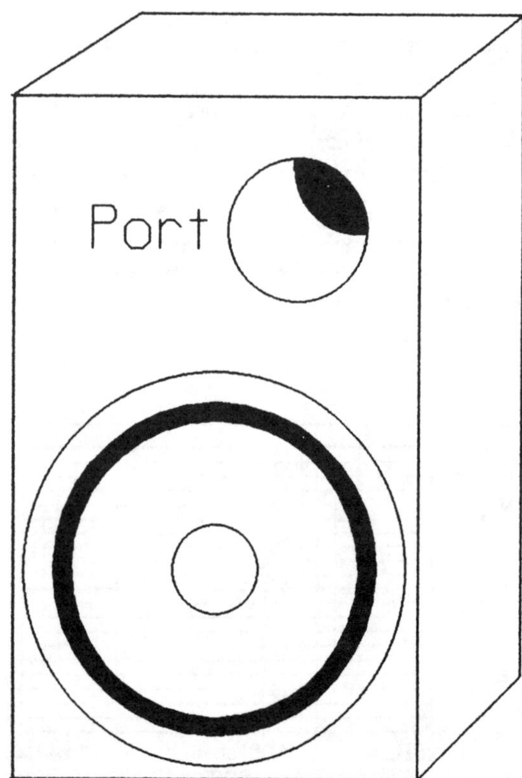

FIGURE 12-7: Vented box (bass reflex) enclosure.

example, a cube or sphere would have one huge resonance and make a very poor enclosure. Note that most commercial speakers have two very similar dimensions, as shown in *Fig. 12-6.*

You can improve a rectangular box by spacing its dimensions 1/3-octave apart. This would produce three small rather than one large resonance and reduce their magnitude. Many small resonances are subjectively less objectionable than one or two large ones.

You can continue the resonance-making process while reducing their magnitude by making the enclosure walls nonparallel. If each dimension covered 1/3-octave, and the walls were nonparallel by a full 1/3-octave, an infinite number of tiny resonances would result.

If you absorb these smaller resonances with acoustic damping material and confine them by very rigid walls, there will be little enclosure resonance to color the sound. Unfortunately, such an enclosure looks strange and is hard to build. It would be aesthetically unacceptable to most homeowners, and too difficult and expensive to build for most commercial manufacturers. Most commercial loudspeaker systems are highly resonant and don't perform to their full potential.

Acoustic-suspension woofer systems are similar to infinite-baffle designs, differing mainly in the way that restoring force is applied to the driver.

In an infinite-baffle system, a relatively stiff spider-and-cone surround restores the cone to its center position. In acoustic-suspension systems, the air in the box acts as a spring to return the cone to its neutral position. The advantage of acoustic suspension is that the woofer's surround and spider may be very compliant. This reduces the spring rate of the system and the fundamental resonance frequency, and allows greater cone excursion.

The long cone excursions made possible by this design allow small drivers to produce high output. The acoustic-suspension design is very popular commercially because it can produce reasonably deep bass and high outputs in a small package.

Acoustic-suspension designs suffer from the same sonic defects as infinite-baffle systems, plus some. The major new problem is inefficiency. The driver must compress the air in the enclosure, which uses a lot of energy. You end up with a small, inexpensive speaker which requires a large, expensive amplifier to drive it to high output levels. The high forces produced by the voice coil when compressing the air cause a large amount of cone flexure. Cone flexure is distortion.

Bass-reflex enclosures (also known as vented-box or ported enclosures) are similar to infinite-baffle enclosures except for a hole which allows air to pass into or out of the box (*Fig. 12-7*). This hole is correctly called a **port**, and is usually made as a tube of carefully selected diameter and length. You'll see why shortly.

The basic idea behind a ported enclosure is to split the single, large fundamental resonance of the infinite-baffle designs into two smaller resonances. These two resonances straddle what would have been the single resonance.

These resonances are produced by a complex set of interactions involving the driver mass, its spring rate, the enclosure dimensions, the air mass within it, and the length and diameter of the port. Its length and diameter are tuned to the size of the enclosure and the mass, spring rate, and other driver characteristics. This tuning process, called **alignment**, defines the magnitude, bandwidth, and resonant frequency of the enclosure.

Only the lower of the two resonances serves the purpose of generating support for the deep bass. The upper resonance often produces nonlinear frequency response in the midbass.

By having two close resonances of different magnitudes, the bandwidth is wider than supplied by a single resonance. This wider bandwidth more accurately produces the mirror-image frequency response needed to compensate for falling radiation resistance. If these resonances have appropriate spacing, location, magnitude, and bandwidth, the bass-reflex enclosure can do a reasonable job of supporting the deep bass.

Designing a bass-reflex system which meets all the above criteria is nearly impossible. Not only do the criteria conflict with each other, but producing the necessary resonant behavior involves very complex interactions with multiple factors.

The complexity of bass-reflex systems prevents trial-and-error design. Computers have become powerful tools in managing the interactions of these systems. However, computers rely on accurate knowledge of many of the driver's physical parameters. These may be difficult to obtain and are prone to inaccuracies caused by variations in individual drivers.

Bass-reflex enclosures have resonances other than those deliberately designed into the system; they have all the enclosure resonances I outlined under infinite-baffle systems. Additionally, vented systems are vulnerable to transient ringing if the design is misaligned.

These systems must have large internal volumes to reproduce deep bass. When designed for use at these frequencies and with large drivers, they are very large.

In summary, bass-reflex systems have higher efficiency, greater output, deeper frequency response, and lower distortion than most closed-box designs. Unfortunately, the large size and complex design process limits their potential.

Passive-radiator systems are modified

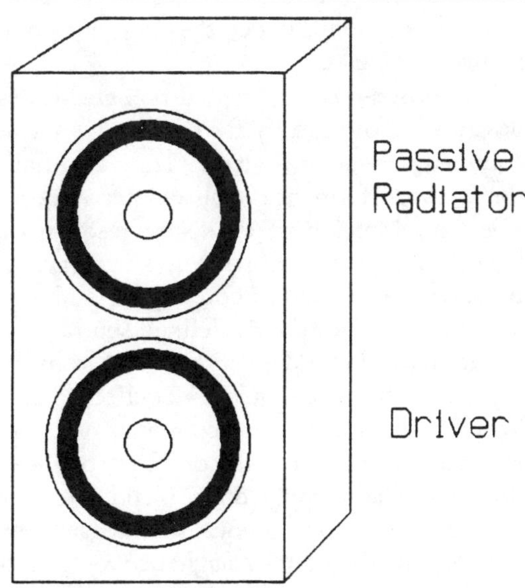

FIGURE 12-8: Passive radiator.

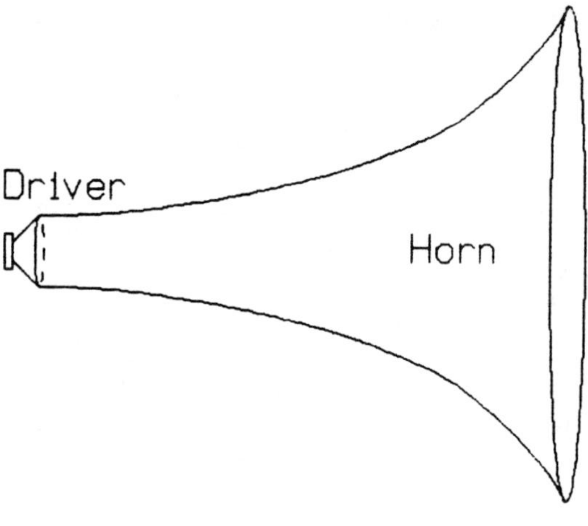

FIGURE 12-9: Horn.

bass-reflex systems (*Fig. 12-8*). The term "passive radiator" is a misnomer, as it is not a radiator of sound. It does not add energy to the system, but is essentially a device that adds mass to a bass-reflex resonant system.

This addition makes it possible to reduce the enclosure size for a given resonant frequency, and also makes it possible to dispense with the tube in a tuned-port system. This, in turn, eliminates air pumping and resonances associated with the tuned-port system.

They not only suffer from the same problems inherent in all bass-reflex systems, they also cost more than a simple port. The manufacturer must carefully control the passive-radiator's tolerances in order to minimize sample-to-sample variability. Other than reduced size, they have little to offer over the basic bass-reflex system, hence they are not commercially popular.

The closed-box and vented-box enclosures described above are by far the most popular commercial woofer systems. They also have the worst performance of all woofer systems. Their combination of enclosure resonances, large fundamental resonances, harmonics of these resonances, overshoot, ringing, and high distortion produce poorly defined sound.

Such flawed sound is very different from the pristine clarity and exquisite detail of an ESL. When these systems are combined with an ESL, there is an obvious discontinuity between the two. They simply don't blend into one smooth, clear, homogenous sound. Is it any wonder that hybrid ESL/magnetic woofer systems have a poor reputation when compared to an all-ESL system?

It need not be this way. Magnetic woofer systems exist which are linear, low in distortion, free of resonances, and support the deep bass. If you use a hybrid system, it will have the same sound quality as a full-range ESL, while simultaneously producing deep bass and high output. There are two types of enclosures which can do this: Horns and Transmission Lines.

Horn enclosures are just what their name implies: by putting the driver in a horn, they effectively couple its relatively small driven area with the air (*Fig. 12-9*). This results in very high efficiency, which not only reduces cone flexure and distortion but greatly reduces amplifier power requirements.

Radiation resistance losses are reduced because the large air mass seen by the horn offers high radiation resistance. This pushes down the frequency where the radiation resistance losses start. If the horn is large enough, deep linear bass can be produced at astonishing output levels.

The operative word here is *large*. You can't beat the laws of physics. The mouth of a horn system must be as large as the wavelength you want to reproduce linearly.

The length of a 32Hz sound wave is about 32'. For a horn to reproduce this linearly requires the horn's mouth be 32-feet wide. Such horn systems have been built into large public buildings. Altec's famous "Voice of the Theater" horn woofer systems require a theater two stories tall for installation. Amateur speaker builders have built such monstrosities. I've seen concrete horns built into the sides of houses, where the horns reside outside in the yard.

Designers have tried to get around the size problem. The most famous of these is the Klipschorn, which attempts to use the intersection of two walls and the listening room floor as part of the horn. This is a good idea, but in practice leaves much to be desired.

Other than their size, horns have few problems. Resonances can exist in the enclosure on the woofer cone side opposite the horn. Careful design can minimize them. The system is so efficient that very little driver excursion is needed, and any resonances will be small. Distortion in the horn's throat is a major problem at high

FIGURE 12-10: Transmission line.

frequencies, but at low frequencies it can be minimized by careful design.

Horn woofer systems, while theoretically excellent, are simply too large to be practical. Additionally, horn design and construction is a formidable task. Regrettably, these problems preclude their use in practical home music systems.

Transmission-line enclosures consist of a long tapered tube behind the driver (*Fig. 12-10*). This tube is called a transmission line or just "line." Its careful design can solve most woofer enclosure problems.

Though the line is open at the end, it is so long it eliminates phase cancellation. A long line has an extended path from one side of the woofer to the other. There will not be enough time for the air to travel this path and cause phase cancellation, which is the same as using a very large, flat baffle around the woofer. In this respect, the transmission line behaves like an infinite-baffle enclosure.

An important difference between an infinite baffle and a transmission line is you can virtually eliminate resonances in the latter. Of the three dimensions, one will be down the length of the line. Since the line is open on the end, there is no wall which can reflect the sound and cause a resonance.

The line walls define the other dimensions. This could cause one or two large resonances, particularly if the cross section of the line is square. You can stop these resonances by making the line walls nonparallel. In other words, taper the line.

If the walls aren't parallel, the line can't generate the large resonances characteristic of and damaging to the performance of rectangular enclosures. Instead, it has a large number of very small ones. By adding damping material to the line, we can absorb these, and for practical purposes have a nonresonant enclosure.

You can see how a tapered transmission line solves the problems of phase cancellation and enclosure resonances. It would be perfect if it also could support the deep bass which is lost to falling radiation resistance, and it can.

It does so by shifting the rearwave phase so it exits the line in-phase with the frontwave. The rearwave is normally 180° out-of-phase with the front, and therefore cancels the output of the frontwave. If the enclosure delays the rear wave by an additional 180°, it will emerge in-phase with the frontwave and add to its output. By carefully selecting the frequencies affected by the phase shift, we can support the deep bass.

Also highly desirable is the fact that phase shift produces a broadband increase in output instead of a narrow peak typical of resonances. This more accurately matches the falling output caused by decreasing radiation resistance. Because the line is one length, not all frequencies experience the same degree of shift. The phase shift will be the maximum 180° at only one frequency. On either side of this frequency the shift will be either more or less than this, and there will be less reinforcement of the sound.

A transmission line also exerts a powerful effect on the woofer's fundamental resonance. It heavily damps it, which smooths the frequency response and reduces cone excursion. Controlling this and other resonances results in well-controlled transient behavior. Ringing and overshoot are minimal and much better than most other enclosure designs.

The line is stuffed with damping material. I don't mean that there is damping material attached to the walls—rather the line is completely full of damping material. This both absorbs and slows sound wave energy.

Technically, damping material interferes with the motion of air molecules in the line. This friction depletes the sound wave energy by converting it to heat. The damping material has several important functions:

- It slows the passage of sound through the line. This reduces the line length required to produce the desired phase shift. You can build the line about 30% shorter than would otherwise be possible.
- It works with the line to suppress the driver's fundamental resonance.
- It absorbs the sound energy radiated from the rear of the cone. This includes not only the energy coming directly from the cone, but also all the tiny enclosure resonances that remain after tapering the tube. The net result is almost magical: the driver's rear wave radiation and associated resonances virtually disappear.

If we make the line long enough, all the sound energy would be absorbed. However, we deliberately make the line length short enough so the damping material cannot completely absorb the deep bass. We want it to release sound energy from the line when the phase has been shifted enough to support the deep bass.

Transmission lines are moderately efficient. While not as efficient as horn systems, they are more efficient than acoustic-suspension designs. Their efficiency is similar to bass-reflex and large infinite-baffle systems.

SUMMARY OF TL CHARACTERISTICS.

- Little or no cabinet resonances
- Well-controlled driver resonance
- High output
- Flat frequency response
- Extended deep-bass response
- Excellent transient behavior
- Low distortion
- Moderate efficiency

Compare these same criteria with an acoustic-suspension system:

- Marked cabinet resonances
- Poorly controlled driver resonances
- Moderate output
- Nonlinear frequency response
- Poor deep-bass response
- Variable transient behavior depending upon system "Q"
- High distortion
- Poor efficiency

Which of these two profiles best matches an ESL? It should be clear why closed-box and vented systems are not good choices for use with ESLs.

Transmission lines have their disadvantages, the main ones being size and construction complexity. The line will generally need to be 6–12-feet long. While you can fold it into a reasonably-sized enclosure, it is still no bookshelf system even when used with a small driver. Additionally, the internal baffling necessary to produce a folded transmission line is more trouble to build than a simple box. This is not an insurmountable problem for the home builder, but is a major cost problem for manufacturers of moderately-priced commercial speaker systems.

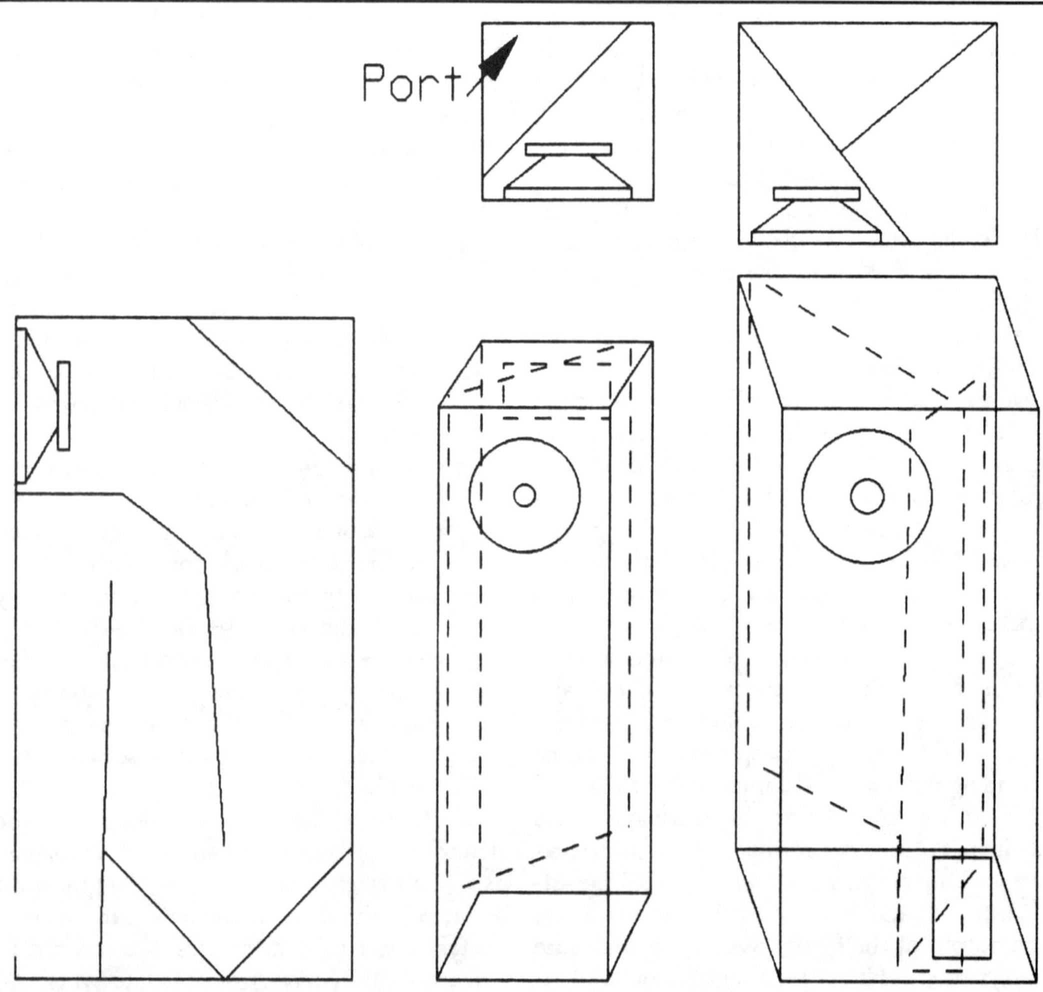

FIGURE 12-11: Various TL designs/layouts.

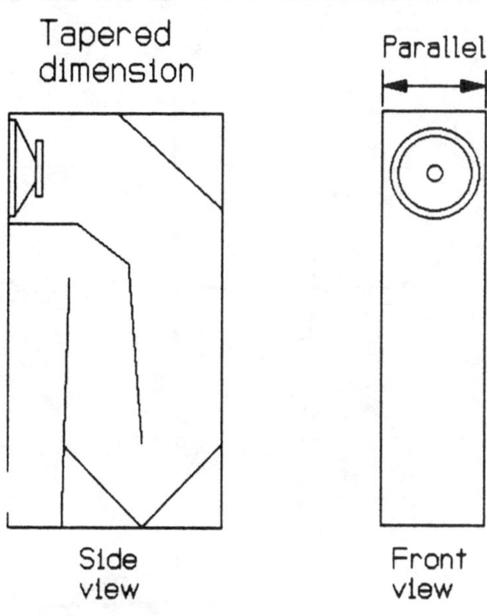

FIGURE 12-12: Rectangular enclosure has parallel walls.

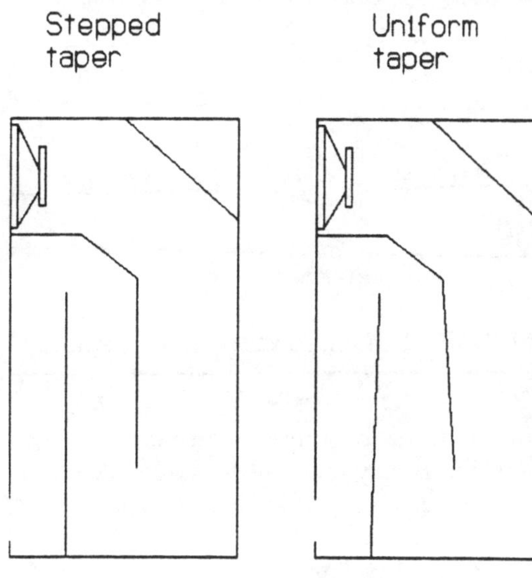

FIGURE 12-13: Angling baffles converts stepped taper into continuous one.

TL DESIGN GUIDELINES. Unlike resonant and horn-enclosure designs, transmission lines are simple. You don't need complex mathematical formulas, computer programs, or trial-and-error tests to design them.

I'll discuss each design parameter in detail and follow with my usual guidelines. You can be confident the guidelines will produce a speaker which performs well.

LINE LENGTH. The line length partially determines the deep-bass response. The standard recommendation for line length is 25% of the wavelength of the driver's fundamental resonance. I think this is a useful guideline, but there is nothing sacred about it. I usually use longer lines, because up to a point the longer the line the deeper the bass.

The line length in most commercial TLs is about 6′, corresponding to one-quarter wavelength at 45Hz. Amateurs sometimes use lengths up to 10′ for very large drivers, which is one-quarter wavelength of 27Hz.

A problem with short lines (6′ or less) is the woofer decouples from the line below resonance and can experience excess excursion. Because of this, and the need for deep bass, your line should be more than 6′.

Unlike resonant enclosures, transmission lines are not tuned, although this term is sometimes applied. Practical experience proves the length isn't at all critical, and that a length between 7–10′ gives excellent results in most systems.

The length is usually a little shorter for smaller drivers, because they generally have higher resonances and greater problems with radiation resistance. The opposite is true for larger woofers.

TL CROSS-SECTION AREA. The commonly accepted rules for line cross-section area are:
- The area behind the woofer should be at least 125% of the woofer's driven area.
- The port should be 100% of the woofer's driven area.

These are good, solid, reliable rules. I call this 125/100 rule the classical TL, and it works very well. But other percentages also work.

I've had equally good results with a 100/70 line. With these more compact lines, you must reduce the amount of damping material to maintain the same performance as in wider lines. The only penalty I've found in narrow lines is the effect of the damping material is more pronounced. This makes the stuffing procedure more critical.

LINE SHAPE. Effective TLs are long. While a 10′ tapered tube protruding from the rear of your woofers might make a splendid conversation piece, in most homes it is neither practical nor aesthetically acceptable.

Transmission lines must be folded or designed to fit into the home environment. Although the integrated, hybrid systems por-

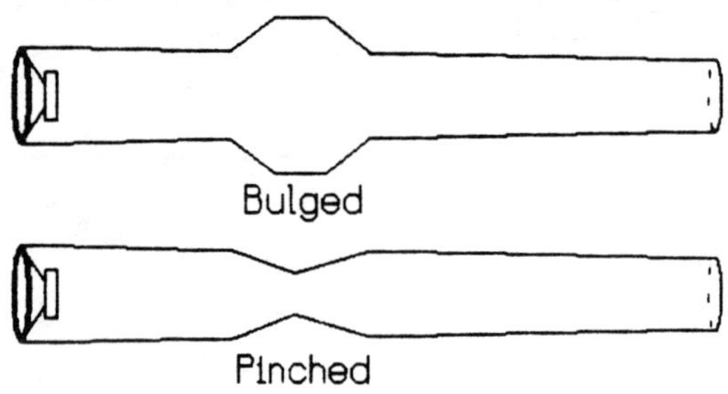

FIGURE 12-14: TL flaws shown conceptually.

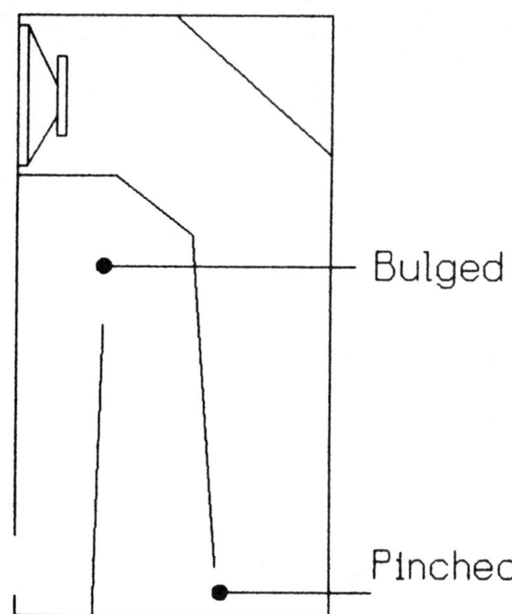

FIGURE 12-15: TL flaws shown in enclosure design.

trayed in this book use an essentially straight line, this is very unusual. Most are folded.

Three elements define line shape:

- Cross section
- Axial folding
- Taper

Fortunately, air is very forgiving of line shape. It doesn't seem to care how it is confined. As long as the cross section is reasonably uniform, any shape works well.

Transmission lines can be folded in any way imaginable, and there have been some very imaginative designs. *Figure 12-11* outlines some possibilities. Your ingenuity is the only limit.

Attaining satisfactory taper should be a major consideration, but this item is overlooked by many designers. Note carefully the types and uniformity of the enclosure tapers in *Fig. 12-11*. They vary from fully tapered to a single step.

It is best to taper *all* dimensions smoothly to minimize resonances. Try to avoid steps; use continuous tapers. In purely rectangular enclosures, this is usually not possible. One dimension will be parallel (*Fig. 12-12*). Despite this compromise, rectangular enclosures work well.

Many designs have steps in the cross section instead of being uniformly tapered. While this works, it is not as resonance-free as a uniformly smooth taper. Often it is easy to change a TL with steps into uniform taper (*Fig. 12-13*).

Finally, pinched and bulged lines should be avoided. *Figure 12-14* shows conceptual exam-

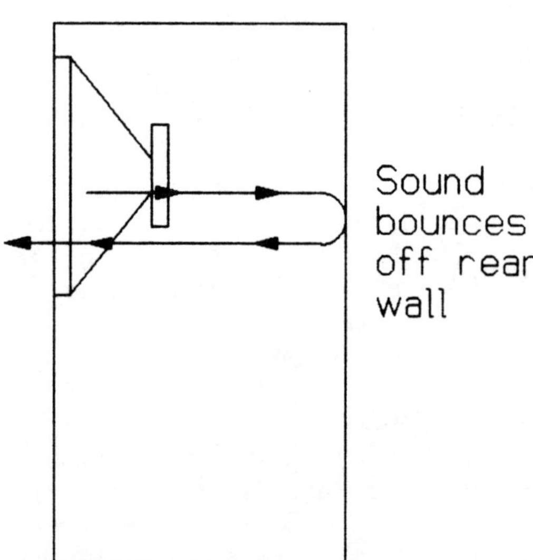

FIGURE 12-16: Sound from rear of woofer reflects off rear wall and radiates through driver.

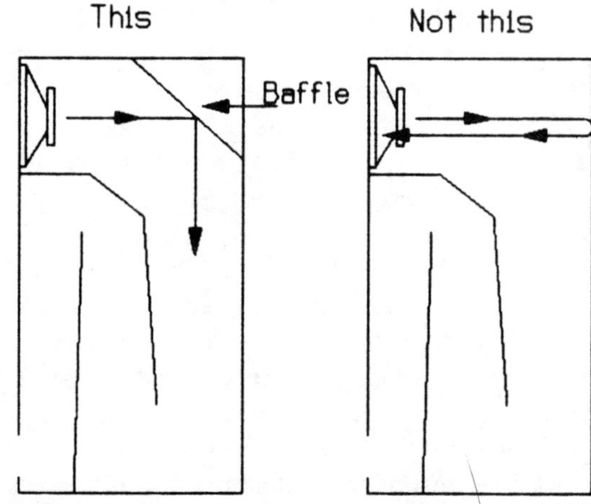

FIGURE 12-17: Baffle corrects rear wall radiation problem.

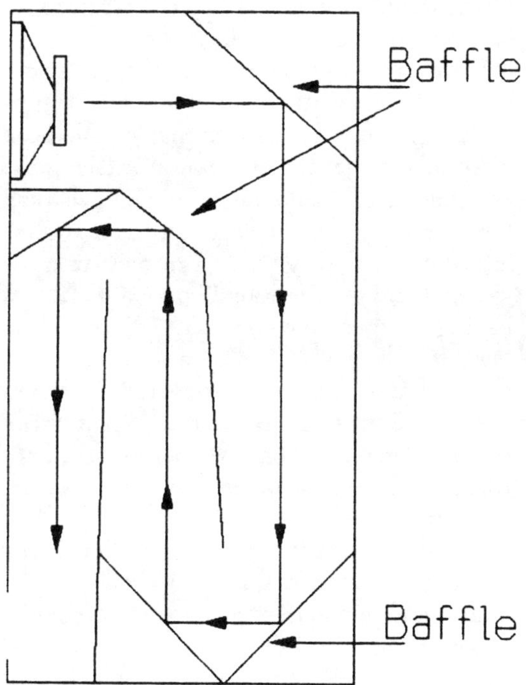

FIGURE 12-18: Baffle corners throughout TL.

GUIDELINE #14:

Magnetic Drivers

1. Use the largest drivers possible consistent with other design parameters.

2. Do **not** use drivers which have hygroscopic (water absorbing) cones, because they change stiffness and mass with humidity. In short, don't use paper-cone drivers. Bextrene, carbon fiber, Kevlar, and other cone materials are better.

3. The frequency response of your driver should extend **linearly** for two octaves above your crossover frequency. If the manufacturer can't supply frequency response curves, find one who will.

4. Buy the highest quality drivers that meet the above criteria. What constitutes high quality? Some clues are: cast magnesium instead of stamped-sheet steel baskets; hexagonal or ribbon voice coil wire instead of round wire; large spiders; synthetic instead of paper cones; flat frequency response; first- class fit and finish.

5. Pay whatever you must to meet the above requirements.

ples of both types. *Figure 12-15* shows these concepts reproduced in real enclosure designs. Like taper, correcting for bulges and pinches is often very simple. Usually, just being aware of the problem breeds solutions.

BAFFLING. One of the major problems with most enclosures is a flat wall directly behind and parallel with the driver. Sound radiated from the rear of the driver strikes this wall and is reflected back out through the cone (*Fig. 12-16*).

In most rectangular-enclosure designs, you can't fix this problem. In TLs, you can, and it's important to do so. The idea is simple: just put an angled baffle behind the woofer (*Fig. 12-17*).

Watch for the other corners as well. In *Fig. 12-18* note that all the corners have baffles.

DAMPING MATERIAL. Many products make good damping material. Common materials include, but are not limited to, polyester (Dacron), fiberglass, wool, and foam. Many builders have strong opinions about damping material. In my experience, they all work well if you get the density right.

You must compensate for different materials by slightly different packing densities. For example, long-fiber wool is the most effective damping material of which I am aware. Under the microscope, its fibers are fuzzy. In comparison, Dacron has smooth fibers and is less

effective as a damping material. Is this a problem? No, just use more Dacron to match wool's performance. Both will do the job.

WOOL. Although natural long-fiber wool is widely believed to have the best damping qualities, it has several problems: it settles; you must support it every 8–12" with dowels or nylon webbing to hold it in position; and, given the opportunity, moths eat it. You must mix it with moth flakes during installation and cover the port with screen or grille cloth to prevent moth entry. It's expensive and not easy to find.

DACRON. Polyester fluff does not have these problems. It's cheap, readily available (as pillow stuffing), doesn't require support webbing, and moths hate it. The preferred form is "stuffing" rather than "batting."

Also, I've noticed that Dacron comes in fine

and medium grades. Salespeople are usually unaware of the two types. If you have a choice, look around and pick the finer grade. This will have smaller fibers which will do a slightly better job of interfering with air movement.

DAMPING MATERIAL PACKING DENSITY. I can precisely define all transmission-line design parameters save one: packing density. While not critical, if the density is way off, performance will suffer.

Some general guidelines exist which will lead you in the right direction. Just be aware that you can do some fine tuning to perfect the sound if necessary.

In the classical 125/100 line, wool is packed at about ½ lb./ft.^{3} of line volume. You will discover that this is not very much wool. You don't actually pack it into the line, you gently put it in the line.

SCREWS
Never drive a screw into particleboard
without first drilling a hole for it.
It will break the wood apart.

Most builders use a constant-impedance line, although some use variable-impedance damping. The line volume is greater at the woofer than at the port. Therefore, to make the line constant impedance, pack the damping material slightly tighter behind the woofer than at the port.

This is not an all-or-nothing procedure, of course. You gradually pack the line more loosely as you move from the woofer to the port, and it's not necessary to put damping material in the last foot or two.

Using wool and a full-size classical line as a reference, other materials and line sizes require slight modifications. Since wool is the most effective damping material, others need to be packed slightly tighter. Exactly how much is difficult to describe. Suffice to say that the difference is slight, probably around 10%.

Lines with smaller cross sections than the classical 125/100 use *less* damping material. The smaller the line, the more critical the damping density becomes.

How do you know when the density is correct? It's not hard to tell. If the sound seems short of bass, particularly deep bass, you have too much damping material in the line. Boomy, loose, muddy, or generally poorly-controlled bass requires more damping material.

If you are going to err, err on the side of too little material rather than too much. If it turns out that there is not enough material, you can often adjust the damping without taking the enclosure apart. Just add more stuffing behind the woofer. You can usually reach far enough into the line with your arm to pack in more material and get the sound you want without having to repack the entire line.

DRIVERS. A poor enclosure can destroy a good driver's performance, but a great enclosure can't make a poor driver perform well. It's Murphy's Law at work again.

There's no way around it, you've got to buy the best magnetic woofers available if you want superb sound quality. I won't specify brands and model numbers because technology keeps improving them. By the time you get this book, newer and better drivers will be available. Instead, let me give you some guidelines to aid you in your search.

One final thought: manufacturers often have regular and high-power versions of the same driver. You might assume that the high-power version is better, but this is not likely to be true.

The high-power versions have heavier voice coils, and perhaps other features like heatsinks which increase their mass. This adversely affects the high-frequency response. Good high-frequency response is much more important than power handling. TLs are reasonably efficient and usually don't need high-power drivers to produce high outputs. Therefore, the regular woofer is likely to be the better choice.

TL CONSTRUCTION TIPS. There are unlimited ways to build speaker enclosures. Your imagination is the only limit. The most popular method is ¾" particleboard that you glue and screw together. Because most of you are likely to use this type of construction, I'll give you a few helpful tips to save you some frustration.

SAWS. Most builders use a table saw to cut particleboard, but it's a poor choice. The problem is that a 4' × 8' particleboard is heavy and awkward. A straight cut is nearly impossible to make with a table saw.

A better choice is the common hand-held circular saw. By itself, it won't cut a straight line either. The trick is to use it with a large metal straightedge.

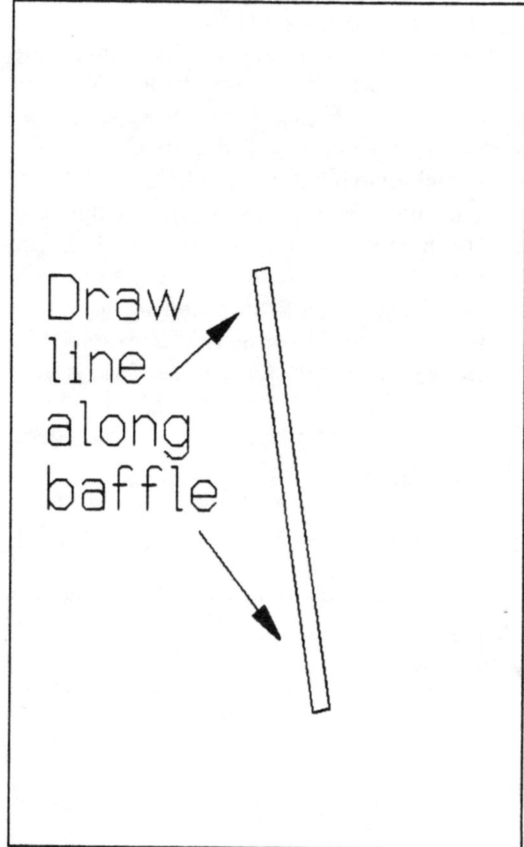

FIGURE 12-19: How to locate screw holes.

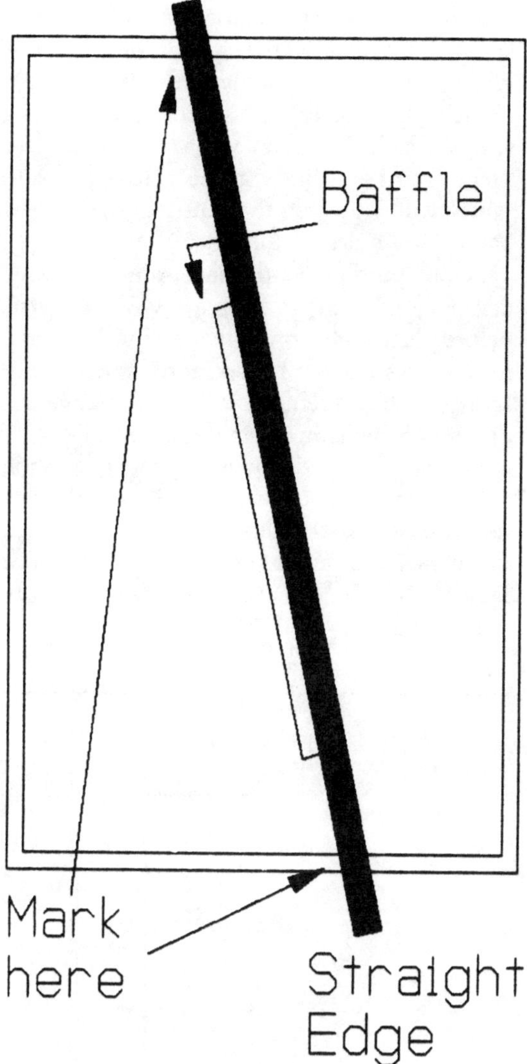

FIGURE 12-20: Identify internal baffle locations on outside edge of enclosure.

Better hardware stores sell aluminum bars 8-feet long. A bar 2-inches wide and ¼-inch thick works well.

To make your cut, take two or more C-clamps and clamp the bar along the cut so your saw's blade is in the correct position. By using the bar to guide the saw, you can make perfectly straight cuts.

SCREWS. Standard wood screws have threads which are too fine to work well in particleboard. It's better to use the extremely-coarse-thread screws called drywall screws, which also have a very deep Phillips head to keep your screwdriver from slipping.

Don't waste your time with slot-head screws. Your screwdriver will slip and cause damage to the particleboard. They require constant adjustment of your screwdriver to drive them, and you can't use a power screwdriver with them effectively. You don't need the aggravation. Use Phillips-head screws.

Place a screw at least every 6″. This means you will drive a surprisingly large number of screws: a power screwdriver is almost essential. If you don't have one, but have a variable-speed drill, you can use it instead. Just get a ¼″ drywall screw bit at the hardware store, mount it in your drill, and go to it. Be careful not to over-tighten the screws and strip them.

If your TL has internal baffling, some tricks will make it easier to build. I find it works best to make the basic box first and assemble everything except one side.

To drill the screw holes in the baffles, put the baffles in the box and draw a line beside each of them (*Fig. 12-19*). Remove the baffles and drill holes 3/8″ to the side of the baffle centerline.

After installing the baffles, you face the

dilemma of drilling the last side. Take a straightedge and lay it along the baffle centerline. Make a mark on the edge of the enclosure which represents the baffle center (*Fig. 12-20*).

After you install the last side, these marks will be visible. You can lay the straightedge on them again, and draw a line across the side which will show you the center of the hidden baffle. Drill holes on this line.

Before putting on the last side, add your damping material. You may wish to wait before putting glue on this side until you have tested the speaker. By mounting it with only screws, you can remove it later to change the density of the damping material.

What glue should you use? White glue is the old standby and still works satisfactorily. But newer glues like the aliphatic resins are better. Of course, you can use epoxy, but the amount required is too expensive when the other glues will do the job.

FINISH WORK. There are many ways you can make particleboard enclosures look attractive. The simplest is the functional finish, which is nothing more than a flat grey or black paint applied with a paint roller.

Before doing any painting, take a putty knife and apply spackle to the screw heads and any other surface irregularities. A large power sander will smooth off the spackle so the edges will match before painting.

Latex paint is very thick, and is the only paint which will hide grain and minor surface imperfections. This feature, and the fact that it's nontoxic, makes it the paint of choice.

Another easy finish method is to wrap the enclosure with cloth. Unless the cloth is very thick, you must paint the enclosure flat black first so you don't see screws and wood through the cloth.

You can get a beautiful and extremely durable finish by covering the enclosure with

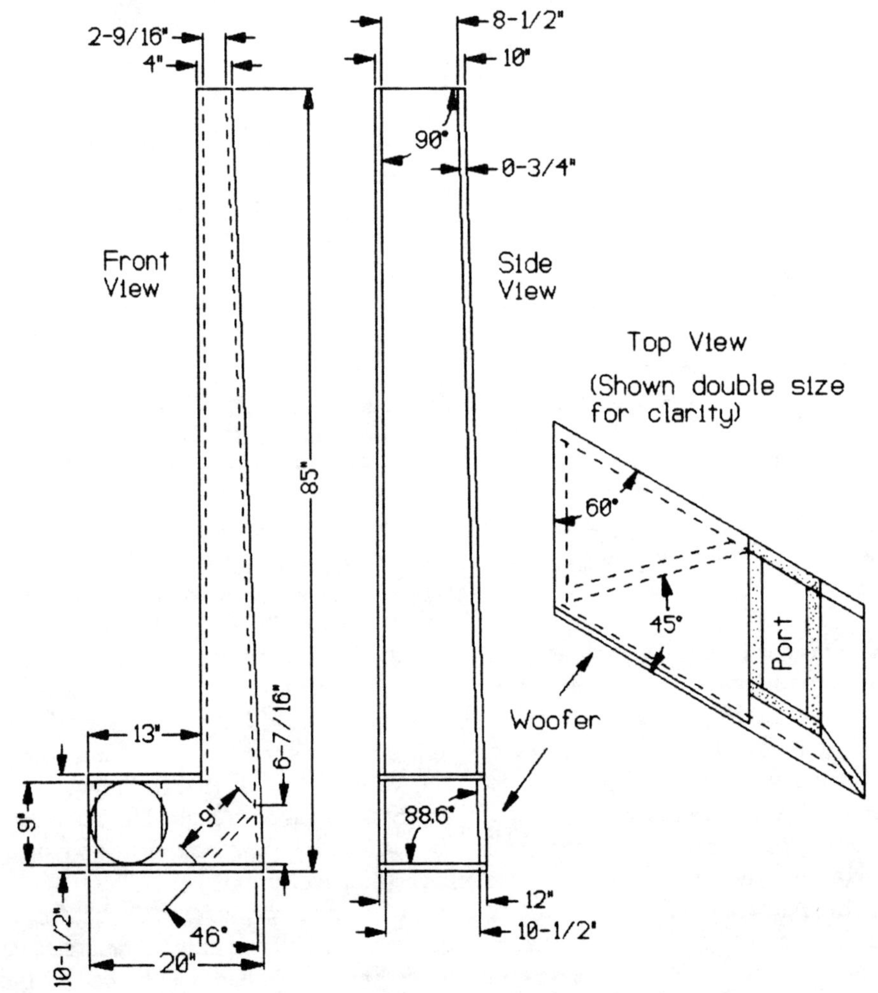

FIGURE 12-21: Compact/Integrated ESL/TL.

Formica or other plastic laminate. These come in almost any pattern and color imaginable. The most popular for speaker enclosures is a wood-grain appearance.

Apply it to the completed enclosure by coating the enclosure and the back of the plastic with contact cement. When the cement has dried, press the Formica to the enclosure and it instantly and permanently bonds to the particleboard. Sand the enclosure so the surface is smooth and the Formica lies flat.

You get only one shot when laminating with contact cement. Be sure you have the plastic in the right position before you push it firmly against the enclosure. Trim the edges of the Formica with a coarse file after you've finished the lamination.

Laminating plastics is not as easy as I make it sound. The material is expensive and there is little room for error. There are special tools for working with these materials that aren't essential, but they sure help. I think that most amateur builders are well-advised to turn this part of their speaker over to a professional.

Finally, you can get particleboard with veneer already applied. I don't recommend its use unless you are a highly-skilled furniture builder. The cutting and drilling is difficult to do without damaging the surface. Even if you build the enclosure without damaging the surface, you still have to laminate veneer along the exposed edges. Finally, you have to put a finish on it.

You can cheat by using an oil finish, but oiled wood finishes aren't durable and dry out. You can reapply the oil to keep it looking good—but who does?

MOUNTING DRIVERS. Consider mounting your drivers on silicone rubber rather than screwing them to your enclosure. When you screw the driver in place, it transfers its vibrations to the enclosure. This vibrates the enclosure and radiates sound into the room. It's better to decouple the woofer from the enclosure in order to minimize this undesired sound.

To mount a driver with silicone rubber, put it face-down and remove any gasket which may be present. Then squeeze a very thick bead of silicone rubber all the way around the driver's mounting surface. This bead should be at least ¼-inch thick, and a ½″ bead is even better.

You should have the enclosure positioned so the woofer cutout is facing upward. Now carefully pick up the driver, and flip it over so it is face-up. Set it carefully in the enclosure

cutout. *Don't press down!* You want a thick pad of silicone under the woofer. If you press down, you will thin the film and increase sound transmission to the enclosure. Also, if you squish the silicone, it may run down the side of the woofer and glue the cone to the frame. Just set the driver in place and let it float on the silicone.

Silicone mounting is semipermanent. In order to remove the driver, you must cut the silicone with a razor blade. For initial testing, you may prefer to mount the driver with screws. After you have everything working to your satisfaction, use silicone.

Pay attention to the connections between the speaker and the hook-up wire. Fixing a failed contact is difficult after you've glued the woofer in place. I recommend you solder the wires to the woofer instead of using plug-in connectors.

SPECIAL CONSTRUCTION METHODS FOR THE COMPACT ESL/TL. The compact-hybrid systems shown later in this book have a very unusual TL design (*Fig. 12-21*). Note from the top view that it is not a rectangle, but a parallelogram. To make this easier to visualize, I have drawn the front and side views as though they formed a rectangular enclosure. When you assemble it, you will cut the parts to the necessary angles to form a parallelogram.

A few comments on its construction will help. Remember to make mirror-image parts for the left and right cabinets. The odd angles require you to make some interesting cuts, but if you pay attention, it is not as difficult as it looks. To make the long vertical cuts at the required angle, clamp a long straightedge to the wood with C-clamps along the cut.

Adjust the foot plate of the circular saw to the required angle. Run the saw along the straightedge to make the cut. Your saw probably will not allow the foot plate to be tilted in the opposite direction to make the mirror-image cuts for the other speaker, so run the saw from the other direction instead.

You will discover that you cannot run the circular saw completely to the right-angle corner without cutting into the face of the enclosure. Solve this problem by cutting close to the corner with the circular saw and finishing the cut freehand with a hand saw. You will have the kerf from the power saw to guide you, so you can easily do an excellent job of finishing the cut.

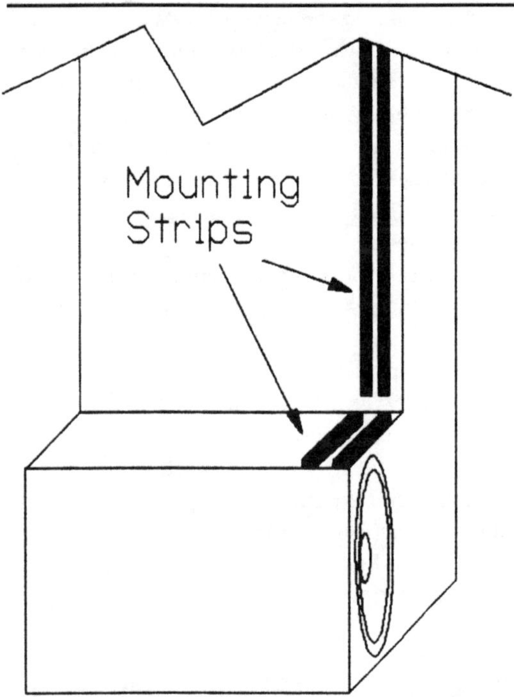

FIGURE 12-22: Use strips of wood to form slot.

Placing the screws at an appropriate angle can be done freehand if you are careful. I find it easiest to drill one screw hole at each end of the enclosure, and assemble the two parts temporarily using only those two screws.

Be gentle to avoid tearing the screws out

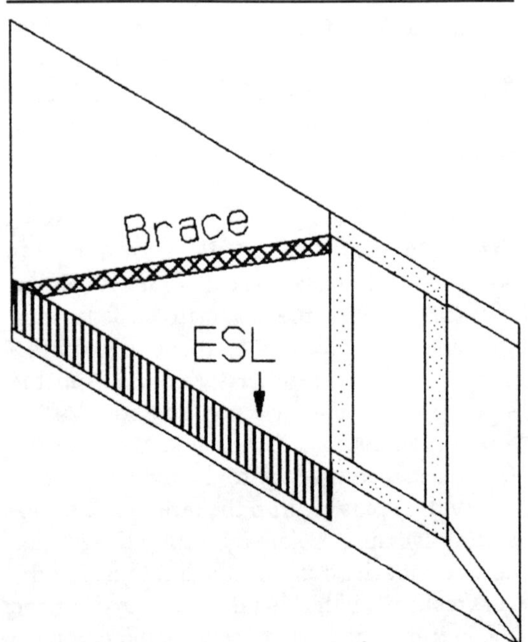

FIGURE 12-23: Aluminum brace to hold ESLs.

while you finish drilling a string of screw holes down the joint. An assistant is essential during this process.

After you drill the holes, disassemble the two parts, apply glue, and reassemble them using all the screws. With all the screws in place, the parts are stable.

You will find it nearly impossible to get perfect joints, as the wood tends to slide due to the angles. It's OK, you can trim up the little overhangs after the glue is dry. A belt sander or plane does the job easily. Don't forget to put in the damping material before you put the last piece in place.

Note that the woofer goes on the side of the enclosure that is tapered. This helps balance the speaker, which tends to be front heavy. Also, it slightly aims the ESLs upward so you are not listening at the joint between the two ESL panels.

Long, tall, skinny TLs like this integrated type are not very stable, particularly on carpet. You can radically improve this with the use of "Tiptoes®."

Tiptoes are steel cones which sit upside down beneath the bottom of the speaker. They penetrate the carpet so the speaker is resting solidly on the floor. If you use only three, the speaker will not rock. They are also very handy when you want to adjust the cabinet angle or correct its tendency to lean.

MOUNTING ESLs. Glue wood strips along the TL to form a slot to accept the ESLs (*Fig. 12-22*). Brace the opposite edge and unsupported top edge with aluminum strips similar

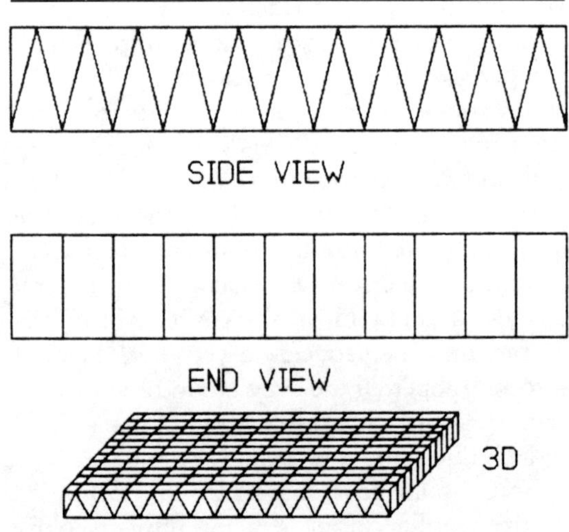

FIGURE 12-24: Foam wedge cutting pattern.

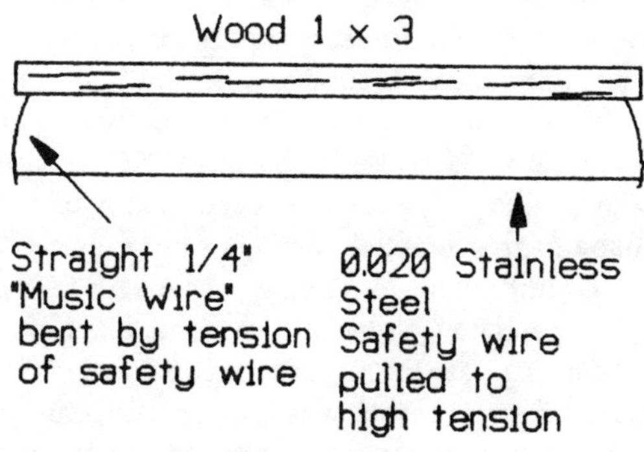

FIGURE 12-25: Hot wire bow.

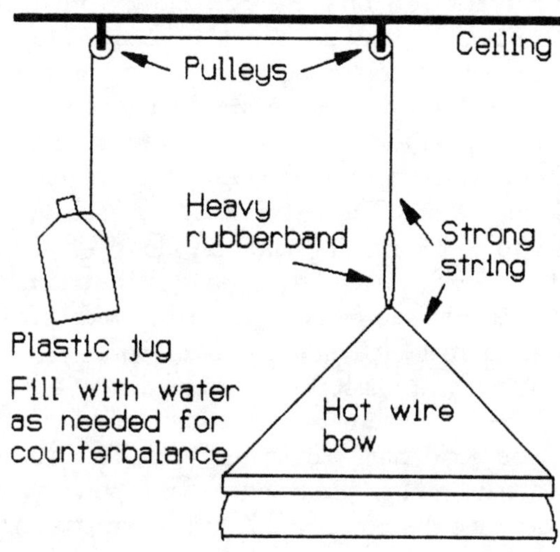

FIGURE 12-26: Hot wire bow suspension system.

to those you used for a cutting guide. I used 1" × 1/8" aluminum which is very soft and easy to bend. Use 8-32 nylon bolts to maintain insulation (*Fig. 12-23*).

When cutting aluminum and other nonferrous metals, use woodworking instead of steel-working tools. For example, your table saw with a wood-cutting blade cuts aluminum quickly and easily without damage to the blade. On the other hand, a steel-cutting hacksaw loads up the teeth with aluminum chips and cuts very slowly.

IN-WALL ENCLOSURES. In *Chapter 4* on frequency response, I discussed building ESLs into walls as a way of enclosing them. You can build TLs into walls, also.

Building the large foam wedges and sheets needed to trap bass energy in large enclosures is difficult. You need special tools and building techniques. Like ESLs, you can build the very-low-cost tools at home. What follows is a discussion of how to make and use these special tools.

BUILDING A HOT-WIRE BOW. Cut the wedges which form the sound absorbers in an anechoic chamber from sheets of polyfoam. This is a yellow/white soft foam often used for mattresses. You can cut small wedges with a *sharp* serrated knife, but an electric steak knife works better. *Figure 12-24* shows suitable cutting patterns.

Knives are slow, imprecise, and too small to cut large wedges. The best tool is a hot-wire bow. *Figure 12-25* depicts one of many ways to build one.

Figure 12-26 shows an optional suspension system for the bow. You can get more precise control if you don't hold the bow's weight while cutting. The suspension is easy to do

and well worth your trouble if you plan to do much work with foam.

The wire diameter, material, and length decide the voltage and current needed to heat the wire. There are too many variables for me to give you precise power supply specifications. To give you an estimate, 0.020 stainless-steel wire requires about 2V and 2A/foot.

The wire temperature is fairly critical. If it's too hot, it glows, stretches, loses its tension, and may break. If it isn't hot enough, it won't cut or cuts agonizingly slowly. The best temperature is just below the point where the wire glows.

A variable autotransformer (Variac) is an ideal power supply. Surplus electronics stores are good places to get these at reasonable prices. You will only use it intermittently, so you can use an underrated one to save money. My bow uses 7–8A, and I've had no trouble with a 5A power supply.

CUTTING WITH A HOT-WIRE BOW. Mark the foam with a felt marker where you want to cut. Place a metal straightedge on the mark on each side of the foam block. An assistant is very helpful. He can hold the straightedge and bow on one side of the block while you do the same on the other side.

Hold the straightedges in place by hand or with small finish nails about an inch long. Drill nail holes in the metal straightedges every 2–3". Push the nails through the holes into the foam by hand.

Place the wire against the straightedges, and

rest it gently on top of the foam block. Turn on the power supply. In a couple of seconds, the wire will get hot enough to start cutting into the foam as it follows the straightedges.

Don't push the bow. Just keep the wire gently in contact with the foam and let the heat do the cutting. Cutting speed is ¼–½″/second. Try to keep it moving smoothly along—if you stop, it will leave a little groove in the foam. This is only a cosmetic problem, but with care you can make it look like a factory job.

When you finish the cut, immediately turn off the power supply so you don't get burned if you accidentally touch the wire. The wire will have molten plastic on it that vaporizes if you leave the power on. A little vaporization occurs at the edge of the cut anyway, but try to avoid massive amounts of smoke.

You should cut in a well-ventilated area. The smoke isn't deadly, but avoid breathing it—besides, it stinks.

COMPOSITE CONSTRUCTION

Composite construction is beyond the scope of this book, but I want to pique your curiosity by giving you some idea of what it can do. You can use hot-wire cutting techniques for cutting all types of foam, both solid and flexible.

You can make composite structures using Styrofoam coated with fiberglass, Kevlar®, or carbon fiber. The idea is to cut foam to the shape of the finished part, lay fiberglass cloth on it, and paint on epoxy resin. After the epoxy catalyzes, you can trim, sand, and paint. Be sure to use epoxy resins. The more common polyester resins will dissolve the foam.

You can make a speaker enclosure by cutting foam into 1 or 2″ sheets, and coating them with fiberglass. The sheets can then be assembled into a box which would be much more rigid and nonresonant than the usual wood box, and would also be very light.

Flat sides on speaker enclosures flex. The sides become speakers themselves and produce sound which colors the speaker-system sound. You can avoid this by using curved cutting templates to cut the foam so the sides are not flat. You can make any shape imaginable, like a cylinder, truncated cone, sphere, or any highly irregular shape.

CHAPTER 13:
SYSTEMS

The purpose of this chapter is to describe the details of designing and building three ESL systems. I want to tie together what I've already presented by showing actual systems and how to complete them. Although much of this information is review, I think you'll find it very helpful to see the *Big Picture*.

I'll depict these as you would build them, with Lincaine perforated-aluminum stators, although you may use a different stator design if you wish. All stators use the same design criteria (D/S spacing, overall dimensions, spacer ratio), so decisions made for perforated metal also apply to the others.

FULL-RANGE SYSTEM. Let's start with the simplest of the three, a full-range crossoverless ESL. You already know it must be big, so let's make it about 2-feet wide. The exact size is influenced by the size of perforated metal. Since Lincaine comes in 2′ × 3′ sheets and other perforated metal normally is 2-feet wide, let's keep it simple by just using whole sheets for this speaker's stators. When you

HOW TO BUILD THESE SYSTEMS

- Full-range ESL
- No-compromise hybrid
- Compact/Integrated hybrid

add the perimeter spacers and some type of frame, the overall width will be around 30″.

The D/S spacing must be relatively wide to produce much bass. Seventy mil is usually too small, and it's virtually impossible to get enough drive voltage for 130 mil if you want high output levels and wide bandwidth.

A reasonable compromise is 90 mil, although this may still be too small if your room has bad bass resonances. You can use 80-mil "unbreakable windows" for the spacers. The other 10 mil of thickness will come from the two glue films.

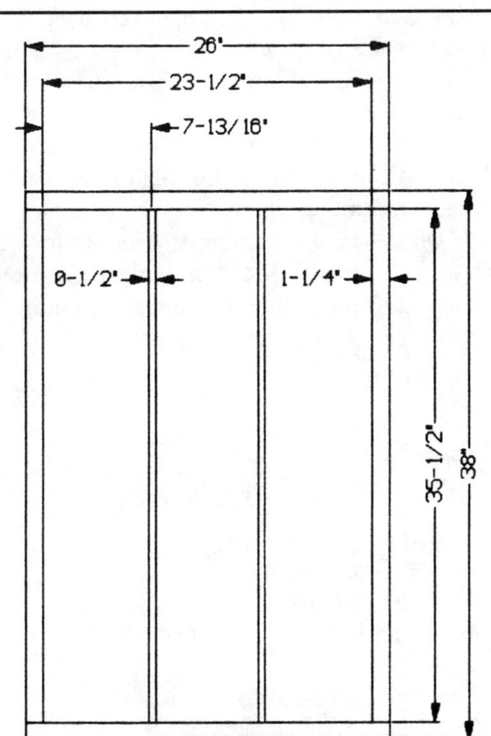

FIGURE 13-1: Spacer pattern for full range ESL.

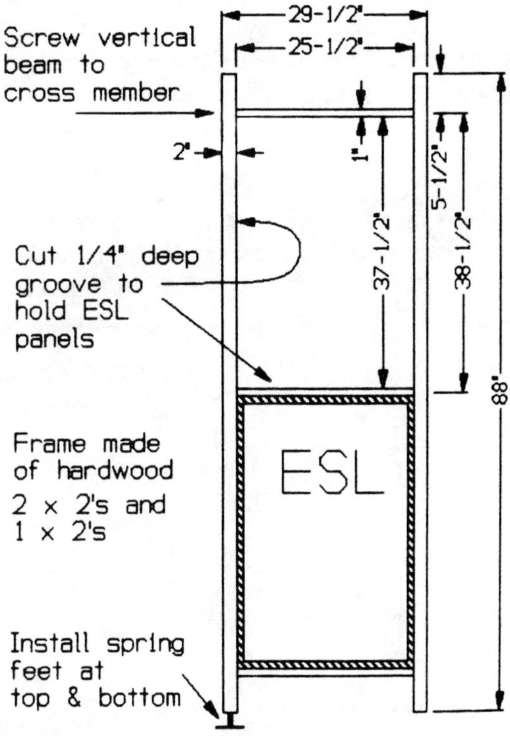

FIGURE 13-2: ESL frame.

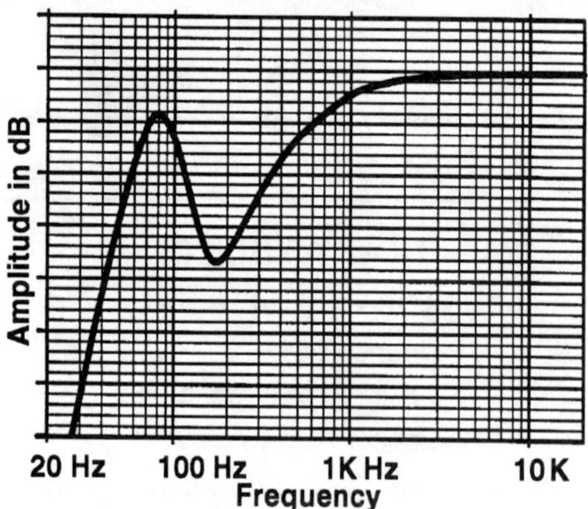

FIGURE 13-3: Unequalized frequency response of large ESL.

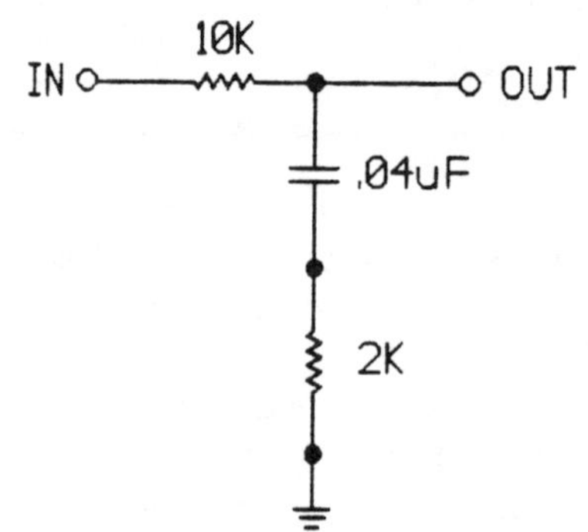

FIGURE 13-4: Passive midrange equalizer.

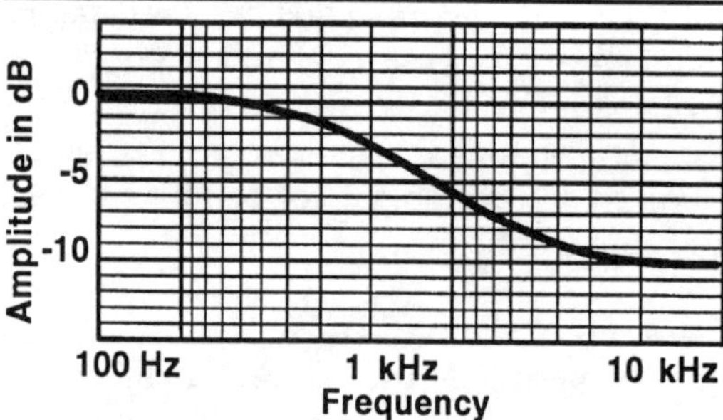

FIGURE 13-5: Frequency response of passive equalizer.

You know that the spacer ratio should be between 50:1 and 100:1. Since the D/S ratio is 90 mil, the free diaphragm distance can range from 4.5 to 9″.

The perforated metal overlays the spacers along the edges by ¼″. This reduces the actual free diaphragm area from 24 to 23.5″. The resulting spacer ratio is 261:1. Since this is too much, we must use internal spacers to break the cell into smaller sections. Let's use two vertically oriented internal spacers to break the large cell into three smaller ones.

If we divide 23.5″ by 3, we get 7.83″ sections. Actually, they'll be slightly smaller than that because the internal spacers have width— let's make them ¼-inch wide. This works out to a spacer ratio of 86:1, which is just about perfect.

How many cells do you want to make? Let's use a floor-to-ceiling line source because it max-imizes output, imaging, and vertical dispersion. Since the typical room is 8-feet tall, this means the speaker must be 6–8-feet tall. Lincaine per-forated metal comes in 2′ × 3′ sheets, so let's make two cells for each speaker. The total height of the two cells will be about 78″, and with some surrounding framework they will be somewhat taller. We can make up the small dif-ference between the typical 96″ between the floor and ceiling of an 8-foot high room with a suitable mounting frame.

Figure 13-1 shows the spacer layout for this speaker. Note that you should place the inter-nal spacers vertically for the least amount of internal spacer area and minimal stray capaci-

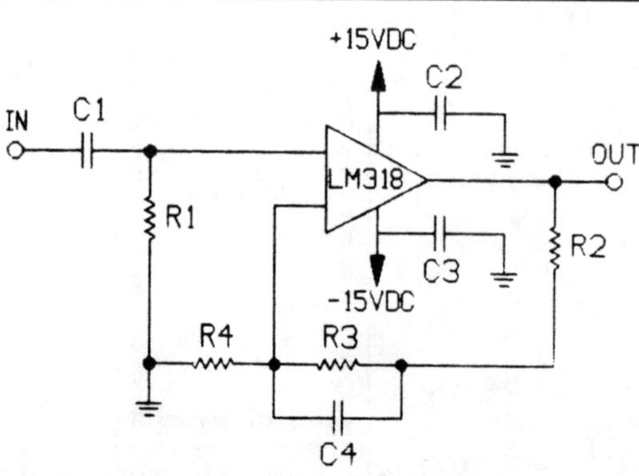

FIGURE 13-6: Active midrange equalizer with gain.

TABLE 13-1

PARTS LIST FOR FIGURE 13-6

C1	0.15µF
C2	0.1µF ceramic disc
C3	0.1µF ceramic disc
C4	8,500pF 2% mica or polystyrene
R1	200k
R2	6.8k 1% metal film
R3	36.5k 1% metal film
R4	7.3k 1% metal film

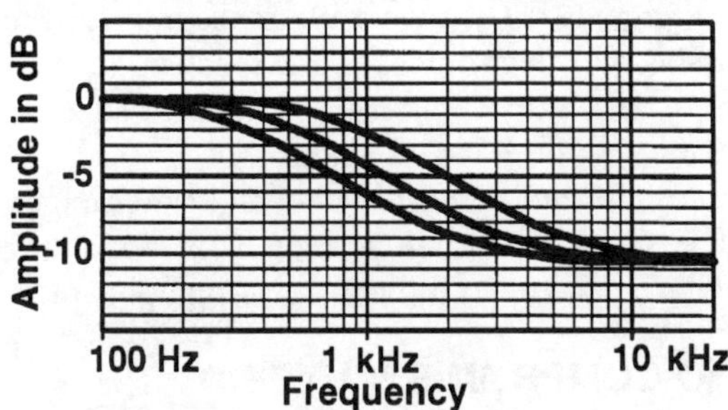

FIGURE 13-7: Affect of changing value of C4.

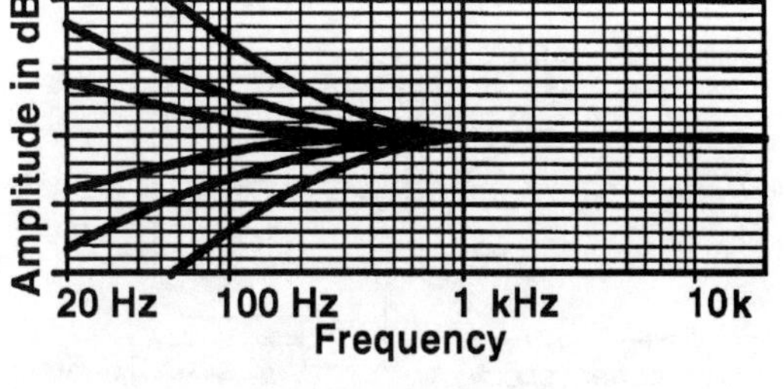

FIGURE 13-8: Typical response of typical pre-amp bass "tone control".

tance. *Figure 13-2* shows the cells in a frame as a room divider.

I have assumed that you are building planar cells. If you want to build curved ones, you must run the internal spacers horizontally and use spacer ratios closer to 50:1 for best diaphragm stability. Breaking each cell into six sections would be best.

FULL-RANGE ELECTRONICS. *Figure 13-3* shows the expected frequency response of the above ESL. We will correct it with equalization.

Figure 13-4 shows a passive-midrange equal-ization-circuit. Its frequency-response curve will look like that shown in *Fig. 13-5*.

This equalizer will have nearly 20dB of inser-tion loss. If you don't have enough gain else-where in your system, or if you want to avoid using a preamplifier, use gain/equalization elec-tronics. *Figure 13-6* shows a suitable schematic.

You may diddle C4 to move the entire fre-quency-response curve up or down, as shown in *Fig. 13-7*. This should not be necessary, however, since the components shown give the correct response and the unit is not affect-ed by impedances as low as 600Ω.

The value of the capacitor C4 is inversely related to the frequency. In other words, a larger value capacitor will shift the curve

downward while a smaller one will shift it upward. Changing this capacitor is an easy way to modify the equalizer for ESLs of differ-ent widths.

These equalizers only compensate for loss-es due to phase cancellation down to around 400Hz. You will need additional equalization below there.

In *Chapter 4* on frequency response, I out-lined the problems inherent with predicting ESL bass response. Therefore, I cannot design an equalizer for you below 400Hz. It may not be necessary anyway, as you will probably find that you can get the bass close enough with the tone controls on your preamplifier.

Standard tone controls usually give a fre-quency-response curve that pivots around 1kHz, as shown in *Fig. 13-8*. When added to the midrange equalizer, you can get a response curve that looks something like *Fig. 13-9*.

This works reasonably well, except that it doesn't suppress fundamental resonance. This resonance will seriously impair the speaker's output and ruin linear bass response. The best course is to suppress it with a notch filter.

The easiest way to get a notch filter is to use an octave equalizer instead of conventional tone controls. You can severely depress the frequencies around and below resonance. This greatly improves output and you can get rea-sonably linear bass.

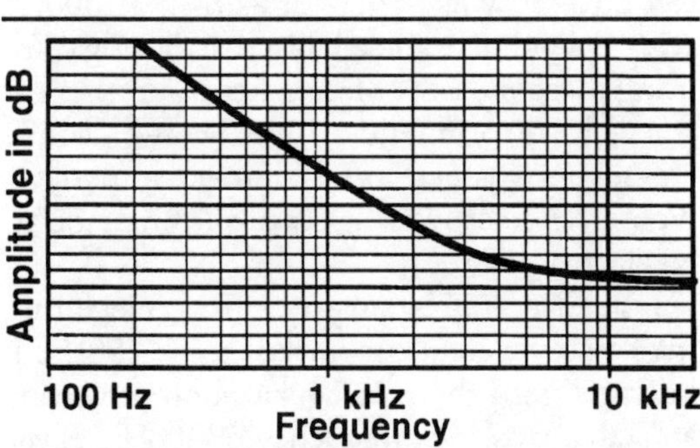

FIGURE 13-9: Frequency response of midrange equalizer plus bass "tone control".

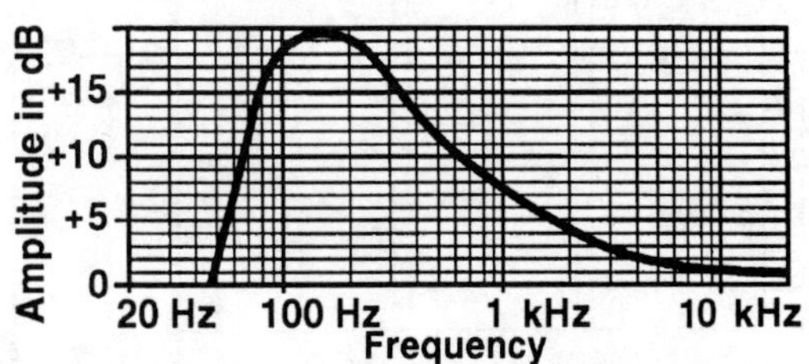

FIGURE 13-10: Frequency response of midrange equalizer plus notch filter.

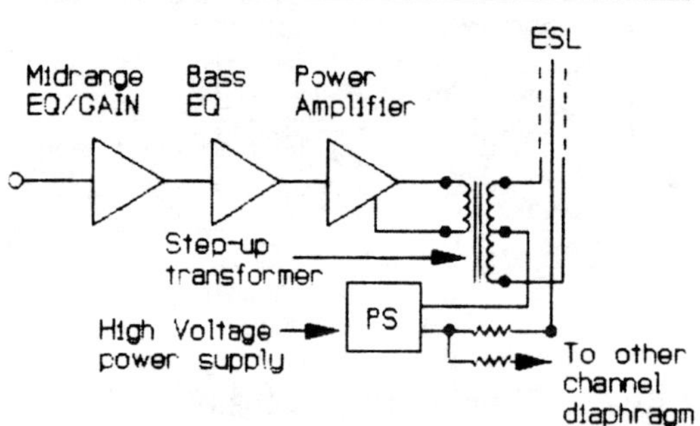

FIGURE 13-11: General schematic diagram of ESL associated electronics.

Figure 13-10 shows the response produced by the combination of an octave and midrange equalizer. This equalizer combination does a reasonably good job of producing a mirror-image response curve of the ESL before equalization.

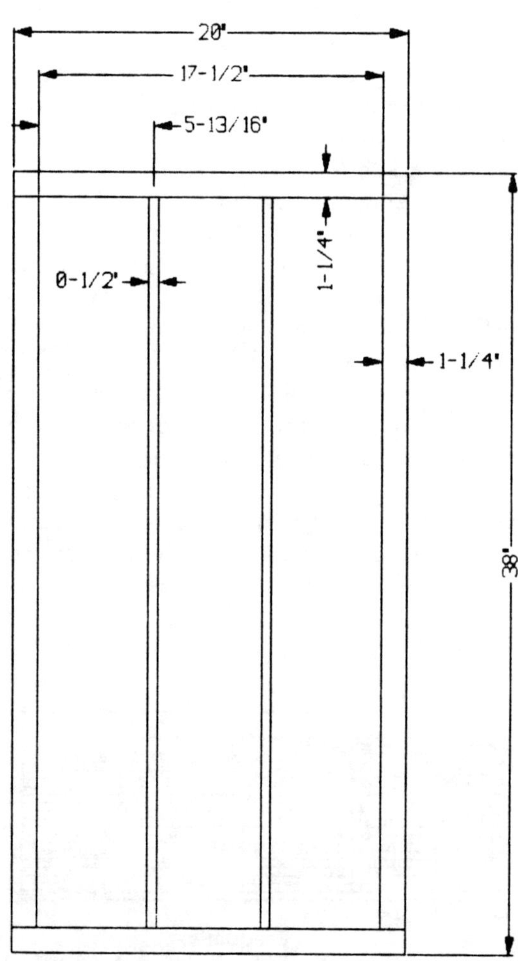

FIGURE 13-12: Spacer pattern for no-compromise hybrid.

Figure 13-11 shows a block diagram of the associated electronics for this speaker. You can find details in *Chapter 7* on associated electronics if you have questions.

You'll need all the drive voltage you can get. The speaker will handle 9 or 10kV. Polarizing voltage will need to be in the range of 4–6kV.

NO-COMPROMISE HYBRID. The full-range system cannot produce high outputs or deep bass. Let's make one which will. To do so, we use a large TL-woofer system to handle the bass.

The ESL still must be very large if we want high output and a reasonably low crossover frequency, but we can make it a little smaller without significant penalty. Lincaine perforated aluminum comes in 3′ × 3′ sheets that we can cut in half to make stators 18-inches wide. The height will remain the same as the full-range system.

We can reduce the D/S spacing to 70 mil. This reduces the demands on drive voltage considerably while maintaining high output.

The spacer ratio demands that the diaphragm be supported every 3.5–7″, so we must still use internal spacers. Let's break the cell into three vertical sections again. Each section will be 5.66″ when the perimeter overlap (¼″) and internal spacer width (¼″) is considered. This gives a spacer ratio of 62:1. *Figure 13-12* shows the spacer pattern. Again, if you want a curved cell, the pattern should be horizontal.

NO-COMPROMISE ELECTRONICS. The midrange equalizer can be the same as the full-range system. I won't repeat the schematics. You may change C4 in the active equalizer to 8,200pF to compensate for the reduced width, but the speakers are similar enough so this isn't really necessary.

You must, of course, use crossovers. You can go as low as 400Hz or as high as 500Hz with splendid results. I discuss the logic for using these crossover frequencies and the reasons I recommend 18dB/octave active crossovers instead of 12dB/octave passive, high-level crossovers in *Chapter 7* on associated electronics.

I prefer a 400Hz crossover. To achieve this, the crossover frequency needs to be 480Hz. It interacts with the gain/equalization producing an actual crossover frequency of 400Hz. *Figure 13-13* shows a composite frequency response of a 480Hz crossover combined with midrange equalization. Because it is difficult to build such

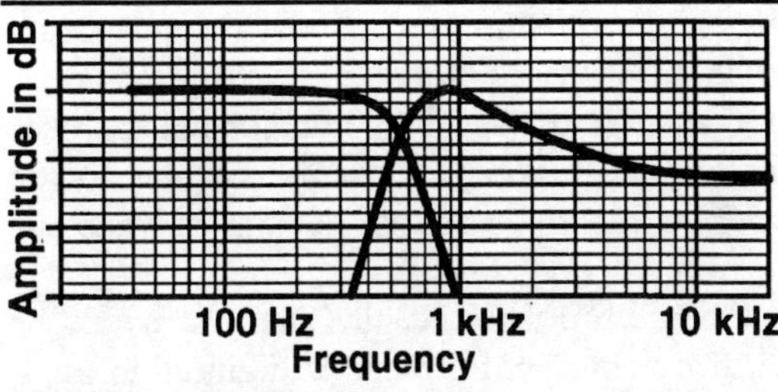

FIGURE 13-13: Frequency response of crossover plus midrange equalization.

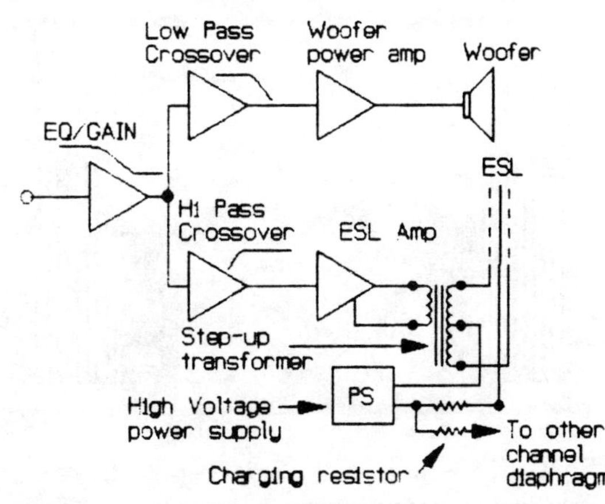

FIGURE 13-14: Block diagram of large hybrid ESL.

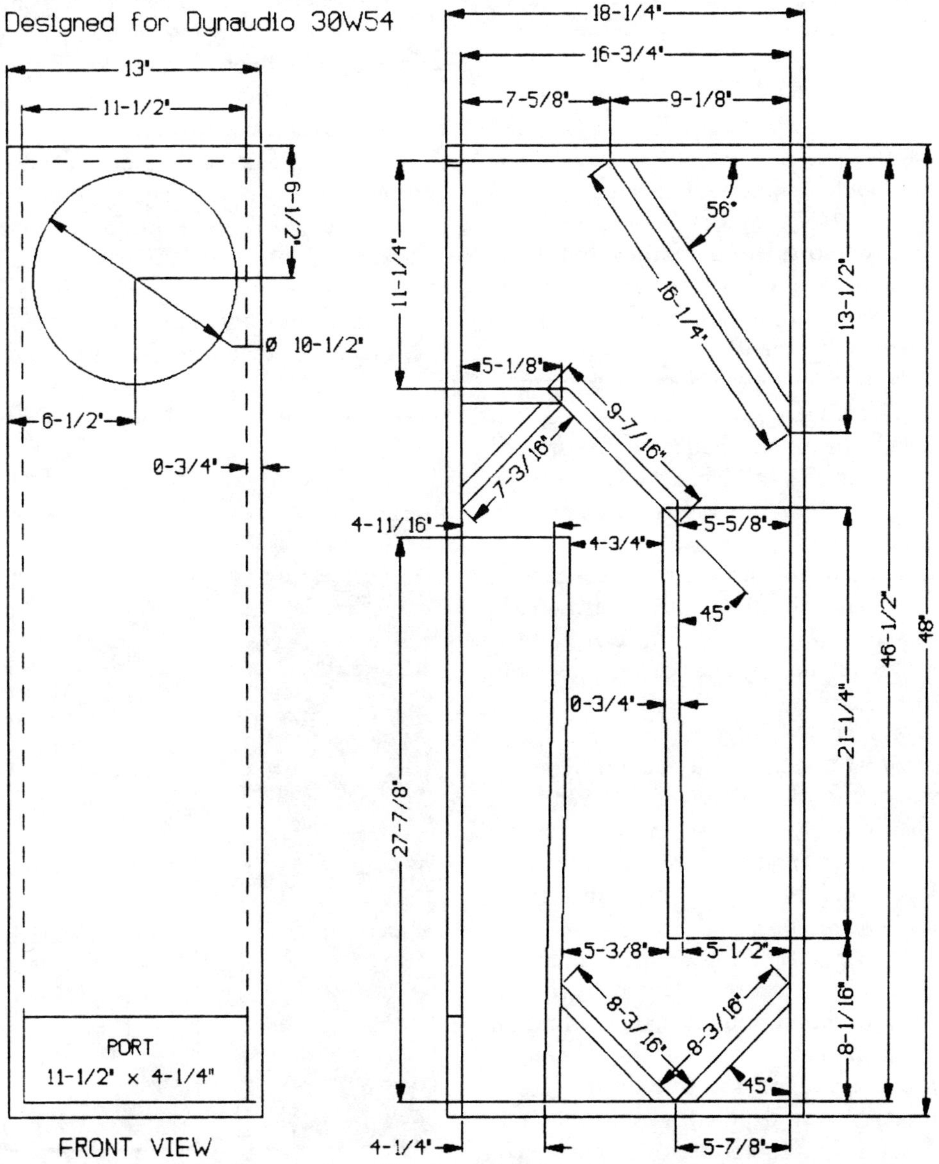

precision crossovers, I'm not including schematic diagrams for them. I anticipate that you will buy commercial ones. See *Chapter 8* on electronic construction for details.

FIGURE 13-15: Large TL.

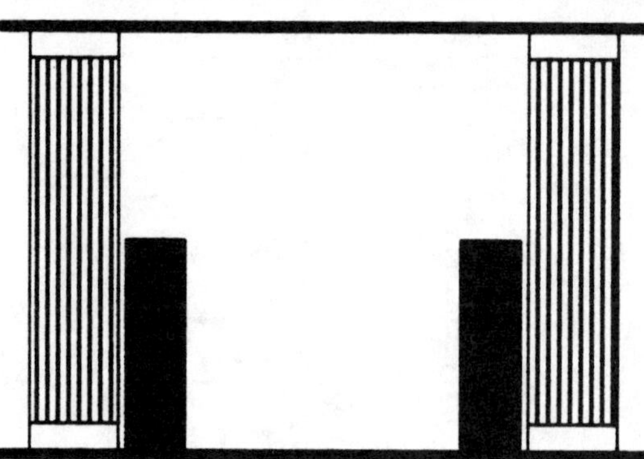

FIGURE 13-16: Hybrid ESL/TL—not integrated.

You can use up to 8kV of drive voltage in this system. The polarizing voltage will be between 3 and 4kV. *Figure 13-14* is a block diagram of the system.

WOOFER SYSTEM. This system needs a no-compromise, TL-woofer system. This means a line nearly 10-feet long with a large (12″) driver. The line should be smoothly tapered.

Keep the woofer well above the floor. Not only will the bass be smoother, but if the woofer is at ear level, you can adjust it fore and aft relative to the ESL for the best phase relationship (alignment).

I haven't found phase alignment to make much difference, but some builders and designers feel strongly about it. If you are one, then you definitely will want the woofer at ear level. *Figure 13-15* shows an enclosure that meets the criteria.

The combination of a large free-standing ESL and a large boxy TL is often aesthetically unsatisfactory (*Fig. 13-16*). You can dramatically improve the appearance by integrating the parts. *Figure 13-17* shows one such hybrid ESL/TL system.

Figure 13-18 shows construction details of the transmission line that will integrate with the ESL. Probably its most unusual aspect is that it is not rectangular in cross section, but a parallelogram. I did this for both performance and aesthetic reasons.

Because a planar ESL is as directional as a laser beam, you must angle it inward at 15–30° for optimum soundstage presentation. If you put the ESL at the front of a rectangular enclosure, you must angle the entire enclosure inward. Most people find this unacceptable (*Fig. 13-19*). To solve this problem, you could recess the ESL from the front of a rectangular enclosure. The ESL would be angled, but hidden from view with a surrounding grille cloth. The resulting enclosure could be put against the wall in the normal fashion. The woofer would fire perpendicular to the wall, while the ESL would aim toward the focal point (*Fig. 13-20*).

Several problems result. The ESL would be far from the woofer plane and aimed in a different direction—not a serious problem, but not desirable either. The ESL would not have free dipole radiation around the back of the TL. Trapping the rearwave in this way will cause resonances and reflections back through the ESL's diaphragm. Aesthetically, it is unattractive because the surrounding grille cloth would make the speaker look like a very large box.

I solved these problems with a design which leaves the sides of the enclosure perpendicular to the wall, while angling the entire face of the speaker. Not only does this make a very attractive enclosure, but it turns the transmission line into a rearwave beam splitter.

By picking the right ratio of width to depth on the vertical part ("chimney") of the TL, and putting the ESL on the inside, I intercepted two-thirds of the rear beam with the TL. I then used different surfaces to reflect parts of the rearwave in two extra directions (*Fig. 13-21*).

A major aesthetic problem is dipole radia-

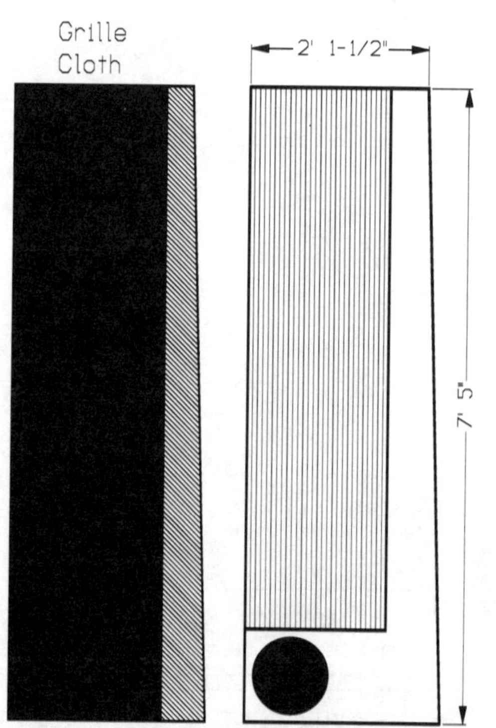

FIGURE 13-17: Integrated ESL/TL.

tors must be freestanding. We usually want our speakers to be out of the way and placed against the wall. The integrated design solves this problem also. It permits the ESL to be away from the wall while seeming to be against it (*Fig. 13-22*).

While this integrated enclosure works very well, it is not perfect. In comparison to the no-compromise enclosure shown above, the integrated one is shorter (about 8'), and it puts the woofer on the floor.

The drawings are difficult to visualize because of the angles. For simplicity, I've drawn the front and side views as though the enclosure were a rectangle rather than a parallelogram. Just make the cuts at the required angle rather than the normal right angle, and you will automatically end up with a parallelogram when you assemble it.

IMAGING CONSIDERATIONS. Recall from *Chapter 6* that unlike magnetic speakers, ESLs sound the same no matter how close you are to them. This makes it possible to have different stage presentations based on the percentage of room acoustics in the image.

If you sit well away from the speakers, they will sound as though the performers are in your room. If you sit very close to the speakers, they will sound as though you are in the concert hall where the recording took place.

Close seating has other advantages as well, including higher SPLs, improved detail resolution, more precise instrument location, and greater intimacy with the performance.

When building integrated TLs of this design, you must give careful thought to the preferred image width and listening location. These fac-

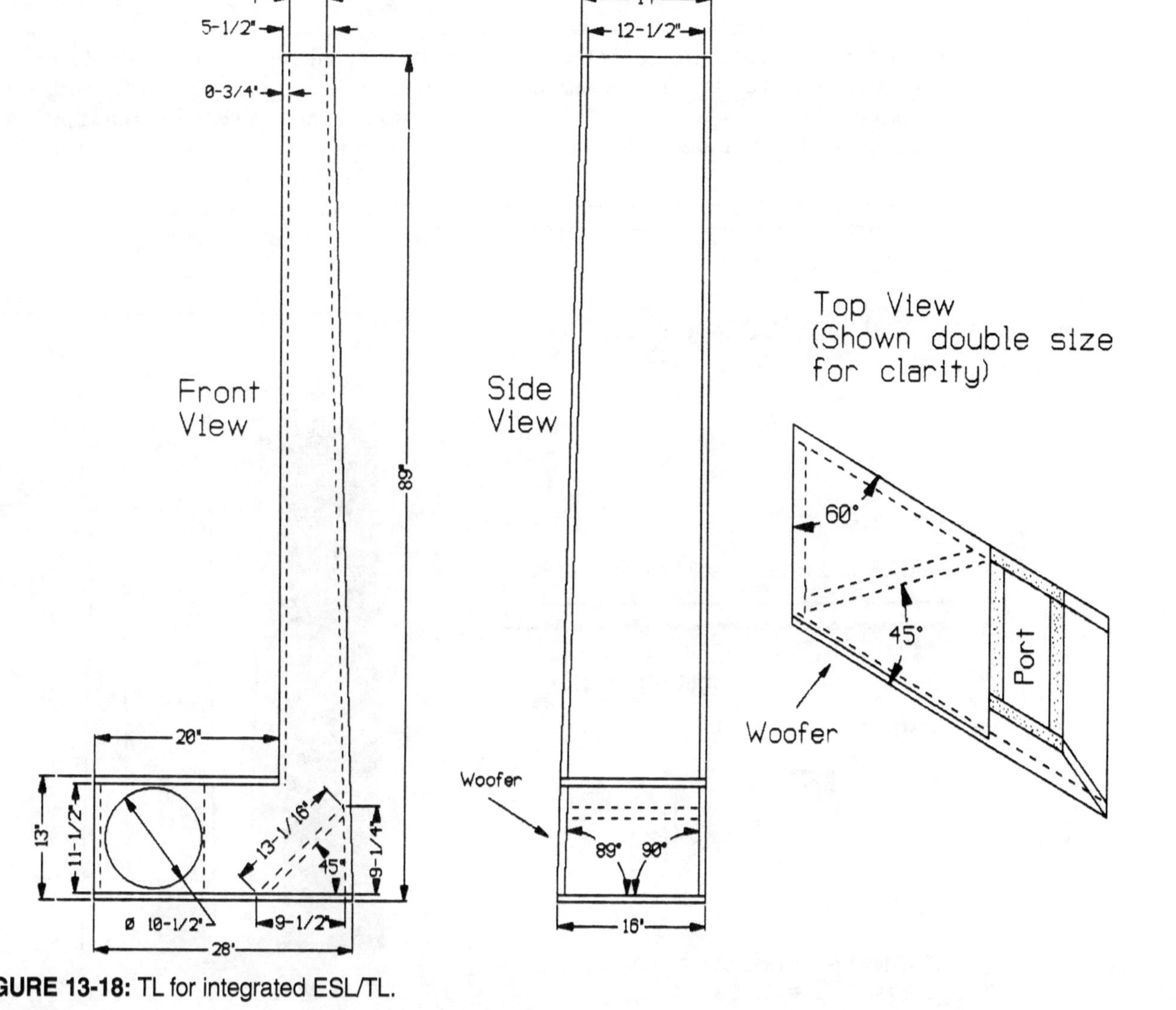

FIGURE 13-18: TL for integrated ESL/TL.

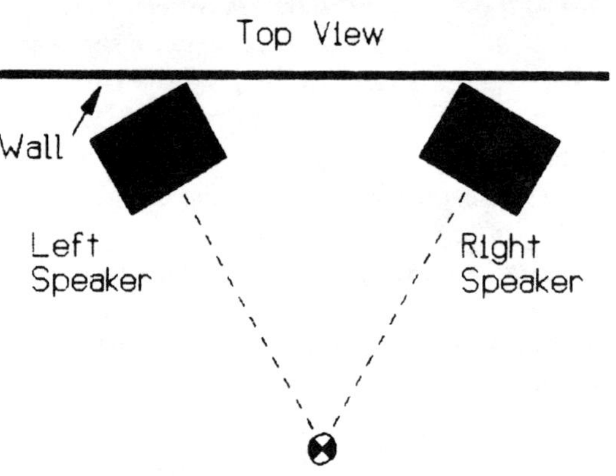

FIGURE 13-19: Poor aesthetics—speakers at angle to wall.

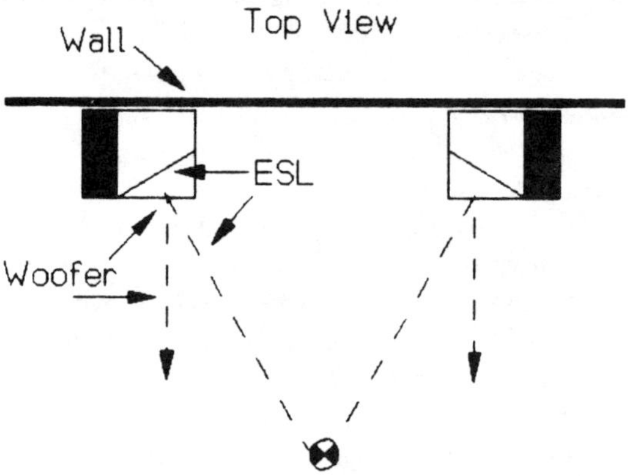

FIGURE 13-20: Enclosure square with wall—Hidden ESL is angled.

tors affect the parallelogram angles you build into the transmission line.

After deciding upon your preferred listening distance, determine the angle you must use to have the speakers face your seat. This may be up to 30° for each speaker. Note that angles less than 30° will reduce not only the sound-stage width but also the effectiveness of the rear beam splitter.

Remember, for optimum performance your

seating position should *not* be against a wall if at all possible. If you are seated near a wall, the sound is smeared by wall reflections immediately behind you. If you must sit near the wall, it is helpful to put some kind of sound-absorbing material on the wall.

If you don't want to have a chair in the middle of your room, you might consider this solution to the problem. Put two tiny marks on the carpet which indicate the proper location for the front legs of the listening chair you normally keep near the wall. Place the chair on the marks for serious listening, but

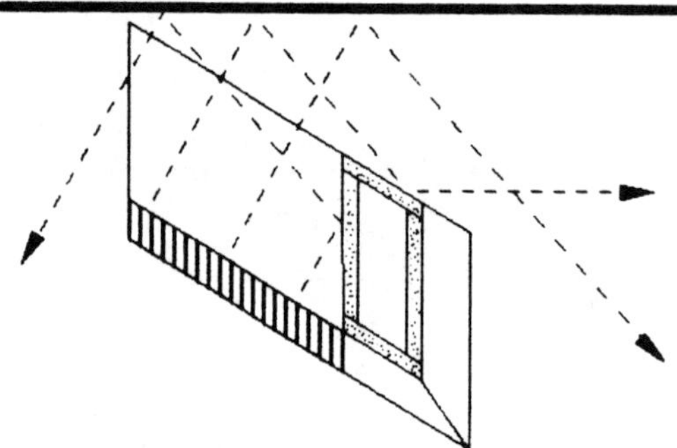

FIGURE 13-21: Beam splitting effect of TL.

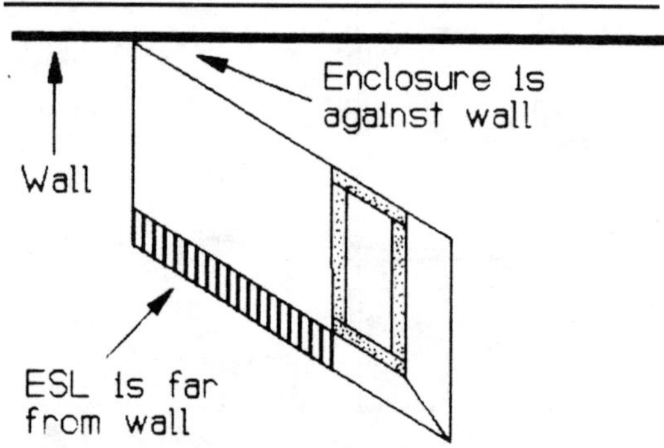

FIGURE 13-22: ESL away from wall but integrated with TL enclosure.

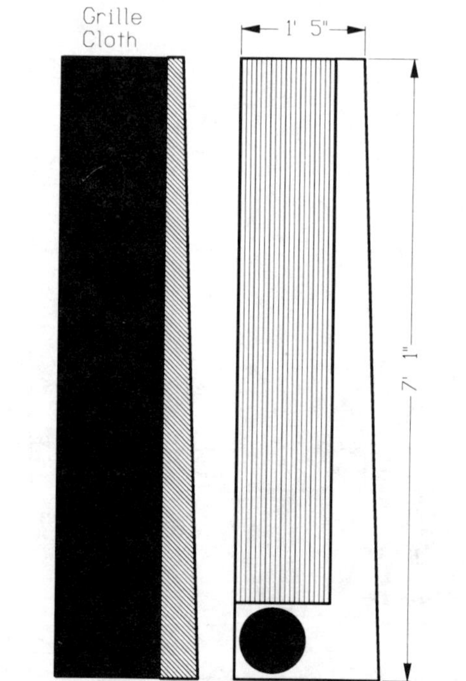

FIGURE 13-23: Compact/Integrated ESL/TL.

otherwise keep it against the wall and out of the way.

COMPACT/INTEGRATED ESL/TL.

The full-range ESL and no-compromise hybrid both suffer from the cosmetic problems of very large size and the need to stand freely out in the room. I dealt with the freestanding requirements with the integrated enclosure, but the large size remains.

A smaller (though by no means tiny) hybrid which performs nearly as well as the no-compromise system is possible to make. By shifting the midrange-equalization curve upward along with the crossover frequency, you can use a narrower ESL. Using these techniques, you can get a 12-inch wide ESL to work very well.

The transmission-line system can be made considerably more compact by using a smaller woofer in an integrated system. An 8 or 9″ woofer can produce very high output levels while sacrificing only a little deep bass.

You can think of the compact system as a miniature version of the no-compromise integrated system. *Figure 13-23* shows the general layout and appearance. *Figure 13-24* is the construction drawing for the transmission line.

The ESL uses the same 70-mil D/S spacing

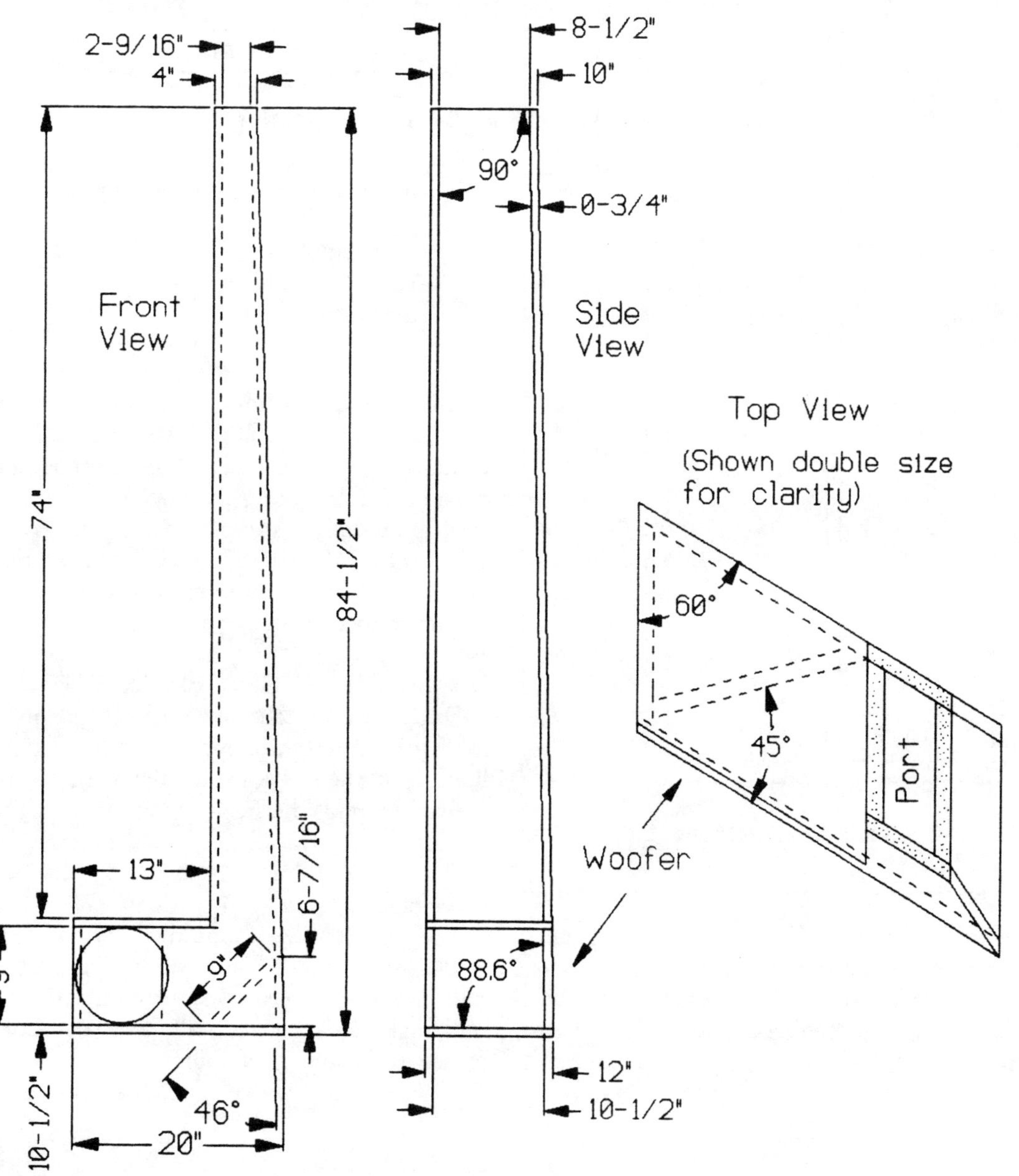

FIGURE 13-24: Construction details of compact TL.

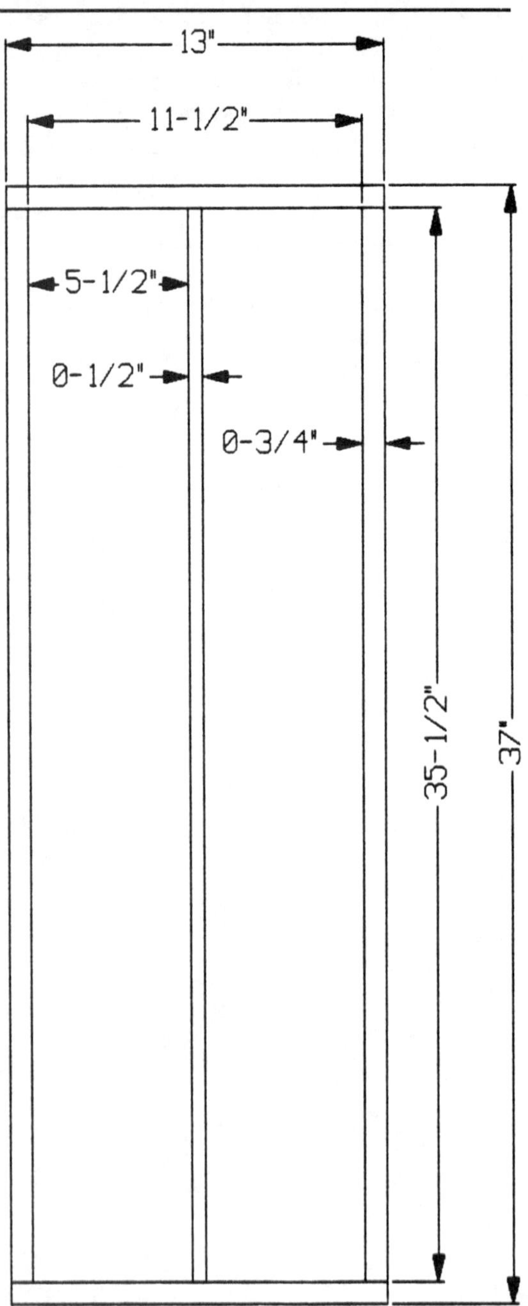

FIGURE 13-25: Spacer frame for compact ESL/TL.

as the large hybrid. The narrow cells only need a single internal spacer to maintain an appropriate spacer ratio. Because space is at such a premium, I've reduced the width of the perimeter spacers to only ¾″. With perforated

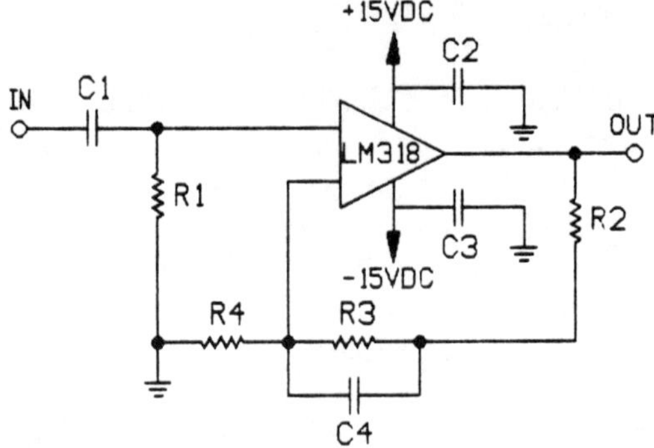

FIGURE 13-26: Active gain/equalizer for compact ESL/TL.

TABLE 13-2

PARTS LIST FOR FIGURE 13-26

C1	0.15µF
C2	0.1µF ceramic disc
C3	0.1µF ceramic disc
C4	7,600pF 2% mica or polystyrene
R1	200k
R2	6.8k 1% metal film
R3	36.5k 1% metal film
R4	7.3k 1% metal film

---

metal, overlapping the usual ¼″, this gives only ½″ of insulator sticking out beyond the conductive part of the stator. You must be cautious in the mounting method you select so you don't short a stator to the frame. *Figure 13-25* shows the spacer pattern. You can buy 2′ × 3′ sheets of Lincaine and cut them in half for the stators.

COMPACT/INTEGRATED ELECTRONICS. The associated electronics are identical to the no-compromise hybrid with two exceptions. First, the crossover frequency should be 550Hz; second, the feedback capacitor C_4 in the gain/equalization electronics must be changed to 7,600pF to push the equalization curve higher (*Fig. 13-26*).

The crossovers should have 650Hz crossover points at 18dB/octave. The combination of the crossover frequency and midrange equalization will give an actual crossover frequency of 550Hz.

CHAPTER 14

SETUP

Where you place the ESL is very important. You can only access its marvelous imaging ability when you abide by certain geometric rules.

These rules apply to all speakers, but conventional ones have such poor imaging characteristics compared to ESLs that proper geometric positioning isn't as important. Also, conventional speakers have wide dispersion, and room acoustics "smear" their sound so badly that precise imaging is impossible. A highly directional ESL lacks these problems, so the effects of improper positioning become obvious.

Having an ESL that is improperly set up is like having an "out-of-tune" race car. Why have one if you don't operate it at its full potential?

Several geometric factors must be considered which I'll discuss in detail. Some of this is review from *Chapter 6* on dispersion. You may wish to reread it before proceeding.

IMAGE WIDTH. With forward-firing, direct-radiating, conventional magnetic speakers, image width depends on the angle formed between you and the speakers (*Fig. 14-1*). This angle is limited: if you increase it beyond a certain point, you will hear the dreaded "hole" in the middle of the sonic image.

The usual limit is about 30°. This translates into a ratio between the speakers' width and the distance from you to the speakers of 2:1 (*Fig. 14-2*). Most listeners prefer wider soundstages, much like listening close to the stage at a live performance: the closer you are, the wider the image (*Fig. 14-3*).

GEOMETRIC POSITIONING FACTORS

- Image width
- Distance from speakers
- Distance from walls
- Angle to wall
- Precision positioning

Speaker manufacturers have tried to increase the width of the soundstage. One way is to mix the two channels together and feed this monaural signal to a center speaker to "fill" the hole formed by widely spaced speakers (*Fig. 14-4*).

This works, but there are two problems: first, few audiophiles are dedicated enough to purchase an additional speaker, amplifier, and mixing/attenuating electronics; second, the sound arrives from two sources of differing distances, which "smears" the sound in exactly the same way as room reflections (*Fig. 14-5*).

Another technique is to have the speakers reflect most of the sound off walls (*Fig. 14-6*). This produces incredibly wide images, but

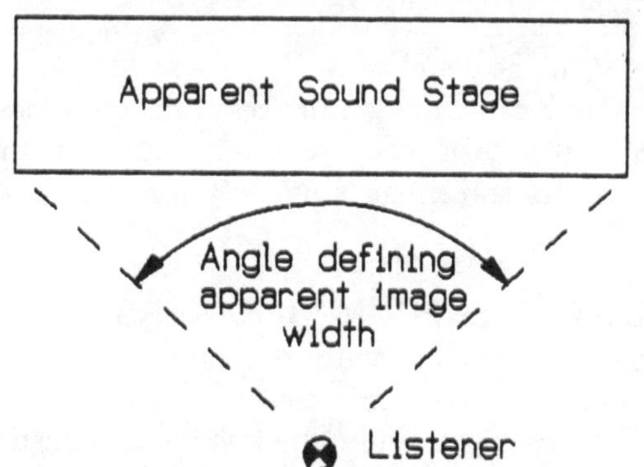

FIGURE 14-1: Listening angle.

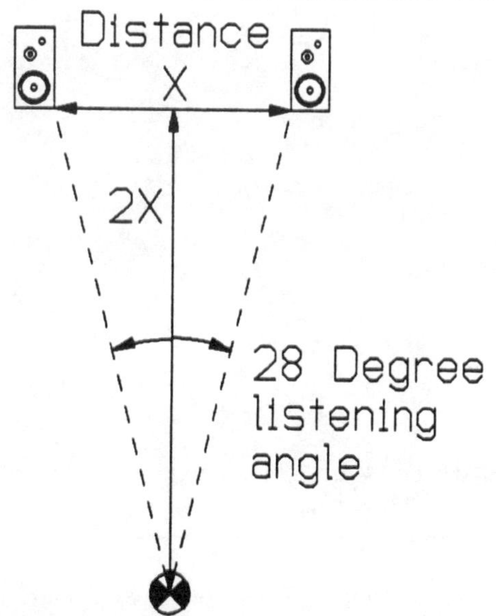

FIGURE 14-2: Typical 2:1 listening angle.

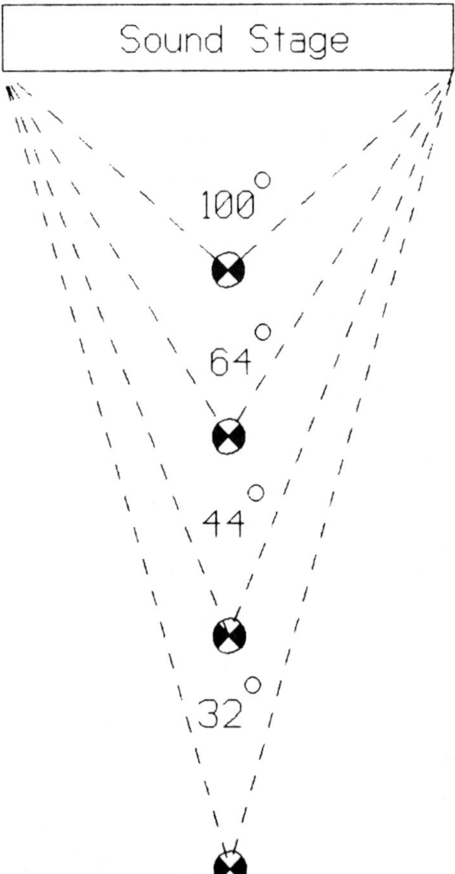

FIGURE 14-3: Distance to speakers affects listening angle.

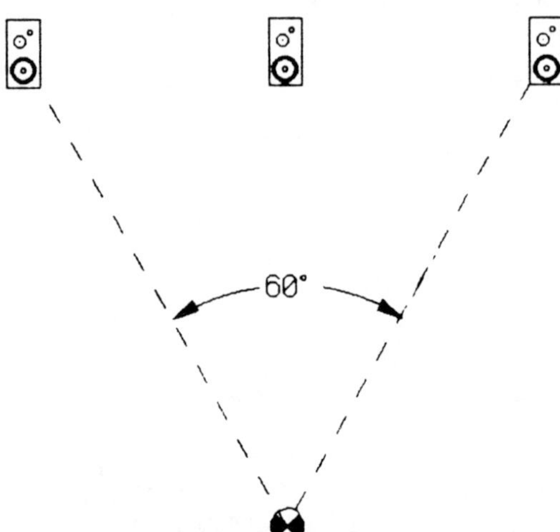

FIGURE 14-4: Center speaker fills "hole" in apparent soundstage.

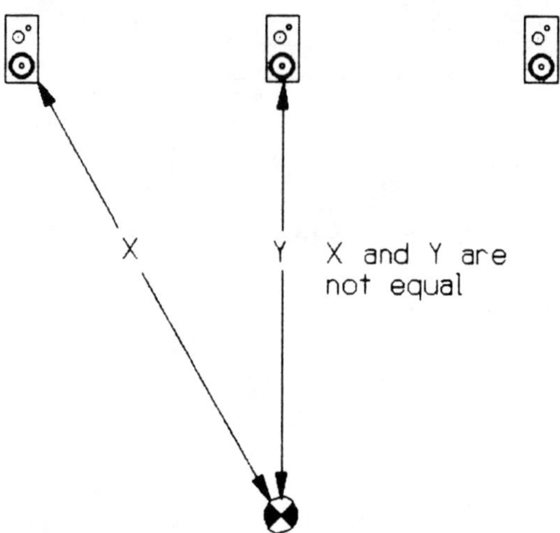

FIGURE 14-5: Unequal distances "smears" detail.

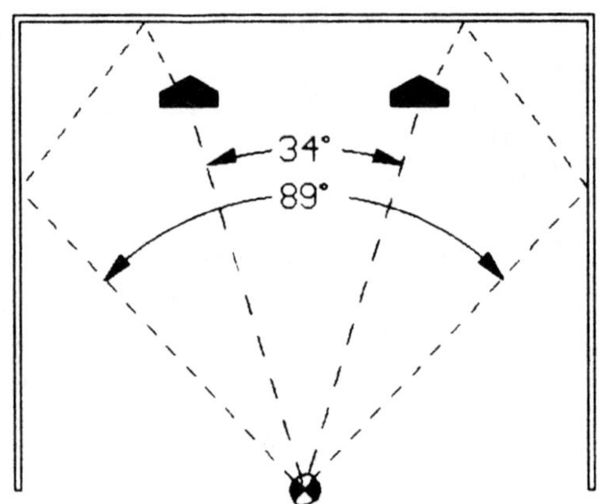

FIGURE 14-6: Direct/reflecting speakers.

again the sound arrives from several sources of differing distances. The image is impressively wide, but diffused and ill-defined. You are effectively trading image width for sound quality.

Planar dipole ESLs solve all these problems. They produce only one sound source and much better phase behavior than conventional speakers. They etch perfect images, so they can present a very wide soundstage *and* stunning image detail.

They can generate flawless images at least 60° wide. You can use a ratio between the speakers and the listening location as high as 1:1 (*Fig. 14-7*).

DISTANCE FROM SPEAKERS. Unlike conventional magnetic drivers, ESLs sound the same no matter how close to them you are. The closer you listen, the less room acoustics affect the sound. The speakers seem to carry you to the concert hall.

If you pick a distant listening location, room

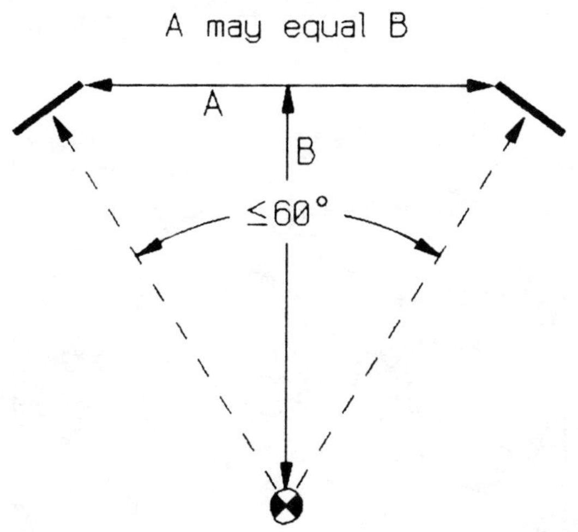

FIGURE 14-7: 60° listening angle using dipole ESLs.

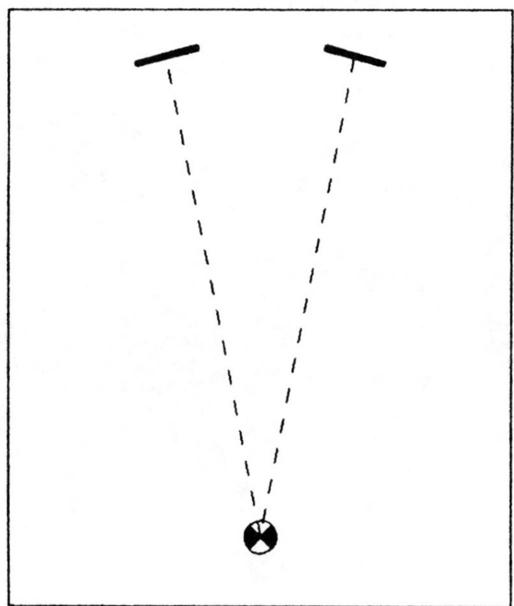

FIGURE 14-8: Musicians appear at far end of your room.

acoustics overshadow the concert hall acoustics. This has the effect of bringing the musicians into your room instead of taking you to the concert hall.

Both locations are equally good. With ESLs, you have a choice, and it is based on personal preference. Remember that dipole ESLs greatly delay and attenuate room acoustics. So, although you may be listening from a distant focal point, the image remains crisp and detailed.

The twin issues of listening distance and image width deserve careful thought. You can literally fine tune your system to produce the type of listening environment you wish. You have several choices of sound-stage presentation:

- A narrow listening angle and distant focus produce an image as though the musicians are at the other end of your room (*Fig. 14-8*).
- A narrow listening angle and close focus produce an image as though you are seated well back in a concert hall (*Fig. 14-9*).
- A wide listening angle and distant focus produce an image of the musicians close to you in your room (*Fig. 14-10*).
- A wide listening angle and close focus transport you to "Row A" in a concert hall (*Fig. 14-11*).

DISTANCE AND ANGLE. A dipole radiator must have "elbow room." It must be free-standing and can't be flat against a wall.

The degree to which this is true is debatable. Some people claim that a dipole works best in the middle of a room, as far from the walls as possible. Others find they can be placed surprisingly close to a wall before problems arise. My experience falls into the latter category.

You'll get no argument from me that a dipole in the middle of a room works beautifully. When in doubt, this is a safe place where

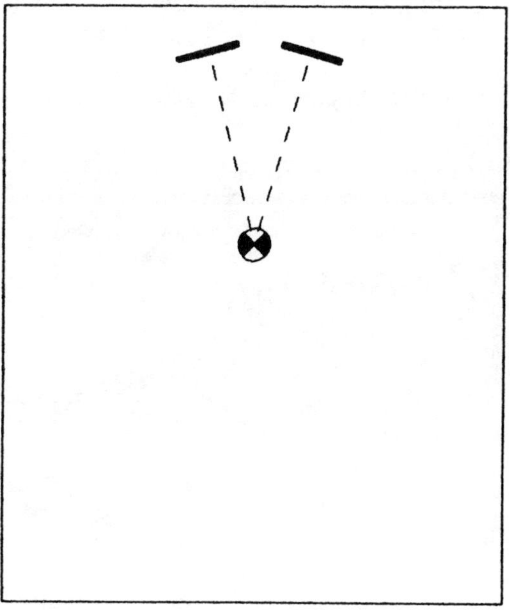

FIGURE 14-9: Musicians appear at far end of concert hall.

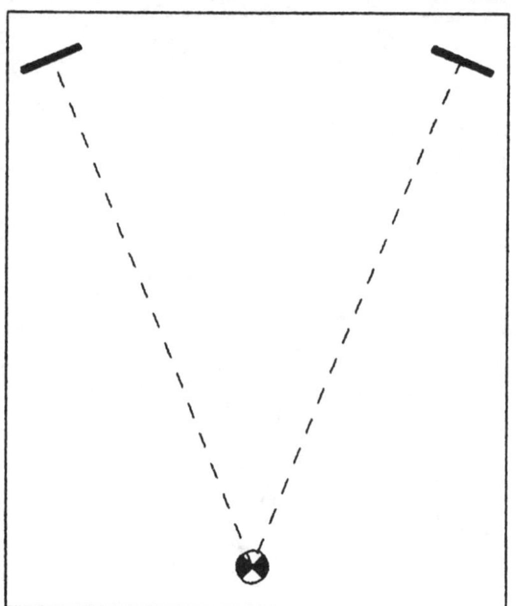

FIGURE 14-10: Musicians appear close to you in your room.

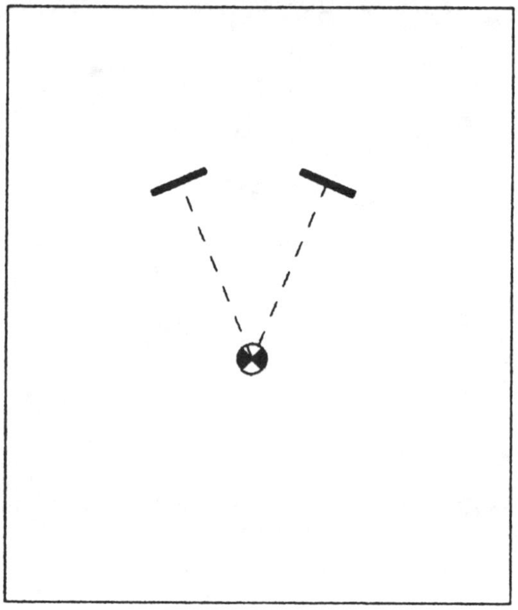

FIGURE 14-11: Musicians appear close to you in the concert hall.

you can be sure you'll have no problems. But I also find that you can put a dipole within a few inches of a wall—if it is at an angle to the wall. *Figure 14-12* shows a planar ESL in a position which usually works well. *Figure 14-13* shows a wide-dispersion ESL which will surely have problems.

When the speaker's rear radiation bounces off the wall and passes back through the speaker, it causes phase interference with the direct sound and peaks and valleys in the frequency response. Also, the sound pressure waves trapped between the wall and the cell

can distort the diaphragm's shape and motion. The curved speaker in *Fig. 14-13* has these problems. By angling the speaker to the wall, the reflected sound is moved away from the diaphragm to minimize these problems. But there is a limit: if the speaker is too close to the wall, even angling won't help.

So the question is, "How close is too close?" This question is not easily answered because it is frequency dependent. The lower the frequency, the further the speaker must be from the wall. As a rule, if your ESL is positioned so the rearwave does not reflect back into the diaphragm, you will have no problems. The degree of angulation and speaker width determine the distance the speaker must be from the wall. *Figure 14-14* shows several examples.

Though this works well, you can often place them closer if you are only producing midrange/highs. Only experimentation will provide the final answer.

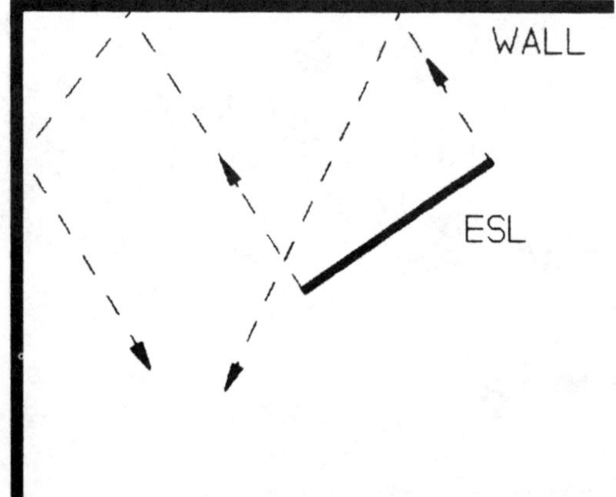

FIGURE 14-12: Good position—rear wave avoids ESL.

FIGURE 14-13: Poor position—Rear wave reflects through dipole.

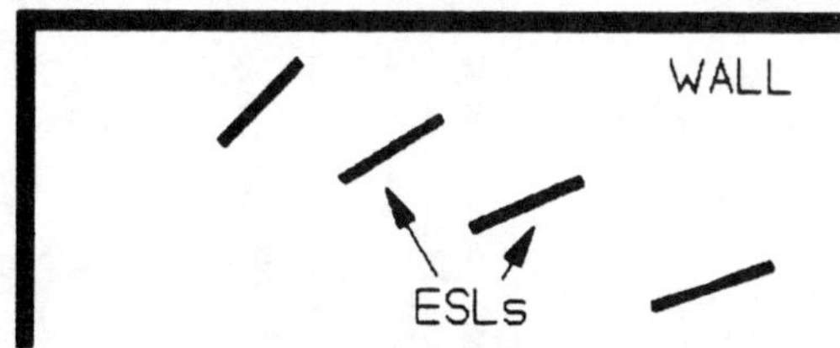

FIGURE 14-14: Greater angles permit ESL to be closer to wall.

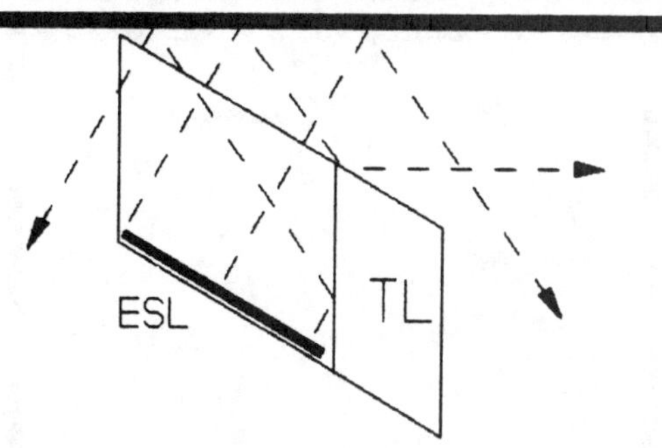

FIGURE 14-15: Rear waves avoid ESL in beam splitter design.

Note that the integrated ESL/TL is a special case. Because of the parallelogram cabinet design, place this type of speaker directly against the wall. In fact, you *must* place them directly against the wall for the beam splitter to work correctly. With my design, none of the rear beams reflect back into the diaphragm (*Fig. 14-15*).

PRECISION DISTANCES. Planar ESLs can produce holographic-quality images. To do so, you must be geometrically precise about their position relative to your seating location.

The distances from the ESLs' diaphragms to your ears must be as equal as possible. The wavelength of a 10kHz tone is only about 1″. Your speaker positioning needs to be at *least* that accurate.

If the distances are unequal by as little as 3″, the image deteriorates. The sound becomes unbalanced, the image no longer appears centered, and adjusting channel balance just does-

n't fix it. The phase relationships are so impaired that you can't even notice the difference when you deliberately put one speaker out of phase! (Normally, an out-of-phase speaker has a diffuse and directionless quality that is very obvious.)

The ESLs are not flawed. After all, the poor sound from an improperly positioned ESL is no worse than the normal sound from a conventional speaker. Once you've heard the image from a properly positioned ESL, you simply won't be satisfied with anything less. The difference is that obvious.

Accurately positioning the ESLs is not as easy as it looks. You must make many equal measurements simultaneously because they interact. Changing one often disturbs another. *Figure 14-16* shows two sets of dimensions which must be equal; *Fig. 14-17* shows others.

If you have an obstruction on the speaker surface (such as a joint between two large cells), be certain you do not listen to this area of the

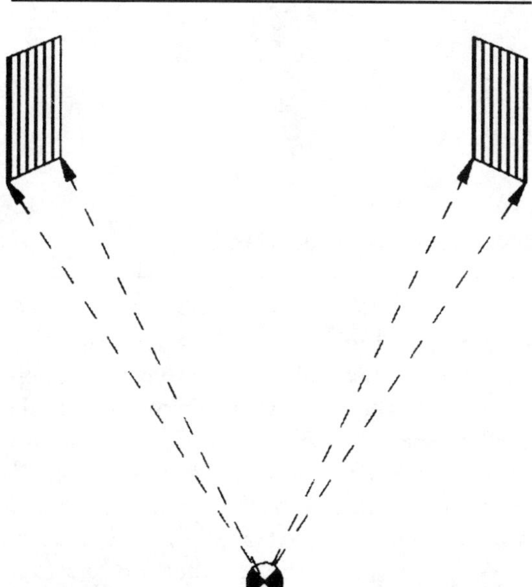

FIGURE 14-16: Distances must be equal.

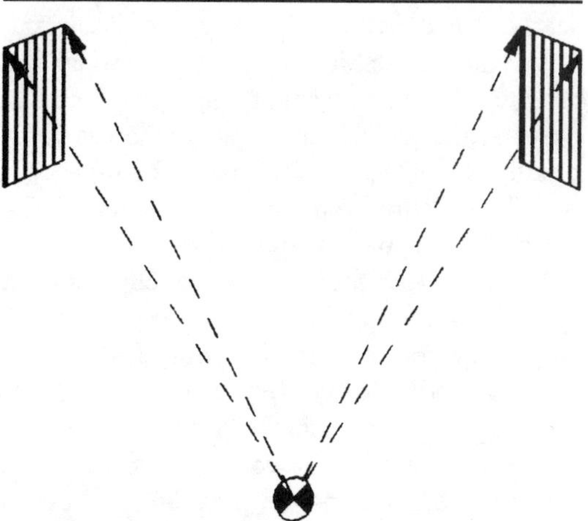

FIGURE 14-17: Distances must be equal.

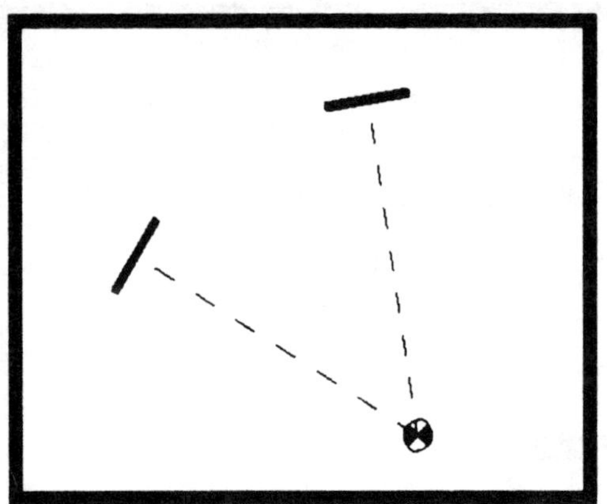

FIGURE 14-18: Asymmetric positioning.

speaker. To prevent this, make the upper and lower dimensions different. If you want the speaker perfectly vertical for aesthetic reasons, these dimensions will also be unequal.

In addition to matching the upper and lower dimensions for aesthetic reasons, you may want the speakers placed asymmetrically, as shown in *Fig. 14-18*, rather than symmetrically, as is usually the case. This complicates positioning.

The speakers must not be twisted or warped, which is a problem if you use very thin frames. *Figure 14-19* shows a twisted ESL.

PHYSICAL SETUP. You can deal with these setup variables in a number of ways. The following is the method I use.

Begin by putting the speakers and your listening chair about where you want them. Remove the grille cloth (if any).

Determine where your ears are in relation to the chair. Usually, this is about 2″ forward of the point where the chair back meets the seat, but it depends on the type of chair and how you like to sit. I suggest you sit in the chair and have an assistant help you. They can use a string with a small weight on it for a plumb bob. Place a small piece of tape or a thumbtack equidistant from the sides of the chair's seat cushion at that point.

Measure from the inside lower edge of each speaker to the tape at the center of your chair. Adjust the speakers and/or your chair so this distance is identical for both speakers (*Fig. 14-20*). To gauge the distance, I use a large tape measure. Alternatively, you can stick a pin in the seat cushion and tie a thread to it. Walk to the various points with the other end so the distances will be identical.

Next, measure from the outside lower edge of

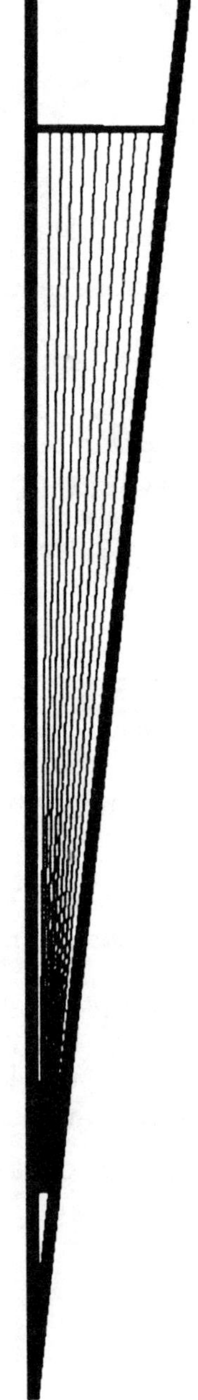

FIGURE 14-19: Twisted ESL—Avoid!

each speaker to the tape. When these are equal to the inner measurement, turn your attention to the top. The measurements from the inside and outside top edges should be equal. An optional aid is a large level. You can put it on each speaker to get them vertical. Note that I said "aid"—it doesn't replace the measurements.

When you have the four bottom and four top measurements equal, the speaker should be directly aimed at your listening location. Make one more set of measurements on a

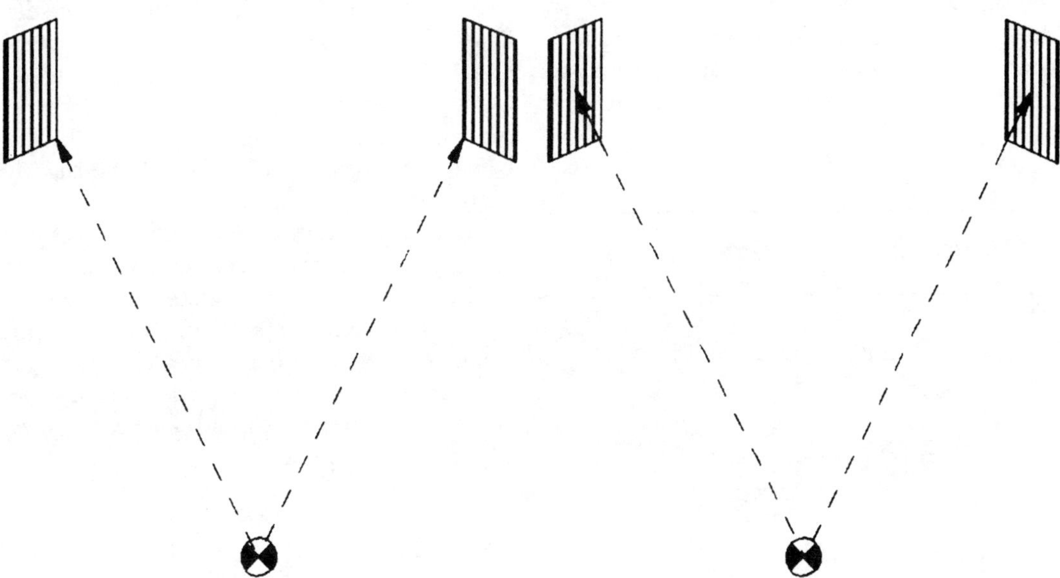

FIGURE 14-20: Equal distances.

FIGURE 14-21: Equal distances.

horizontal line from your ears to the center of the stator—this is the critical one. Since this is the speaker part that actually fires sound directly to your ears, this dimension should be as close to identical as possible for both speakers (*Fig. 14-21*).

If your speakers' frames are perfectly true, getting the top and bottom measurements right will automatically ensure the center is right. However, most of us don't build that accurately. Even if we do, frames sometimes warp. Therefore, it's imperative you check the center measurement, which is more critical than the others.

Now comes the acid test. Sit in your listening chair and look for your face's reflection in the diaphragms. If the lighting is good, it will be easy to see. If necessary, you can hold a flashlight just above your head to make it obvious.

Your reflections must be centered between the vertical borders and in the same relative vertical position on both ESLs. If not, move your chair and/or the speakers to get it right (*Fig. 14-22*).

If you needed to move something, you must go through the measurements again, as they will have changed. Just because you have the reflections centered, doesn't necessarily mean the distance from each diaphragm to your ears is identical. The object is to get the reflections centered left to right *and* vertically equal *and* the chair-to-speaker distances equal. This can be a tedious process, but undeniably important.

The type of mounting system you use can make this process either easy or difficult. Probably the easiest to adjust is the "room divider" ESL which has spring-loaded feet to hold it against the ceiling. Shifting any corner is easy.

The integrated hybrids have narrow bases and are relatively unstable on a soft carpet. They may lean in any direction and be difficult to position. You can shim their bases or adjust "Tiptoes" to position them.

If you are using a separate freestanding

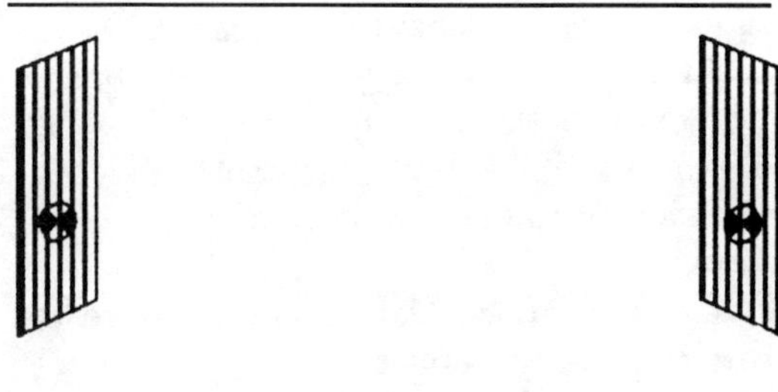

FIGURE 14-22: Reflections must be centered and identical for both channels.

woofer system, you should adjust them so they are also equidistant from your ears. A common question (and one that is not easy to answer) is, where should the woofer cone be relative to the ESL's diaphragm?

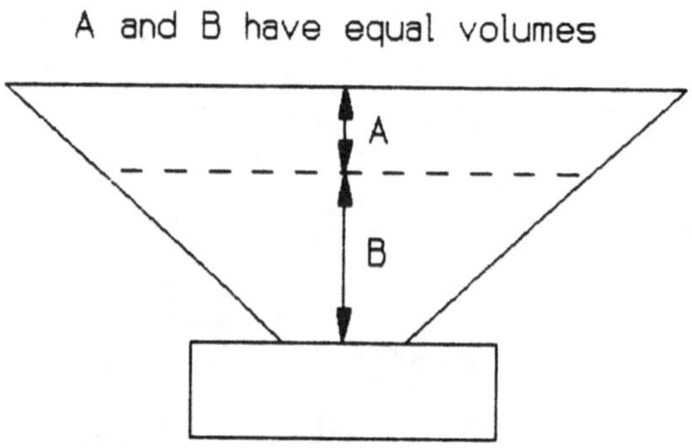

FIGURE 14-23: Align center of cone volume with ESL diaphragm.

I place the woofers in a position where the cone's 50% volume is even with the diaphragm. Generally, that point is about 30% of the way down the cone (*Fig. 14-23*).

When set up correctly, the phasing/imaging of planar ESLs is so precise that moving your head to the left will not shift the image to the left, but to the right. In other words, your brain prefers it over loudness when determining source position!

If you are using an integrated ESL/TL, a hard surface must be behind the speakers for the beam splitter to work well. A bare wall or window is best, as heavy drapes, cork walls, or other sound-absorbing materials will reduce rearwave output and dispersion.

ELECTRICAL SETUP. You *must* have the correct phasing between the two channels. Failure to do so will ruin the image.

If your electrical connections for both channels are identical, the phasing will automatically be correct. When in doubt, reverse the connections to one ESL while playing a monaural signal source. When the speakers are in phase, the source should appear to float in the room between the speakers. When they are out of phase, the sound will have a diffuse and directionless quality. You should have no trouble telling which is correct, but if you can't, you don't have your speakers precisely positioned.

Determining the correct phasing between

each ESL and its respective woofer is more difficult. Do this after the system is fully operational. The effect will be subtle, but you can usually detect a slight increase in fullness in the upper bass/lower midrange when the drivers are in phase. The sound will be slightly thinner when out of phase.

In my experience, this is audible only if you use high-quality, odd-order Butterworth filters in the crossovers. I can't detect a difference using 12dB/octave crossovers. Evaluating this is difficult if a delay occurs while you reverse the leads to your woofers. You can make an instant A-B tester to tell with authority. *Figure 14-24* shows a suitable schematic.

CRITICAL ADJUSTMENT. In a hybrid system, you *must* match the woofer level to the ESL level. The correct frequency balance will make or break your system. I can't overemphasize how critical this adjustment is.

Unfortunately, without sophisticated instrumentation you can't objectively adjust the frequency balance. You will have to do it subjectively, but without master tapes of live concerts what do you use for a reference standard? None. You must use your best judgment. This can be tricky, but some suggestions will help guide you.

Unlike most electrostatics, yours will not

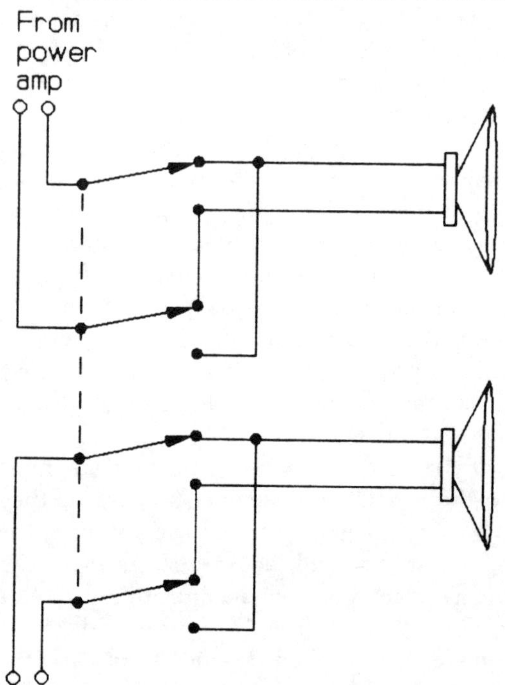

FIGURE 14-24: Test circuit for phase tester.

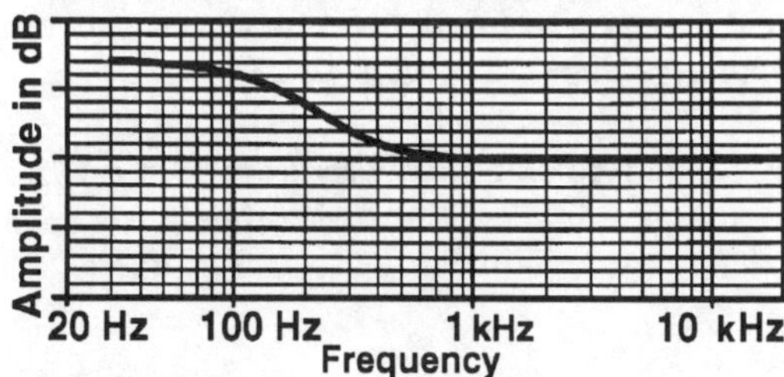

FIGURE 14-25: Excessive bass.

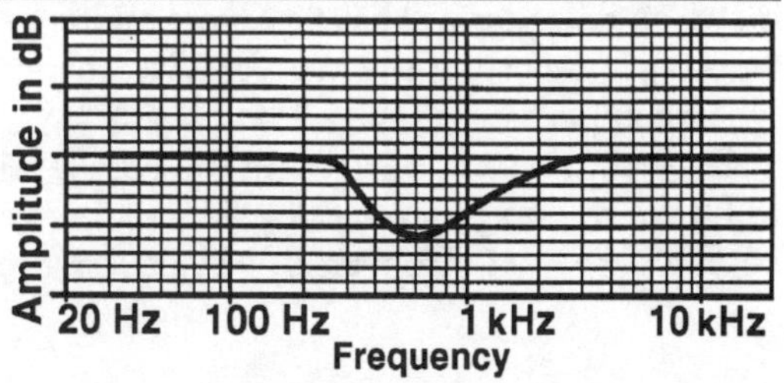

FIGURE 14-26: Midrange "suckout."

sound bright and thin if you have equalized them properly. If you built transmission-line hybrids, the bass will be neither boomy nor thin. Your system will have the full, rich sound of the best magnetic speakers, but with the detail, imaging, and delicacy of electrostatics.

Although you are adjusting the bass, the trick is to listen to the *midrange* when making the adjustment. Adjust the woofer level so it is full, but completely clean and clear.

If there is lack of definition in the midrange, the woofer level is too high. If it is thin, the woofer level is too low.

If the bass is excessive when the midrange seems properly full, try adding stuffing to the TL or placing the speaker in a different room location. It might seem as though you can't increase the woofer level enough to get ample bass without the midrange becoming muddy. Try removing stuffing from the TL or a different room position.

Sometimes this isn't a fault of the TL. Keep in mind that you probably have been listening to a woofer system with a pronounced resonance. A TL system won't have this quality. If you try to adjust a TL so it sounds as "bassy" as your previous highly resonant woofer, you are asking for trouble.

The problem is with our "audio memory." After we listen to an inaccurate system for a while, our mind thinks that's how the music should sound. When you hear a linear system, it sounds wrong.

Audiophiles commonly have this problem. I even found myself caught in this trap until I took objective measurements and discovered my error.

If you find the system sounds bass-shy when the midrange is full, and you've adjusted the stuffing in the TL, simply adjust the woofer level so the midrange is clean—even if the bass sounds inadequate. Listen to it for a while. Unless the stuffing is way off, you will soon realize that the present sound is right and the previous sound was wrong.

A good test of your TL's stuffing and performance is the very deep bass. A properly operating transmission line will *not* sound "bassy" until it encounters some really deep bass. Then it will amaze you with its power, but it won't do this if it's "overstuffed."

If the deep bass seems excessive when the mid-bass sounds right, it's a sure sign that you've been listening to an inaccurate resonant-woofer system prior to your TL experience.

All this assumes you have a variety of first-class source material, which is a major problem since very little commercial material is of reference quality. Be particularly careful when using the male voice as a reference. Good recordings of male voices are rare, because cardioid microphones exhibit **proximity effect** when used at close range. This phenomenon produces exaggerated bass and is largely responsible for the unnatural quality of male voices and guitars on commercial source material.

A recording engineer may try to correct this problem with his equalizer. It then becomes his judgment of what the voice should sound like, and that's hardly what you should use as a reference standard! The solution is to use sensitive omnidirectional microphones placed several feet from the performer, but recording studios just don't do this. Consequently, I discourage you from using male voices as a test of driver balance.

Recording studios never record "pop" music naturally, so it makes a poor reference standard as well. Use complex, classical orchestral music—recorded naturally, if possible. Don't put all your eggs in one basket either. A variety of recordings from different manufacturers will help you eventually find an average balance which works well. If you set the balance with classical music, you should find other types are automatically properly balanced.

This discussion assumes you are using appropriate crossover points and equalization. If you use excessively low crossover points with the attitude that no woofer system can match an ESL, you will have a severe midrange "suckout." Adjusting the woofer/ESL balance won't compensate for this.

On such a system, increasing the bass level to fill the midrange produces wildly excessive bass (*Fig. 14-25*). Conversely, adjusting the bass to an appropriate level produces a severe midrange depression (*Fig. 14-26*).

If you find yourself in this predicament, there are two solutions. One is to increase the midrange equalization. The other is to raise the crossover point, usually to 400–500Hz.

CHAPTER 15

TESTING

The most frustrating aspect of high-quality sound reproduction is the controversy surrounding many aspects of equipment performance. The lack of clearly defined, objective, reproducible testing procedures results in a wide diversity of inaccurate, confusing, arbitrary, and misleading opinions.

Most listeners, and all would-be speaker builders, wish to know for a fact what a component will sound like *before* they sink considerable time, effort, and money into a project. High-fidelity sound reproduction is not an art; it's a science. We should have meaningful scientific testing methods. Unfortunately, advertising hype and consumer ignorance have turned equipment evaluations into a three-ring circus.

Most audiophiles have strong opinions regarding the sounds of various components. Everyone can recall personal experiences where a component sounded "veiled" or "shrill," or some other equally nebulous, subjective term. We can point out that we have heard a different component which sounded better.

We often try to explain the difference with some unproven technical rationale. Examples include: one is tubed, while the other is transistorized; one has capacitors in the signal path, while the other does not; one operates in Class A, while the other is Class AB.

This type of component evaluation is aggravating because different listeners reach different conclusions. Since the results are unreproducible, you can't tell which is right.

Listeners argue endlessly over unproven subjective impressions. One will never convince the other he/she is wrong. After all, they both heard what they heard. For one to agree with the other implies that he/she can't hear well, or that his/her opinion is less important. Human nature revolts against this.

This whole problem can be put into perspective once everybody understands that *both* listeners are telling the truth and neither is wrong. When I hear an audiophile describe a sound as "veiled" or "shrill," I believe him/her. I only have trouble with the conclusion that the transformer, amplifier, stator, or whatever, was the cause.

To determine a problem's true cause, you must eliminate all possible factors which could account for any differences you hear. In scientific circles, this is termed "isolating the variables." When you are certain there is only one variable, only then can you be certain it is responsible for the sound you hear.

This concept is fundamental to all testing, and everybody accepts it. For example, when testing amplifiers, audiophiles understand that you can switch only from one amplifier to another. You can't switch amplifiers and loudspeakers at the same time and draw any valid conclusions from the sound differences you will surely hear.

What I find amazing is that most listeners ignore this basic principle, even when it is so widely recognized. Typically, an audiophile will eliminate some variables but will ignore others, such as power output, internal gain, phase inversion, loudness differences, psychological variables, and clipping behavior.

TESTING FLAWS. I will explore and demonstrate different testing procedures, using as examples two controversial topics. One is direct-drive amplifiers versus hybrid amplification. The other is perforated-metal versus wire-stator construction. Builders argue endlessly about the merits of each and how different they sound, yet controlled scientific testing proves there is no difference.

Let's look very closely at amplifier and transformer testing to see if we can determine why listeners arrive at different conclusions. I'll use a friend and fellow ESL builder, Barry, who wanted to find out if transformers in a hybrid (conventional amplifier plus step-up transformer) amplifier system degraded the sound compared with a D/D (direct-drive, high-voltage) amplifier.

He tested transformers in two ways. First, he listened to a D/D amplifier for several minutes. Then, he turned off the system, disconnected the D/D amp, hooked up a hybrid amplifier, and listened again. He concluded that the sound he heard from the hybrid amplifier was "veiled" compared with the D/D amplifier.

Second, he listened for several minutes to a moving-coil phono cartridge through transformers, then stopped the music and switched to an electronic "pre-preamp." After listening again for several minutes, he found the cartridge sounded "clearer" with the "pre-preamp." He blamed these flaws on the transformer.

He concluded from these tests that transformers are bad and that D/D amplifiers are the best way to drive ESLs. Should you believe Barry? Is he right or wrong? Can you trust his conclusions? Remember that he heard what he said he heard, and there is no reason to doubt his observations.

Consider the following factor in the D/D amplifier test: was the transformer the only variable? Clearly, it was not. Many other *uncontrolled* variables were involved, not the least of which was an entirely different amplifier.

That may have caused the difference he heard. How could he know whether the amplifier or the transformer was at fault? He couldn't. Furthermore, many more uncontrolled variables were present in his test. Psychological variables are extremely important, and are such a common cause of error that they are carefully controlled in valid scientific testing. Yet audiophiles nearly always ignore them.

In this example, they include prejudice, time delay, memory error, and placebo effect. Specifically, Barry had to *remember* the previous sound to compare it. He had no immediate reference, and audio memory is notoriously inaccurate.

A very significant **time delay** occurred while he hooked up the other system during which he forgot to some degree what the previous sound was like. He also admits to being **prejudiced** in his expectations that the D/D amp would be better.

Not only is it possible, it's highly likely that his memory recalls the D/D amplifier as essentially perfect, while his prejudice easily finds imperfections in the hybrid system.

The **placebo effect** is also operable here. He *expected* the D/D amplifier to sound better so it did.

These psychological variables are very real. The effects they produce are not imaginary, and you don't have to be crazy to be affected by them. Prejudices still work at an unconscious level no matter how objective a person is. These effects are so important that they have forced scientists to test in the "blind" mode.

Valid testing would demand that Barry not know which amplifier system he was listening to so he could not prejudge its performance. Somebody else would switch the amps so Barry would not know which one was playing.

Scientists usually go one step further and use the "double-blind" testing procedure. Not only is the listener "blind" to what he is hearing, but the person switching the amps also does not know which amp is operating. This way, he can't give the listener any clues, conscious or unconscious.

Other variables exist as well. For example, was the hybrid amp clipping, while the D/D was not? If so, Barry would be comparing a distorted to a clean amp. Hardly a conclusive test of an amplifier, much less a transformer.

What about playback levels? If one is even slightly louder, it will sound better. What about source material? Was it identical for both tests?

I could go on, but my point is that many uncontrolled variables could have been responsible for the sound Barry heard. Again, I stress Barry's sincerity when he says that the sound he heard was "veiled." We have no reason to doubt what he heard; only his conclusions are suspect.

What about his test with a moving-coil cartridge and transformer? Again, there were many uncontrolled variables, including an amplifier present in one test which wasn't present in another.

You cannot judge ESL transformers based on listening tests (uncontrolled or otherwise) of moving-coil cartridge transformers. The two aren't similar enough to draw conclusions.

Specifically, one is a transducer driving a transformer that is driving an amplifier, while the other is an amplifier driving a transformer that is driving a capacitor. Another difference is that one is a high-voltage, high-power device, while the other is extremely low voltage and low power. One is very small, the other is large; the loads are totally different.

TEST EQUIPMENT AND PROCEDURES. Well-controlled testing procedures are readily available. They don't require much equipment and are very accurate, reliable, and reproducible. They do require extra effort, but this is necessary if you wish to draw valid conclusions.

Such testing is not based on instruments, because while they are accurate, we all recognize that they do not define everything we hear. The tests I will describe rely on the most sensitive instrument of all—the human ear.

Controlled testing requires you to isolate all the variables and have an immediate reference. This may or may not be easy depending upon what you are testing.

The easiest and most precise of such tests is the bypass test. This compares the output of the device under test to the output of a straight wire, and is superb for testing signal processing devices and low-level amplifiers such as tape recorders, equalizers, crossovers, transformers, preamplifiers, and mixers.

Required test equipment includes your basic sound reproduction system, a DPDT switch, an audio voltmeter, and an audio generator. If you don't have an audio voltmeter, you can use the analog meters on a good tape recorder or oscilloscope. However, considering the investment you have in your sound system, the cost of an audio voltmeter is insignificant and you really should get one.

If the meter has a very high input impedance (on the order of $10M\Omega$) so it does not load sensitive circuits, you can use it for all sorts of electronics work. Suitable audio meters are often just your basic FET VOM or DVM, but with a decibel scale and a specified frequency bandwidth listed in the specifications. I much prefer analog meters for this function, but you can also use digital types.

You do not need a full-blown, precision, sine/square-wave audio generator. However, you need at least a steady-state, midrange sine wave for setting levels. Tuners or tape decks often have a built-in tone generator which will work, and some test CDs have steady-state tones.

Mount the switch in a box with whatever jacks are necessary to hook up your equipment. You will test in monaural, as you should have only one variable, so switch only one channel. Send the output to both loudspeakers if you wish.

Let's assume you are testing a linear amplifier such as the high-level section of a preamplifier or a microphone mixer. Connect a high-quality signal source (usually a tape recorder or CD player) to the switchbox so you can direct the signal alternately through the device or through a short, straight piece of wire (*Fig. 15-1*).

The reference wire is usually soldered directly to the switch terminals, but a short connector cable works fine and jacks offer flexibility. The wire must not be long and curly, or it can act as an inductor with potentially adverse effects.

The output from the switchbox feeds the rest of your audio system so you can listen critically. The output level of the device under test must be *perfectly* matched to that of the

wire, as the ear is amazingly sensitive to slight differences in level.

Measure the wire's output level, then match it with the device. Be as precise as you can, preferably within a tenth of a decibel.

You must not drive the device under test into distortion. Although this is usually not a problem with low-level electronics, if you have a distortion analyzer available, you should use it.

The source material you use is important. Simple, isolated sounds such as triangles, percussion, and solo instruments are not very challenging. Complex material such as a full symphony orchestra more readily detects problems.

Testing with white noise (a combination of all frequencies such as the sound of a waterfall) is a must. A good source is a tape recorder playing blank tape at a high level with any noise reduction turned off. Another good source is FM interstation hiss. Turn off any muting and pick a frequency that has no station. White noise is the best test for evaluating frequency response differences.

Perform the test by listening for *any difference* in sound as you flip the switch from the device to the straight wire. The comparison is immediate, so no memory is involved. Your reference source is essentially perfect.

You must not know which switch position represents the wire and which the device under test—this is **blind testing**. Have an assistant connect things up for you, or at least unplug the output jacks after you have set everything up. Then randomly switch them around and plug them in again so you do not know which is which.

You can do the switching, but it is better if your assistant does it. Ideally, he/she shouldn't know what the switch positions represent either, which is called **double-blind testing**.

Have a panel of listeners judge rather than just yourself. You may think you hear a dif-

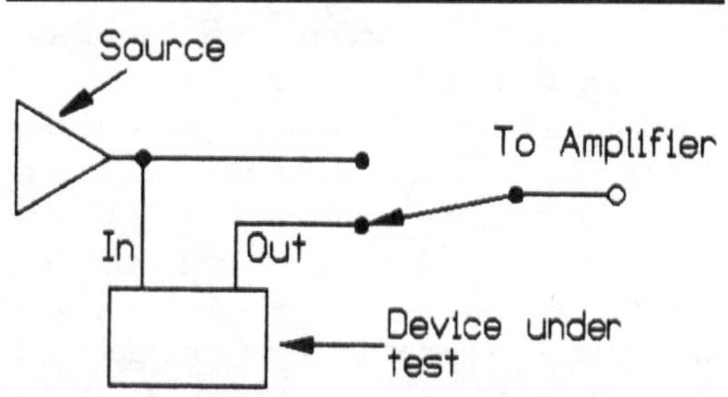

FIGURE 15-1: Switch box for bypass test.

ference and be wrong. If there is a difference, most of the panel should be able to hear it and agree.

The listeners need not be "golden ears," since they only listen for any sound difference. I like to think I fall into the "golden ears" category, but I've found that anybody can hear a difference when I do if he/she gets a little instruction and practice.

Well-controlled tests include tally sheets and statistical analysis; however, most panels readily detect differences and agree on them. If differences are so subtle that there is significant disagreement, the difference is probably either nonexistent or insignificant.

Again, let me stress that the only thing you must determine is whether there is any *difference* in sound. Trying to subjectively describe any difference you might encounter is neither necessary nor desirable. If there is any difference, the item under test is distorted in some way. If several test items are distorted, the one with the least difference is the most accurate, although it is not perfect.

One more thing must be kept in mind if you hear a difference, and that is phase. Some electronics invert the signal phase, while others don't. This will cause the drivers to move in opposite directions when switching during the test. The effect is surprisingly subtle, but when I hear differences, I routinely reverse the phase to see if they disappear.

You must be creative to test some components. For example, to test one which has equalization, such as the RIAA equalization in a phono preamp, you must build a passive reverse RIAA equalizer. Mate the equalizer with an attenuator to reduce the tape recorder source output to the level and equalization characteristics of a phono cartridge. You can thus get linear output from the phono preamp and avoid overloading it.

Crossovers require reverse equalization unless you limit the input signal to their linear pass band. Alternatively, you can mix the high- and low-pass sections to see how well they match the straight wire.

Mixing the filters introduces two variables in that you have two different amplifiers at work (unless the crossover is passive). However, this does not necessarily void the test, because you will be using them together in your speaker system and can consider them one component.

You must be very careful when drawing conclusions from this or any other testing. The tests are valid only for the specific situation; you may not infer performance for different situations.

For example, I have done considerable testing of phono preamps using reverse RIAA equalization and attenuation, while feeding master tapes through the preamps. I have found preamps which did not sound different from the straight wire (although they were rare).

When two of these "perfect" preamps were tested in a direct AB comparison against each other using a cartridge playing a disc, however, there were obvious differences. Clearly, they behaved differently when driven with a highly reactive phono cartridge than when driven by line-level electronics. In other words, perfect performance with the tape recorder source did not predict the disc performance.

The bypass test is extremely sensitive, and it is not unusual to detect differences, although you will be surprised by their subtlety. Such reliable, reproducible testing allows you to look for a problem's source and attempt to eliminate it through a circuit change or other modification.

Testing power amplifiers is more complicated, since you no longer have a perfect reference source. A fundamental need for power gain exists which a wire alone cannot supply.

Therefore, you can't say that a given amplifier is "perfect." You can only say it sounds different from, or identical to, another amplifier. While not ideal, this is nevertheless a very good test if properly done, and particularly when trying to answer questions such as, "Which is better, a D/D amp or a hybrid?" Or, more accurately, "Joe claims amplifier A sounds `better' than amplifier B. Let's do an honest test and see if a panel of listeners hears any difference between them."

Amplifier testing is similar to bypass testing. Double-blind, immediate AB switching comparisons are made with complex source material while simply listening for any differences. But the setup is different.

You will need a switchbox that will handle significant power and, if testing ESL amplifier systems, it must handle high voltages. Also, you *absolutely must* do distortion measurements!

It has been my experience that all sonic differences in properly operating, high-quality amplifiers are due to overload conditions which drive them into high distortion. To deal with this, you must know for a fact that you never drive the amplifiers into any type of distortion. Obviously, an amplifier that is grossly distorting is going to sound different from one that is not.

You will be surprised by how easily (and often) you distort amplifiers at the levels at which you listen. You will be amazed by how quietly you must listen to moderate-power amplifiers to ensure that you are truly comparing amplifiers and not distortion products.

LOUDSPEAKER TESTING. Loudspeaker testing is difficult to do in a scientifically valid manner because the only "perfect" reference is a live performance. But between the sound of live performers and the sound from a loudspeaker are vast numbers of variables that cannot be isolated. While no objective test exists that defines absolute loudspeaker performance, there are tests that can define certain aspects. You can compare ESL-stator-construction techniques, for instance.

Many ESL audiophiles believe that different stator-construction techniques produce different sounding ESLs. Theory suggests that the stator should not affect the sound if there is adequate open area.

To settle this question, I devised a test to compare a perforated-metal to a wire stator using double-blind, immediate-AB-comparison techniques. The cells were identical in every way except for material.

The test was difficult, although theoretically it should have been easy. I needed a switchbox which could directly switch the high-voltage drive from one cell to the other, while the cells were hidden from the listening panel by either a curtain or darkness.

When I did this, the panel immediately noted obvious differences. Analysis showed several causes. First, the perforated cell was less efficient than the wire one. Only a little, but even a fraction of a decibel is easily detectable in AB testing. Therefore, I had to add an attenuator.

Using an attenuator on only one cell introduced another unpermitted variable, so I had to add an attenuator to both cells. To control the variables, it was necessary to use the same attenuator, which was then switched to different levels when I switched from cell to cell.

This required a relay and fixed resistors at the amplifier input, as high-voltage-drive-signal attenuation was not practical. The high-voltage drive also had to be switched with relays so the entire change could be instantly made. In short, the switchbox became complicated and expensive.

Additionally, output levels had to be taken with an SPL meter rather than with an audio voltmeter. This could not be casually done, because room acoustics caused marked output differences in different locations at any steady-state frequency. I found it necessary to use white noise with the bass filtered out ("A-weighted" frequency response) to obtain a steady reading which was not significantly influenced by room acoustics. The SPL meter had to be placed in the listening location, untouched, with the person reading it always standing in the same position.

Further testing still showed obvious sonic differences which were easily detected by all members of the listening panel. Now, most audiophiles would stop here, describe the sound in some nebulous, subjective way, and draw conclusions about which type of stator construction is best.

This would be a serious mistake, for although the test appears to be valid, it is not. I still did not control all the variables! Yes, there was a sound difference, but it was not caused by stator construction.

The true cause was easy to find. Since the cells could not occupy the same physical space, the room acoustics were different for each.

I tried to solve the problem by putting one cell immediately in front of the other but differences persisted, although to a lesser degree. This led me to believe that I was at least on the right track.

Finally, I decided the test would have to be conducted outdoors in an open field. To blind the listeners, it would have to be conducted in the dark.

This was quite a task. I had to string power cords and haul tape recorders, switching networks, speakers, chairs, speaker stands, and amplifiers out in a field. Some light was needed after sunset while everybody found their seats and began their duties, so the speakers had to be hidden from view until the actual testing commenced. I had to wait for suitable weather and a new moon. Arrangements were made with some neighbors to turn off outdoor lighting. Are we having fun yet?

There were still differences! But this time, at least, they were very slight.

The test was repeated with the cells separated rather than with their edges close together. I thought that putting the cells together effectively put a baffle on one side of one cell in one switch position, which would have been reversed on the other. Sure enough, with that correction the sound was finally identical.

I realize you may not believe what I am saying. You are certain you have heard differences in components and see no reason to go through a scientifically controlled testing program to pick a good one. You are sure I'm exaggerating the problems. You are sure I'm wrong. You are thinking, "All amplifiers sound the same! Who does he think he's kidding? What nonsense!"

Good, that's what I want you to think! I don't want you to believe my conclusions. I want to turn you into a skeptic. I want to plant the seed of doubt in your mind so you understand that you cannot trust uncontrolled testing. The responsibility is now yours to prove me wrong—but not with assumptions and guesses. You have no choice but to devise and use controlled-testing techniques.

Once you've discovered its accuracy, you will refuse to believe any other results. When everybody refuses to place credibility in uncontrolled tests, controlled testing will become the norm and we will be free from the scourge of inaccurate, misleading, and confusing opinion that plagues us.

CHAPTER 16

FURTHER RESEARCH

SANDWICHED CELLS. Much work is required to fully understand the operation of membrane speakers. The most pressing need is to develop larger excursions so we can build speakers with smaller radiating areas which produce high output. One approach I don't think has been fully explored is stacking (sandwiching) cells.

Consider an ESL that is six-layers thick, as shown in *Fig. 16-1*. You can use common stators if alternating diaphragms have opposite polarizing supply voltages. Compared with a single-diaphragm ESL of otherwise identical design, the sandwiched cell would have ten times the output for the same drive voltage.

Here Murphy's Law nails us: the speaker would also have ten times the excursion. We will need to increase the D/S spacing to compensate, which will reduce output substantially. To find out how much, we must experiment.

Increasing the D/S spacing will reduce the output for a given set of voltages. Having more diaphragms, however, would only increase the output more than the reduction of the electrostatic force would decrease it.

The same is true of the speaker's capacitance. For a given D/S spacing, the capacitance will increase ten times. However, increasing the D/S spacing will decrease it. Again, not as much as six diaphragms will increase it, but it will help keep it within reason.

If a multi-diaphragm ESL can produce high outputs, it could be made much smaller than the usual ESLs. I envision a multi-diaphragm ESL perhaps 12″ across in a transmission line. This would settle the issue of hybrid versus full-range ESL operation once and for all.

BETTER TRANSFORMERS. Step-up transformers that could maintain linear frequency response at higher turns ratios would be very helpful. They would also need to sustain higher voltages.

Current transformer design is apparently incapable of better performance. Perhaps a radically new transformer design that can solve this problem has not yet been discovered.

The basic problem is that the wires making up the windings are physically separated. A promising solution is in the area of high-temperature superconductors. This would permit extremely small wire and windings as well as

very-low-leakage inductance, the main cause of high-frequency resonance that limits response.

D/D AMPLIFIERS. A breakthrough in direct-coupled amplifiers could relegate step-up transformers to the junk pile. But will anyone ever design one? Serious barriers exist to making them acceptable to most ESL users.

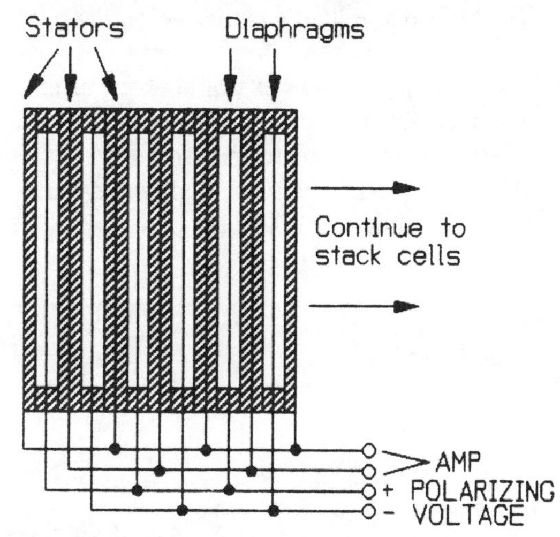

FIGURE 16-1: Compound ESL.

As a start, an acceptable amplifier can't use vacuum tubes. Tubes deteriorate, are large and bulky, and produce much waste heat. In addition, the multiple power supplies they require drive up the amplifier's cost. A satisfactory D/D amplifier must use transistors.

Furthermore, it must be efficient. A Class A amp which dissipates several kilowatts of power is totally unacceptable for most ESL owners.

I am optimistic, however. I believe a suitable D/D amp design is possible. All we need is sufficient demand for a manufacturer to put the necessary resources into designing one.

HYBRID AMPLIFIERS. Better step-up transformers and D/D amplifiers involve major engineering problems that make their development impractical or impossible. But one solution to audio-drive problems that is well within the grasp of current technology is the hybrid-amplifier system. A conventional power amplifier and step-up transformer could be specially modified to perform ideally with ESLs.

The hybrid would be a conventional-design Class AB power amplifier modified to produce higher than normal voltage. Instead of the 100–200V power supply as in the typical high power amplifier, it would use a 500–1kV power supply. The current capability could be reduced to make the amplifier smaller and more economical.

The step-p transformer would have a *lower* turns ratio than we are now forced to use, perhaps 20:1 instead of 50:1. This would greatly improve high-requency response and bandwidth.

The hybrid amplifier's higher voltage would more than compensate for the transformer's lower step-up ratio. For example, if the amplifier had a 500V power supply, it could drive a 20:1 step-up ratio transformer to 10kV. If the amplifier had a 1kV power supply, the transformer would deliver 20kV!

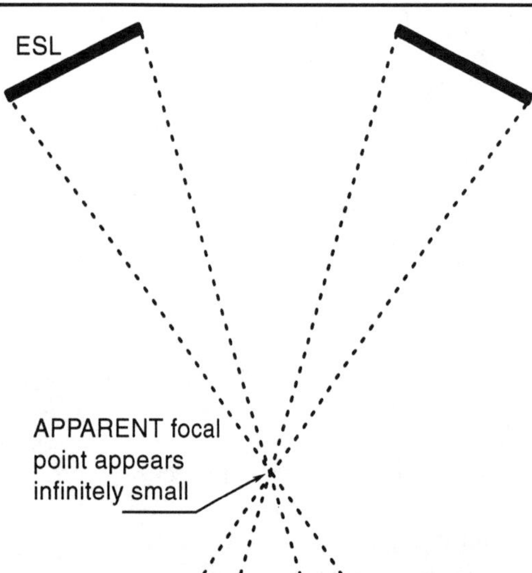

FIGURE 16-2: Focus is infinitely small despite wide beam.

This represents many kilovolts more drive than the best existing amplifier/transformer systems, and it would have better frequency response as well. The ESL could use wider D/S spacing to take advantage of the additional voltage.

Drive voltage of this level might make electrostatic bass a practical proposition. It might be possible to develop a full-range electrostatic of reasonable size that could produce high out-put. Combining this hybrid amplifier with a sandwiched-cell design could produce fabulous results. The hybrid amplifier is an exciting idea. No technological breakthrough is needed—just a designer who will sit down and do it.

RUGGED CELLS. Current construction techniques are barely adequate. ESLs should be more rugged. Another great disadvantage is they are not portable. Better cell construction would help.

DIAPHRAGM EVAPORATION. With time (fortunately, several years), ESL diaphragms develop large holes. The cause is unclear, although it seems as though the diaphragm material literally evaporates. The process may start with a tiny hole, but I haven't seen pin holes grow larger. Even if a pin hole is the initial cause, why would it continue to enlarge?

More study into this problem could lead to diaphragms which last indefinitely. The same material or techniques that resulted in permanent diaphragms could be used to make them thinner and lighter.

BEAM SPLITTER. Beam-splitter technology is only in its infancy. The idea has great untapped potential, but needs much more refinement. Experimenting with beam splitters is relatively cheap and easy. An industrious amateur could make significant contributions to the art and science of ESLs in this area.

PSYCHOACOUSTICS. While physical parameters are important, I find the subject of psychoacoustics intriguing. Several observations have no clear-cut scientific answers, but I would like to know more about them:

- A single planar speaker has a tweeter beam as wide as the speaker. When you cross the beams from two ESLs during stereo listening, they subjectively become infinitely narrow (*Fig. 16-2*).
- Planar ESLs sound subjectively louder than wide-dispersion ones, although they are radiating identical energies.
- Phase relationships seem different between planar cells, curved planes, and domes/cones.

If we could understand why crossed beams sound infinitely narrow, perhaps we could develop a speaker which *seemed* to have wide dispersion while keeping the beams confined enough to minimize room acoustics.

If we could understand why narrow-dispersion speakers seem louder than wide-dispersion ones, perhaps we could make a wide-dispersion speaker that seems to have more output. In other words, we could improve speaker efficiency, which would reduce both distortion and amplifier requirements.

If we could understand the phase behavior differences between different types of radiators, perhaps we could modify inferior drivers to make them sound better. Few would complain if we could make a conventional magnetic speaker sound like an ESL.

APPENDIX A:
SOURCES

Acoustic Instruments, Inc.
1730 Hamlet Dr.
Ypsilanti, MI 48198-3609

- ESL electronics

B & F Enterprises
119 Foster St.
Peabody, MA 01960

- Various high voltage power supplies and associated parts

ESL Clearinghouse
Roger R. Sanders
PO Box 647
Halfway, OR 97834

- Genuine Mylar film, ESL step-up transformers, and contacts through the ESL Clearinghouse for information, parts, and exchange of ideas

H & R Corp.
401 E. Erie Ave.
Philadelphia, PA 19134-1187

- High voltage power supplies, transformers, and associated parts

Old Colony Sound Laboratory
PO Box 243
Peterborough, NH 03458-0243

- Crossover & amplifier kits

APPENDIX B:
SUMMARY OF GUIDELINES

GUIDELINE: Charging resistor value
20–200MΩ. Use the higher values for small cells and low frequencies. The value is not critical. 20MΩ works well in large ESLs.

GUIDELINE: Fundamental resonant frequency
100Hz (±50) for large ESLs in midsized rooms—higher for small ESLs and lower for very large ones.

GUIDELINE: Maximum polarizing voltage
50V/mil of diaphragm-to-stator spacing.

GUIDELINE: Spacer ratio
60:1 to 100:1

GUIDELINE: D/S spacing
- *Tweeters: 30–40 mil*
- *Midrange/tweeters: 70–90 mil*
- *Woofers: 120–260 mil*

GUIDELINE: Hole size limit
Perforations should be at least twice the stator thickness.

GUIDELINE: Open area percentage
More than 40%

GUIDELINE: High field density stators
- *Use thin stators with tiny openings.*
- *Openings should be no larger than the D/S spacing.*
- *Maintain at least 40% open area.*
- *Make openings at least twice the stator's thickness.*

GUIDELINE: Stator opening size
Perforations should be no larger than the D/S spacing.

GUIDELINE: Audio drive voltage
100V/mil of diaphragm-to-stator spacing.

GUILDLINE: ESL capacity
200–250pF/ft²

GUIDELINE: Stator current

$$I = \frac{V \times C}{165}$$

Where:
I = Current as mA
V = Drive voltage as kV
C = Capacitance as pF

GUIDELINE: Crossovers
- *Use as few as possible.*
- *Use them only below 600Hz.*

GUIDELINE: Magnetic drivers
- *Use the largest drivers possible consistent with other design parameters.*
- *Do not use drivers which have hygroscopic (water absorbing) cones, because they change stiffness and mass with humidity. In short, don't use paper cone drivers. Baxtrene, carbon fiber, Kevlar, and other cone materials are better.*
- *The frequency response of your driver should extend linearly for two octaves above your crossover frequency. If the manufacturer can't supply frequency response curves, find one who will.*
- *Buy the highest quality drivers that meet the above criteria.*
- *What constitutes high quality? Some clues are: Cast magnesium instead of stamped sheet steel baskets. Hexagonal or ribbon voice coil wire instead of round wire. Large spiders. Synthetic instead of paper cones. Flat frequency response. First class fit and finish.*
- *Pay whatever you must to meet the above recommendations.*

INDEX

INDEX

INDEX

INDEX

SUPER SOFTWARE FROM OLD COLONY SOUND LAB

AIRR SOUNDBLASTER SOUND CARD SPEAKER RESPONSE SOFTWARE
SOF-AIR1B3G **$49.95**
Julian J. Bunn

AIRR (Anechoic In-Room Response) offers a cheap alternative to loudspeaker measurement for owners of PCs equipped with a SoundBlaster (this brand only) sound card. Based on the principle of pulse mode excitation of the loudspeaker followed by Fast Fourier Analysis, AIRR features full color real-time plotting of both the digitized pulse received by the microphone and the frequency response of the loudspeaker system. Requires PC with SoundBlaster 16 sound card; VGA screen with at least 640 x 480 resolution, preferably color; and microphone (usually supplied with the sound card).

BASSBOX 5.1 FOR WINDOWS
SOF-BAS1B3G **$99**
Harris Technologies

BassBox 5.1 is a great program that aids in the design of bass loudspeaker enclosures in two ways. First, it models how a speaker will sound in a variety of boxes, and second, it models the maximum loudness of the loudspeaker/box combination. Once these tools help you decide what size and type of box to use, BassBox will then help you calculate the dimensions of its port, if the design is vented, and the box itself. Many other capabilities. Windows 3.1+; DOS 5.0+ recommended; IBM 386SX+; 4Mb RAM; mouse; 4Mb hard disk space (cannot be run from floppy). Also available:

SOF-BAS1B3G/X BASSBOX 5.1 PLUS XOVER 2.03 (below) $118

XOVER 2.03 FOR WINDOWS
SOF-XOV1B3G **$29**
Harris Technologies

This package helps in the design of 2-way and 3-way passive dividing networks, as well as simple load compensating circuits, and calculates the component values for many common first-, second-, third-, and fourth-order networks. Two-way designs include Bessel, Butterworth, Chebychev, Gaussian, Legendre, Linear-Phase, and Linkwitz-Riley, while 3-ways include 3- and 3.4-octave spreads between crossover frequencies. Windows 3.1+; DOS 5.0+ recommended; IBM 386SX+; at least 2Mb RAM (4Mb recommended); mouse; 1.5Mb hard disk space.

CALSOD 1.30 STANDARD WITH GRAPHICS
SOF-CAL2BXG **$69.95**
Witold Waldman/Audiosoft

From Australia, CALSOD (or "Computer-Aided Loudspeaker System Optimization and Design") is one of the world's most famous software packages in the field of crossover network design. It combines the transfer function of an LC network with the acoustic transfer function of the loudspeaker, by using some form of iterative analysis. CALSOD creates, through the process of trial-and-error curve fitting, a suitable transfer function model which it can then optimize. The program differs considerably from other software in that it models the entire loudspeaker output of a multi-way system, including the low-end response, as well as the summed responses of each system driver. For PC/XT/AT and PS/2 with 512K of free RAM and DOS 2.10 or higher. Hard disk with 1.6Mb free space recommended (except for demo). 8087/80287/80387 coprocessor recommended but not necessary. CGA, EGA, VGA, or Hercules graphics card required. IBM—PLEASE SPECIFY DISK SIZE.

CALSOD 3.00 PROFESSIONAL WITH GRAPHICS
SOF-CAL3BXG **$269**
Witold Waldman/Audiosoft

Although CALSOD Standard 1.30 is an excellent and popular package which more than does the job for most people, CALSOD Professional 3.00 differs in that it is more extensive and aimed (as CALSOD was originally intended) at professional engineers. Additional features of 3.00 include double ported bandpass enclosure models; vented-box, closed-box, and passive-radiator simulations; active crossover network simulation; impedance optimizer for designing zobels; SPL optimization; importing of SPL and impedance data from MLSSA, SYSid, System One, LMS, and IMP loudspeaker test systems; optimizing Thiele/Small

parameter estimator; piston approximation for driver models; simulation of the effects of room gain; and much more. IBM—PLEASE SPECIFY DISK SIZE.

PXO PASSIVE CROSSOVER CAD
SOF-PXO3B5G **$50**
Robert M. Bullock III, Robert White

New PXO version adds improved ease of use, faster performance, and high resolution graphics to six great programs: PASSIVE TWO-WAYS, PASSIVE THREE-WAYS, EQUALIZER UTILITY, RADIATION PATTERNS, EX(CURSION)-LIMIT, and CROSSOVER TRANSFER FUNCTION. Also has transient response capability for combined crossover sections. 1 x 5-1/4", DS/DD. IBM only.

THE PRIVATE FILES OF G.R. KOONCE
SOF-KPFNBXG **Each $14.95**
G.R. Koonce

At long last, we have persuaded Mr. Koonce—one of America's great computer/loudspeaker gurus—to make available to the public this 1.7Mb, five-disk set containing 19 unique programs from his own personal library. As wide-ranging in purpose as they are fun to use, these handy IBM files developed over the years for Koonce's personal use can now benefit you! Disk #1: Vented Box Design. Disk #2: Closed Box Design/Testing. Disk #3: Crossover/Padding Design. Disk #4: Miscellaneous Box Design/Testing. Disk #5: Driver Evaluation/Overshoot Function. PLEASE SPECIFY IBM DISK SIZE AND REPLACE "N" IN THE PRODUCT CODE ABOVE WITH THE DISK NUMBER YOU DESIRE. Also available:

SOF-KPF6BXG KOONCE PRIVATE FILES, 5-DISK SET $69.95

THE LISTENING ROOM FOR IBM
SOF-TLR1BXG **$47.50**
Sitting Duck Software

This interesting program predicts standing wave modes in small rooms and is designed for positioning speakers—and the listener—in such a way as to minimize standing wave effects and other room-generated influences. It allows for a full range of speaker and listener movement in 3D space and continuously updates a standing wave Pressure Versus Frequency display. IBM 256K RAM; DOS 2.11+; CGA, EGA, VGA, or Hercules graphics required. Also available, by Ralph Gonzalez:

SOF-TLR3M3G THE LISTENING ROOM FOR MACINTOSH
$47.50

BOXMODEL VERSION 3.0 WITH GRAPHICS
SOF-MOD4BXG **$59.95**
Robert M. Bullock III, Robert White

This worldwide best seller's regular capabilities and features have included sealed, vented, and passive radiator designs; four concurrent design capability; on-line help; 8088 through Pentium support; no coprocessor required, but recommended; DOS real-mode application; auto-detect CGA, EGA, VGA display; pull-down menu; inclusion of atmospheric effects on design; port length calculations; inclusion of box, port, and absorption losses in analysis; filter-assisted alignments; equalized alignments through 8th order; and automatic file save/recall. Graphing capabilities in BoxModel have included maximum and relative SPL; voice coil impedance magnitude and phase; acoustic phase response; vent air speed; driver piston excursion; and passive radiator excursion. Now, transient response and group delay have been added as new graphs. In addition, other new features in BoxModel include series and parallel compound and isobaric designs; graphics printer dumps, support for over 300 printers; generation of optimized flat alignments; display of coefficients of transfer function; and redesigned user interface. IBM—PLEASE SPECIFY DISK SIZE. Also available:

SOF-BPB1BXG BANDPASS BOXMODEL WITH GRAPHICS **$50**
SOF-TLB1BXG TRANSMISSION LINE BOXMODEL WITH GRAPHICS **$50**

CALL OR WRITE TODAY
FOR YOUR FREE COMPLETE CATALOG OR MORE INFORMATION ON ANY OF THESE PROGRAMS!

MACSPEAKERBOX — SOF-MSB1M3G — $39.95
Eldon Sutphin

This software allows you to design and examine the low frequency characteristics of bass reflex, closed box (acoustic suspension), and infinite baffle types of enclosures. Thiele/Small driver parameters are used to calculate the response for various design tradeoffs. Macintosh 512K, Macintosh Plus, or Macintosh SE required. ImageWriter or ImageWriter II recommended for hard copy.

BOXRESPONSE WITH GRAPHICS — SOF-BOX2B5G — $50
Robert M. Bullock III, Robert White; Glenn Phillips

Very straightforward, menu-driven, and flexible, this package provides model-based performance data for either closed-box or vented-box loudspeakers with or without a first- or second-order electrical high-pass filter as an active equalizer. It can be used for designing closed, vented, passive radiator, and electronically augmented vented boxes. The disk also contains seven additional programs: AIR CORE, SERIES NOTCH, STABILIZER 1, OPTIMUM BOX, RESPONSE FUNCTION, L-PAD, and VENT COMPUTATION. IBM 5-1/4", DS/DD.

LDCAD — SOF-LDC2B5G — $11.95
Madisound Speaker Components

Produced as an innovative supplement to the The Loudspeaker Design Cookbook, LDCAD is a special collection of programs and resource lists on a single 360K, DS/DD floppy disk designed to introduce speaker enthusiasts to the kind of work possible on IBM-type personal computers. The disk includes four design programs, including BOX PLOT and EASY CROSSOVER DESIGN, as well as information on Madisound's audio bulletin board and its contents. IBM only.

LOUDSPEAKER DESIGN POWERSHEET PROFESSIONAL — SOF-PSH2BX — $69.95
Marc Bacon

The Professional version of this program covers 19 different kinds of bass loading with extensive graphing capabilities; volume calculation for 5 different enclosure shapes; evaluation of cavity resonances, rectangular panel resonances, and the coincidence effect; 24 different types of crossovers; 10 miscellaneous programs for shaping circuits, zobels, room interaction, and coil design; 8 programs for evaluating driver parameters and losses; electrical laws; conversion factors; room acoustics; and more. A Basic version which includes 41 of the above programs is also available. IBM PC or compatible with 640K of memory, preferably a hard disk, and Lotus 1-2-3, Quattro Pro, or another spreadsheet which can use Lotus *.WK1 files. PLEASE NOTE THAT SPREADSHEET SOFTWARE IS NOT INCLUDED, AND PLEASE SPECIFY DISK SIZE. Also available:

SOF-PSH2M3 LDP PROFESSIONAL FOR MAC — $69.95
SOF-PSH1BX LDP BASIC FOR IBM — $49.95

WOOFER-SATELLITE OFFSET — SOF-WSO1BXG — $34.95
Sitting Duck Software

When, due to aesthetic considerations, woofer systems are placed at distances from the listener which are different from that of the satellites, serious dips in frequency response may result. The magnitude, width, and frequency of the dips are a function of the distance differential and the crossover network in use. This program plots the frequency and phase response curve which results from user-determined offset differentials and network configurations. 256K RAM; CGA, EGA, VGA, or Hercules graphics required. IBM only.

QUICK BOX — SOF-QBX1BXG — $34.95
Sitting Duck Software

QUICK BOX allows you to rapidly design closed, vented, and fourth-order bandpass boxes. The program's Library Manager, which allows "quick preview" of driver parameters and box possibilities, comes complete with data for 38 common drivers—more are easily added by the user. Supports CGA, EGA, VGA, or Hercules; coprocessor; LaserJet or dot matrix. IBM only.

LMP LOUDSPEAKER MODELING PROGRAM FOR IBM — SOF-LMP3BXG — $49.95
Ralph Gonzalez, Bill Fitzpatrick

LMP models multiway loudspeaker systems, with the resulting frequency and phase response curves predicting the on-axis SPL produced by your choice of crossover, drivers, and enclosure design. The Macintosh version adds visual and audible square-wave prediction using the internal speaker or audio output jack. CGA, EGA, VGA, or Hercules graphics capability required. PLEASE SPECIFY DISK SIZE. Also available:

SOF-LMP3M3G LMP FOR MACINTOSH — $39.95

QUICK & EASY TRANSMISSION LINE SPEAKER DESIGN — SOF-QET1B5 — $8.95
Larry D. Sharp

This unique new booklet starts with the basics: what a TL is, where it came from, and how it evolved over the years. Then it lays out a step by step process for designing a TL system that will sound good every time. And for those of you who own a personal computer with Lotus 1-2-3 or equivalent on it, there's also a computer worksheet diskette included that does the math for you and prints out your system design information. 1993, 22pp., 8-1/2 x 11, spiralbound; IBM 5-1/4" DS/HD. Booklet is easily usable without software, but PLEASE NOTE: LOTUS 1-2-3 OR EQUIVALENT SPREADSHEET SOFTWARE IS REQUIRED TO RUN WORKSHEET DISK AND IS NOT SUPPLIED WITH THIS PACKAGE.

ROOM DESIGN POWERSHEET — SOF-RDP1BX — $59.95
Marc Bacon

This program covers a wide range of knowledge in an easy-to-use spreadsheet format. Working from a main menu, the user can access programs dealing with room resonances, reverberation, boundary augmentation, wall diffuser design, resonance traps, and so forth. Requires an IBM PC or compatible with 640K of memory, hard disk, and Lotus 1-2-3, Quattro-Pro, Excel, or any other spreadsheet which can use Lotus *.WK1 files. COMPATIBLE WITH EXCEL AND LOTUS 1-2-3 FOR WINDOWS; NOT USABLE WITH QUATTRO PRO FOR WINDOWS. PLEASE NOTE THAT SPREADSHEET SOFTWARE IS NOT INCLUDED. PLEASE SPECIFY IBM DISK SIZE DESIRED.

SPEAKER SYSTEM DESIGNER 4.2+ FOR WINDOWS — SOF-SSD2B3G — $279
Bodzio Software

SSD4.2+ enables you to create, evaluate, and then optimize 2-, 3-, 4-, or 5-way loudspeaker systems prior to starting enclosure assembly. You are also able to model the behavior of the crossover when loaded by the driver in an enclosure and observe the dramatic effect on the frequency response curve of the crossover which the driver may have. Functions available in this package include driver reference library creation; loudspeaker enclosure design and optimization; compensation of the driver impedance or amplitude; crossover filter design and optimization; system frequency response evaluation and optimization; frequency response of the system "in room"; L-pad, series LRC, and zobel network calculators; impedance peak suppressor; and much more. IBM 286+; 2Mb RAM min.; min. 2.5Mb hard disk plus 1Mb to install; Windows 3.1; SVGA with 800 x 600 pixels, 16 colors.

TERM-PRO LOUDSPEAKER DEVELOPMENT — SOF-TRM2BXG — $399
Wayne Harris

High-resolution graphics, on-line help, and a menu-driven format that makes this program extremely easy to use. Features common to TERM-1 and TERM-PRO include a 10,000-driver-capacity database with multiple library support; enclosure design capabilities for sealed, ported, and isobarik sealed and ported; predicted enclosure response and SPL plots; port and enclosure layout design functions, including wedge, rectangular, or bandpass designs; passive crossover design for 1st, 2nd, and 3rd order HP, BP, LP, notch filters; L-R, Bessell, BEC, Butterworth, Chebychev design; and acoustic curve overlays with crossover enabling toggle. TERM-PRO (only) also includes single reflex bandpass (4th); isobarik SRBP (4th); SRBP with coil (5th); isobarik SRBP with coil (5th); dual reflex bandpass (6th); isobarik DRBP (6th); DRBP with coil (7th); and isobarik DRBP with coil (7th). Originally intended for auto sound use, but of universal value. IBM XT/AT, 640K RAM; MS-DOS 3.0+; CGA, EGA, or VGA. PLEASE SPECIFY DISK SIZE DESIRED. Also available:

SOF-TRM1BXG TERM-1 LOUDSPEAKER DEVELOPMENT — $199

DISCOUNT AND SHIPPING TABLE

ORDER VALUE	<$50	$50-$99.99	$100-$199.99	$200+
Take Discount	0%	5%	10%	15%
Add Shipping				
USA Surface	$3	$4	$5	$6
USA Air	$7.50	$15	$20	$30
Canada Surface	$5	$7.50	$15	$25
Canada Air	Same as USA Air			
Other Surface	$10	$20	$30	$40
Other Air	Add $10 to Other Surface			

We accept Mastercard, Visa, and Discover, as well as money order or personal check in US dollars drawn on a US bank.

OLD COLONY SOUND LAB
PO Box 243, Dept SND
Peterborough NH 03458 USA
24-Hr. Tels. (603) 924-6371, 924-6526
Fax (603) 924-9467

Vacuum Tubes for **Audio** Are Back!

Glass Audio brings together yesterday's tube with today's improved components, voltage control, and the exciting new Soviet tubes, to make smooth sound in your livingroom possible again!

Get your first issue FREE to examine, without obligation.

YES!

Please send me an issue of ***Glass Audio*** to examine FREE for 30 days. When I choose to subscribe, I'll pay just $23.00 for six issues (1 year) of the best information on tubes to be found anywhere! If I decide that ***Glass Audio*** is not for me, I'll write "cancel" on my invoice and owe nothing.

Name ____

Street & Number _________________________________

City _______________________ State _________ Zip _________

In Canada, add $6 per year in postage. Overseas Rates: $45 for one year. Remit in US $ drawn on a US bank only. Rates subject to change.

Glass Audio

PO Box 176, Dept. ELS5, Peterborough, NH 03458-0176 USA
(603) 924-9464 or FAX 24 hours a day to (603) 924-9467